practical computing for biologists

practical computing
for **biologists**

Steven H. D. Haddock
*The Monterey Bay Aquarium Research Institute,
and University of California, Santa Cruz*

Casey W. Dunn
*Department of Ecology and Evolutionary Biology,
Brown University*

 Sinauer Associates, Inc. • Publishers
Sunderland, Massachusetts U.S.A.

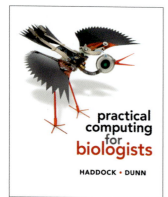

ABOUT THE COVER

Ann P. Smith created the sculpture on the cover, incorporating many electronic components used for scientific data acquisition. We are fortunate to have this image, which projects our belief that the intersection of technology and biology does not have to be intimidating. The body is an Arduino microcontroller board, and other parts include thermistors and a 555 digital timer. Some of the pieces came from special items we sent to her and others were from a wonderful box of parts she has collected through the years. She dumped them out on the table at Julian's in Providence and let us paw through them during lunch. More of Ann's artful creations can be seen at burrowburrow.com.

Notice of Liability

Due precaution has been taken in the preparation of this book. However, information and instructions described herein are distributed on an "As Is" basis, without warranty. Neither the authors nor Sinauer Associates, Inc. shall have any liability to any person or entity with respect to any loss or damage caused or alleged to be caused, directly or indirectly, by the instructions contained in this book or by the computer software and hardware products described.

Notice of Trademarks

All product names are trademarks or registered trademarks of their respective owners. In lieu of appending the trademark symbol to each occurrence, the author and publisher state that these trademarked product names are used in an editorial fashion, to the benefit of the trademark owner, and with no intent to infringe upon the trademarks.

Library of Congress Cataloging-in-Publication Data

Haddock, Steven H. D. (Steven Harold David), 1965-
 Practical computing for biologists / Steven H. D. Haddock, Casey W. Dunn. -- 1st ed.
     p. cm.
 Includes index.
 ISBN 978-0-87893-391-4
 1. Biology--Data processing. I. Dunn, Casey W. II. Title.
 QH324.2.H33 2010
 570.285--dc22

                          201003948610

 9 8 7 6

This book is dedicated to Erika and Lucy

CONTENTS IN BRIEF

CONTENTS

PART I   Text Files   7

Chapter *1*
GETTING SET UP   9

Chapter *2*
REGULAR EXPRESSIONS: POWERFUL SEARCH AND REPLACE   17

PART II   The Shell   45

PART III   Programming   103

PART IV   Combining Methods   243

PART V   Graphics   321

Chapter **17**
GRAPHICAL CONCEPTS   323

Chapter **18**
WORKING WITH VECTOR ART   345

Appendices   449

ACKNOWLEDGMENTS

Erika Edwards was supportive in every way at every stage of writing this book. She was always ready to enable writing time, and helped talk through key issues of content and presentation. Lucy's patience and imaginative suggestions for plot twists were very generous.

C.D. thanks Steve Dunn for teaching him to build logic circuits and free dive in kelp forests at an early age, and Karin Dunn for encouraging disassembled household appliances. S.H. thanks his parents for providing an environment (toolboxes, 65-in-1 electronics kit, hatchet, etc.) where learning and experimentation were fun and unconstrained. We couldn't have completed this book without the support of our friends and our families.

The members of our labs have contributed many recommendations and examples for us to draw upon, and their questions and adventures while learning new computer skills were direct inspiration for writing this book. They have also been very understanding during our periods of distraction.

We thank Andy Sinauer for taking this book under consideration in the first place, and Bill Purves for again bringing about life-changing opportunities.

The staff at Sinauer Associates provided expert support and found a way to bring our vision of the book to life, especially Azelie Aquadro, Joan Gemme, Chris Small, Marie Scavotto, and Andy Sinauer.

Critical review of early drafts was supplied by many readers, who flagged confusing sections and weeded out misteaks. These include Randy Burgess (our copy editor), Julie Stewart (our #1 supporter), Katie Mach, Ashley Booth, Kevin Raskoff, Joan Gemme, Dan Barshis, Susanna Blackwell, Carl Haverl, Michael Alfaro, Dan Riskin, Chris Perle, Carolyn Tepolt, Kevin Miklasz, Jason Ladner, Megan Jensen, Lou Zeidberg, Katherine Elliott, Alana Sherman, Tim McNaughton, Stefan Siebert, Henry Astley, and Rebecca Helm. Several classes served as beta testers, including the students of Topics in Ecology and Evolutionary Biology at Brown in the spring of 2009 and the Hopkins Marine Station graduate seminar.

While we were writing, a welcoming environment was furnished by the Helen Riaboff Whiteley Center, and Pat Lundholm and Ted Johanson of the Vinalhaven Writers' Sanctuary. Jim Edwards created writing time in Providence by always being ready to play with Lucy. Mark Bertness helped guide and support C.D. in balancing book writing with starting a faculty position.

The highlight program (www.andre-simon.de) allowed us to rapidly format our example code, and iAnnotate made it possible to review chapters on the fly.

Jean-Philippe Rameau and Philip Glass provided the soundtrack for critical writing periods when the fate of the world hung in the balance.

We are grateful to Ann P. Smith (burrowburrow.com) for creating an inspirational cover sculpture, incorporating many components that had special meaning to us.

STEVEN H. D. HADDOCK
CASEY W. DUNN

BEFORE YOU BEGIN

Introduction

The goal of this book is to show you how to use the many tools you already have on your computer to make it do more work for you. You will learn skills for shaving minutes or hours off many small tasks—and because these skills are scalable, you will also be able to use them to accomplish things that would take you many weeks or even years to do by hand. The focus is on general solutions applicable to a range of problems, rather than on cookbook recipes not useful in any other context. In addition to learning new skills, you will begin to recognize when to apply these tools to make your analyses easier. Once you are familiar with the basic tools, you will learn how to combine them to improve the efficiency, flexibility, and reproducibility of your overall workflow. In some cases, you will even be able to ferry your data through a series of analyses without touching a keyboard.

The few computer books that have been written with biologists in mind focus mainly on programming, in particular as it is relevant to bioinformatics. This is not a book on bioinformatics; it takes a broader approach, both in the range of tools and the variety of applications. As you will see, effective computing can benefit many more aspects of science than just molecular sequence analysis.

Why this book?

There are already many books dedicated to each of the different subjects we cover here, including programming, databases, graphics, and Unix commands. Most of these texts aim to be comprehensive, providing a thorough background on the concepts, theory, and history of the technologies they cover. However, we know many frustrated biologists and eager students to whom these in-depth books are of little use without an initial grounding and a practical overview. In our experience, 90% of the computational problems biologists face can be addressed with less than 10% of the commonly available tools, but a typical scientist has no idea which 10%. This is why we decided to write a problem-centric book to complement the many technology-centric books that already exist—we wanted to create a resource from the perspective of the life-scientist user base.

In the short term, writing a special program to help with your data analysis may seem like a big investment in time and effort, especially when the choice is between spending six or so hours to write a script, or just working through the dataset in two hours of copying and pasting. In the long run, however, an investment in script-writing quickly pays off, since it subsequently takes just minutes to process hundreds of files. You also have the benefit of rapidly being able to redo your analyses when more samples are added to the dataset.[1] The process of generating tools for analysis becomes easier each time you do it, and as you become more proficient with the languages and tools, eventually you will be able to write a script faster than it would take to process even a single file by hand. You also quickly accumulate a set of programs that take relatively little time to adapt to new purposes.

Learning to write programs is only one aspect of being able to use computing for your research. The more general challenge is to figure out when and how to apply each of the many tools available as you move from data collection to analysis and presentation. Recognizing that a nail needs a hammer and a bolt calls for a wrench is just as important as having these tools in your toolbox and knowing how to use them. This broader perspective is addressed with a flowchart that helps you with the tool-selection process, as well as in a chapter on organizing your dataset before you even start collecting it.

Why biologists?

We chose to focus on biologists rather than scientists-at-large for a few reasons. We ourselves are biologists who also happen to have backgrounds in computing. This book grew out of years of developing tools for our own research and helping other biologists with their computational problems. Although the methods and background we provide would be just as useful to scientists in other disciplines, as well as to engineers and other professionals, we think it best to stick to the audience we've been serving all along. We also want the examples to represent familiar problems, so that the reader will have a concrete idea of how the material applies to issues they routinely face, even before they have to consider technical details.

Biology is becoming far more computationally intensive, yet undergraduate and graduate biology curricula have not kept up with this new reality. As a result, many biologists find themselves lacking the training and information-processing skills needed to tackle exponentially expanding datasets and more complex analyses. This text aims to fill that gap by providing training that is increasingly important for scientific success. Ultimately, we hope it will encourage biology departments to develop standard computing courses for their students.

---

[1] As our colleague Sönke Johnsen likes to say, "There's a reason they call it re-search."

Is this about using a particular computer or program?

This book is about using computers practically. We focus on general, flexible, scalable tools, rather than on telling you how to click through the menus of a particular version of a program to accomplish a given task. We concentrate on the technologies used most frequently within the biology community, in order to maximize your ability to understand, recycle, and reuse tools that have already been developed by other life scientists. We also consistently lean towards open source software, largely because of its availability, flexibility, and transparency.

Some of the biggest data analysis challenges faced by biologists are in figuring out how to organize their data and workflows, and these issues have nothing to do with the type of computer you use. However, it would have been too cumbersome for this book to walk through each example with specifics for more than one type of computer or operating system. We have therefore used Apple's OS X as the primary setting for our examples and procedures. OS X is an easily accessible Unix variant. Unix itself has been around for forty years now, and is here to stay. It has long been the standard for demanding scientific analyses, and can now be used on any personal computer.

In addition, Appendix 1 provides extensive detail on how to install and use the tools presented here on computers running either Microsoft Windows or Linux. Microsoft Windows is not based on Unix, so many details are slightly different and some topics—for example, using the command line—require work-arounds. Linux, like OS X, is based on Unix, so in this case, the differences relevant to this book are minimal.[2] Appendix 1 also explains how to install Linux on a Windows computer, which can be a good option for those who have a Windows computer but want to test the Unix waters.

One of the most flexible and powerful skills you will learn in this book is how to write a program. There are two principal aspects of programming: the conceptual building blocks (loops, decisions, arrays, input, output), and the language-specific syntax (the words you type to embody those concepts). Although most beginning programmers are concerned about the syntax, the building blocks are far more important: once you become comfortable with them, you can quickly adjust to any programming language. For this book we have chosen Python as our demonstration language, for several reasons. It is powerful, yet easy to read. It is growing in popularity, and is rapidly being adopted as the standard for bioinformatics, Web development, and introductory programming courses. While your fundamental programming ability will never go out of style, the Python syntax you learn should also serve you well for many years to come.

---

[2]Unix operating systems are used "behind the scenes" in many household devices, including DVRs and the iPhone.

To readers who will use this book on their own

We expect that many biologists will use this book to improve the efficiency of their research, help scale up existing projects, or develop the skills needed for new types of studies. The book is designed to serve these independent readers, and in fact, our own graduate students and colleagues have used chapters of the manuscript in this way. Much of what we ourselves use in practice was garnered through self-directed experience, and we have tried to collect this knowledge in one place to make it easier for other scientists to do the same.

Some recommendations for working through the book: Try out as many examples as you can and explore beyond their immediate scope as described in the text. Also, start keeping a file on your computer of useful commands and things that you discover, to help you generate your own quick reference card.

When something doesn't work as expected, it can be difficult to troubleshoot in isolation. If you come to a point where you can't sort things out on your own, take advantage of our online community at `http://practicalcomputing.org`. By reading related posts and asking other readers for advice, you can gain a sympathetic ear and get a quicker resolution for your setback. You can provide feedback about what you find most versus least useful in the book, and let us know of errors that may have crept into the text. Finally, you can exchange ideas about applying your new skills to your research.

To teachers using this book

Although the book is written to stand on its own outside the classroom, it is also designed to be used as a primary or secondary text within the classroom. We use it as a textbook by having students read thematically connected chapters ahead of time. Then, in class, we walk through examples in more detail, and expand on the content in the areas the students find either most confusing or most relevant to their own interests. It helps to keep in mind that students who are knowledgeable and confident when discussing biology may feel intimidated or frustrated by computing concepts, and they may interact differently because they are sensitive to being found "ignorant."

In a classroom setting where students have different types of computers, you will also have to determine how you want to achieve operating system independence. Many of the lessons are platform-independent, but Chapters 4 through 6 assume a Unix command line is available. Windows users might at first feel slighted by these sections, but some general solutions are provided in Appendix 1. Ultimately, you will have to choose whether you conduct the class in a lab with OS X machines available to all; request that students with Windows computers install Linux (which is much simpler now with the widespread use of virtual machines, as described in Appendix 1); or have students install a software package (Cygwin) that gives them some Unix functionality (also described in Appendix 1).

The book is organized into sections, but only parts I , II, and III need to be covered in sequence; the skills taught there not only build on each other, but provide a foundation for the entire book. Beyond that, you are free to jump around as you see fit. For example, Chapters 4–6 on the Unix shell fit well with more detailed shell material from Chapters 16 and 20. The three chapters on graphics can be covered at any time.

While we give examples and opportunities for practicing new skills as they are developed, we do not provide formal exercises or test questions. There are so many diverse data-analysis issues in biology that it should not be difficult to generate exercises which are especially relevant to your students' subdisciplines, whether molecular, ecological, or biochemical.

Beyond this book

We hope that this book helps you start solving many problems immediately, as well as showing you how to do things you didn't know how to do before. Beyond that, it is also intended as a stepping-stone to more advanced topics. The long-term goal is to empower you in the use of your computer in such a way that you know how to continue growing, training yourself with both other books and Internet resources. We have found that not knowing where to get started is a major roadblock for many biologists facing new computational challenges, and that once they are pointed in the right direction, they quickly gain traction and expertise. Advice on additional reading is provided at the end of many chapters.

It is our hope that by showing biologists how to apply the powerful tools already installed on their computers, they will be free to spend more time on the pleasures of scientific inquiry and less time on the frustrating aspects of data processing.

How to use this book

You can, of course, pick and choose the parts of this book that you feel are most relevant. However, we recommend that you follow the chapters of Parts I through III in the order that they are presented. The later chapters within those sections assume understanding of the concepts and skills (and in some cases, the system settings) described in earlier chapters. Part IV takes a broader view of ways to combine tools into processing pipelines and techniques for working with larger datasets. Part V, on graphics, is most relevant when preparing figures for publication or presentation. Part VI includes more specialized topics, which may or may not be relevant to your particular needs. The appendices serve as reference material, largely for the tools described in Parts I through III.

Throughout the book, graphical elements help to highlight certain sections. Icons in the margin point out passages of the book which may be of special interest or importance:

A tip to make your work easier

A potential pitfall that might trip you up

A pitfall that might cause widespread failure

Sections that vary for Windows users. Refer to Appendix 1 or a marginal note

Sections that vary for Linux users, including Linux via Virtual Box within Windows

Text is also formatted to indicate its usage:

System items include files and folders, program names, and menu commands

`Code characters` indicate programming language and terminal commands. The background color indicates whether they are from a `terminal window` or part of a script in a `text document`.

We have customized the `Courier` typeface used to represent code characters to more clearly distinguish between the number `1`, lowercase `l`, and uppercase `I`, as well as between the number `0` and the capital letter `O`. At places in the text, a tab character is represented by the symbol △. Further information can be found at the companion site `http://practicalcomputing.org/`.

PART I

TEXT FILES

## Chapter 1

GETTING SET UP

Before setting out to learn new computing skills, you will need to take care of a couple of things. Here we describe how to prepare your computer so that you can easily follow along with the examples and apply the new tools we cover. We also explore some fundamental concepts about text files.

An introduction to text manipulation

A variety of seemingly complex and disparate computational challenges routinely faced by biologists can be reformulated as simple text manipulations that can be addressed with a common set of tools. These tasks include aggregating data from several sources that are each formatted differently, or reformatting text files from an instrument or database so they can be further processed by another program. Learning how to automate such tasks can save time and allow you to broaden the scale of your work. There is a good chance that the skill set you build addressing one challenge with general text manipulation skills will be directly applicable to an entirely different challenge in the future. Focusing on general approaches to text handling, therefore, is often a better use of your time than learning the idiosyncrasies of how to get a particular program to accomplish one specific task.

Plain text files are the common currency of many programming environments and their accompanying data. Even for software packages or instruments which store data in special proprietary file formats, there is usually an option to export or import data as plain text files. This provides a way to do things with data beyond what the people writing the software might have intended. You can, for instance, export data from a program frequently used in one discipline and then import it into an analysis program used in a different discipline. The trick to this, though, is that the text file generated by one program is rarely suitable for import as-is into another program; usually it must first be modified.

Learning how to do this will empower you to perform fundamentally new analyses and help your research break new ground, and is a primary goal of this book.

What are text files?

All computer files consist of a linear sequence of binary numbers. Text files are just a special type of binary file where those numbers correspond to human-readable characters such as digits, letters, and punctuation, and white space such as spaces, tabs, and line endings.

REPRESENTING TEXT WITH NUMBERS The correspondence of particular numbers to particular characters follows agreed-upon conventions, the most common of which is the American Standard Code for Information Interchange (ASCII). For instance, in ASCII the number 65 (`0100 0001` in binary) represents the uppercase letter A, and the number 49 (`0011 0001` in binary) represents the digit 1. Notice that in this last example, the digit 1 is not stored as a direct binary representation of the number 1 (that is, `0000 0001`). Nor is a multidigit number such as 12 stored as a direct binary representation of the number 12 (that is, `0000 1100`); instead, it is stored as two sequential codes for the digits 1 and 2. Text files are inherently more human-friendly than computer-friendly—a person can read them as-is, but a computer must interpret them to use the data for calculations or other analyses.

Appendix 6 shows the complete set of characters encoded by the ASCII standard. An extended character set, including accents and special symbols, is supported by another standard called Unicode, which offers encodings such as UTF-8 and UTF-16. International readers are especially likely to encounter Unicode.

Although all files are inherently binary in nature, a file is usually referred to as being either a **binary file** (a file that doesn't simply encode a series of text characters) or a **text file** (a binary file that does encode such a series). There are many sequences of binary information that don't represent text and can't even be displayed as such. It is more compact, and faster for the computer, to store numbers and other data as direct binary encodings, rather than as the letters and digits describing their values. These performance considerations were more important in the past on slower computers with less memory.

For most biological data, a text representation of data works perfectly well, and is in fact desirable. Why? It comes down to transparency and portability trumping size and speed. Transparency means that the contents of text files are readily accessed. With a binary file, you need a map of how the data are stored to be able to extract them; with a text file, you don't. It can be very difficult to peer into a binary data file and see what's there without thorough documentation of the file structure. A JPEG file, for example, contains compressed binary image data. Try opening it in a text editor, and you'll see that it looks like gibberish. Portability becomes an issue if the program that generated certain files stops being supported or distributed. It is not uncommon to see an ancient computer limping along in the corner of a lab, just to be able to run a 15-year-old program that is the only means of extracting data from an obsolete file format.

In contrast, the organization of data within a well-conceived text file should be entirely self evident to anyone who opens the file in any text editor. Even if the program that generated a particular type of text file goes away entirely, the ability to just look inside such files and see what data they contain will allow you to rearrange, extract, and continue to utilize those data. So storing data in a well-organized text file greatly extends the lifetime of the data and increases the number of things you can do with them.

The organization of data within a text file

The most common way to organize data in a text file is as **character-delimited text**. The layout is two-dimensional, with columns and lines. The fields within each line are separated by a particular character—the delimiter—which is usually a comma, space, or tab. If the character used to delimit the text occurs within one of the fields, quotation marks are usually placed around the entire field. This indicates that all the characters within are part of the same datum, and that if the delimiter character occurs within the quotes, it does not imply a separation of values as it normally would. The first line in character-delimited text files often consists of a series of descriptions of the columns, delimited by the same character as the data themselves. This description line is called the **header**.

Character-delimited text is an excellent way to handle two-dimensional data, such as a series of measurements taken through time or across multiple samples. Many data loggers and instruments can be set to store and export their data in this way. All spreadsheet programs can export delimited text, usually offering you options for which character to use for separation. However, character delimitation is not the best option for all types of data. If the data are hierarchical in nature, for instance with varying degrees of nestedness and a different number of nested elements at each level, it is a very inefficient way to store information. A phylogenetic tree is an example of this category of data. Character delimitation also breaks down if the fields are very large and vary in length from sample to sample, as is the case with molecular sequence data. Large data values are more conveniently stored across several lines, rather than in a single column in a table.

For these and other reasons, a wide array of formats is available for organizing different types of biological data within text files, often with many formats for the exact same type of data. For example, there are FASTA, GenBank, and NRBF formats for molecular sequence data, and SDTS, Google Earth, and GRASS formats for spatial data.

We will show you how to extract data from one type of text file format, reorganize it, and store it in another type. Later we'll guide you through some of the steps involved in identifying why a file that is supposed to be understood by a particular program isn't, and then using these same reorganizing skills to fix the problem.

Text editors

Given the importance of text files for science, one of the applications that you'll spend the most time interacting with will be a text editor. Although you might be tempted to start with a word processor you are already familiar with, word processors don't actually save plain text. Their files include formatting and layout information in addition to the text to be displayed. Not only is this formatting information unnecessary, but it will corrupt your files by filling them up with unexpected characters. You will almost never use a word processor for any stage of manipulating text data or scripts. Instead, there is a wide variety of text editor programs that focus exclusively on the raw characters found within a text file. These editors show you exactly what is in the file, character by character, and provide a range of useful tools that make it easier to view and edit these characters.

Installing TextWrangler

Although OS X comes with a text editor (appropriately called TextEdit, and located within your Applications folder) we do not recommend that you use it for the tasks we describe in this book. This is because it is actually a hybrid between a word processor and a text editor, sometimes leading to confusion about when you are seeing formatted text versus raw text, and it isn't optimized for displaying and editing data and software. Instead, we recommend that you download TextWrangler. This is the free version of BBEdit, and includes most of that editor's important features. There are many great text editors available, but TextWrangler is very powerful and distributed free of charge. All the editing in this book can be reproduced using this program.

OTHER TEXT EDITORS TextWrangler is used for the examples throughout this book. If you are using Windows, we suggest you use Notepad++ as an equivalent. For Linux, we recommend either gedit, or the popular text editor jEdit (consult Appendix 1 for details). There are many text editors, and each has its camp of advocates. Most commercial products are reasonably priced. If you also want to try other editors, some popular alternatives for OS X include BBEdit and TextMate, each of which has its strengths, or Emacs, which is popular among programmers. Because programmers spend so much time doing repetitive tasks in text editing programs, an extensive feature set has been built into many text editors to streamline operations and check errors as you go. These features include coloring text based on the context, auto-completing the names of items in your scripts, filling in templates with commonly used program snippets, and reformatting your programs for clarity.

You can download TextWrangler from the BareBones Software Web site, `www.barebones.com/products/textwrangler`. We used version 3.0, and there may be minor differences in other versions. TextWrangler shows multiple open documents in a single window, organized in a drawer to the right of the editing window that you can use to toggle between them. After you install TextWrangler, it won't necessarily open all the appropriate files by default when you double-click them. You may need to drop some files onto the program's icon or use the Open dialog box within TextWrangler to access them.

Optimizing text appearance within a text editor

Although you will be dealing mostly with text files that contain no information on formatting, you still have control over how the text is displayed on the screen through the text editor preferences. You can, for instance, set the display font. For data and programs, it is usually desirable to use a fixed-width font (also called a fixed-space or monospaced typeface), such as Courier. The advantage of these fixed-space fonts is that data and program code will line up better from line to line since each character has the same width. You can do a quick test to see if your typeface is monospaced by typing the letters iiii and OOOO on consecutive lines:

Proportionally spaced	Fixed-space
iiii	iiii
OOOO	oooo

If the left and right edges of the sets of characters line up, you probably have a monospaced typeface.

Most text editors also provide the handy feature of displaying line numbers to the left of the text. This is helpful for navigating large files and going straight to problems in the file that were identified during testing by other programs. To enable line numbers in TextWrangler, open an existing or blank document, click on the Text Options icon at the top of the window, and select Show Line Numbers in the pull-down menu.

Remember, these display settings do not alter the text file itself—they are just editor settings that control how file contents are displayed. You can have different display settings for different files, but these are not stored in the file as with word processors. The text editor is separately keeping track of your preferences for each file.

Line endings

The characters used to designate the ending of a line—that is, the symbol inserted when you hit the [return] key—were never standardized across different operating systems. This may seem like an esoteric technical issue, but it is one of the most common causes of problems when transferring text files from one computer to another, or even between programs on the same computer. It is likely that you have

come across this problem before, such as when opening a result file from a remote computer and finding that all the contents are shown on a single line. It is also one of the most common reasons that a program will not accept text data that appear to be correctly organized. Many programs that can't handle different types of line endings do not give you an error telling you this—they just crash. If you can't get something to run, it is a good idea to first check the line endings of your input files.

There are two line ending characters, **carriage return** (abbreviated CR or \r) and **line feed** (abbreviated LF or \n). Unix systems, including OS X, use a line feed, some older Macintosh programs use a carriage return, and Microsoft Windows uses a carriage return followed by a line feed! When it is within your control to decide which line ending to use, or when you are making a first stab and don't know what to expect, go with a line feed. It will generally give you the fewest problems across systems.

You can check which line endings are used within a file by opening it in TextWrangler and taking a look at the status bar at the bottom of the window. Next to the horizontal scroll bar there is a drop-down menu that shows the current line ending character for the file. To change this character, just click the drop-down menu, select a different option, and save the file.

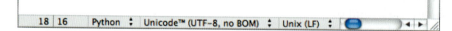

Unlike cosmetic changes, changing the line ending character *does* modify the file by altering which characters are recorded within the file itself.

The example files

You will be working with a variety of example files throughout this book. These are available for download as a zip file at `http://www.practicalcomputing.org`.

Installing the example files

After downloading the zip file, place it in your home folder. This is the folder where your Documents and Pictures folders are stored by default; it is also the folder that typically opens if you create a new Finder window on OS X.[1] Double-click the zip file to extract its contents; this will generate a new folder called pcfb. The pcfb folder has three folders within it. The first folder, examples, has the data files you will need to follow along with the book. The second, scripts, has programs and scripts you will use and modify. The third, sandbox, is an empty folder

---

[1] If you don't know what your home directory is, see the explanation accompanying Figure 4.2. For Windows and Linux users, put the folder in the directory named with your username, and see Appendix 1 for more information on the file structure.

that provides a convenient place to work with the files you will generate. Many of the examples will assume that these folders are located in the specific location described above—so if you put it somewhere else, you will need to adapt the examples accordingly.

Exploring the example files

Go ahead and use a text editor to explore some of the files in the examples folder you installed earlier. First, from within TextWrangler (or the alternative text editor for your computer) open ctd.txt. This file contains columns of comma-delimited measurements of salinity, temperature, oxygen concentration, and various other measurements at different depths off the coast of California. Now, open ctd.rtf from within TextWrangler. This file contains the exact same data but has been edited in a word processor and now contains hidden formatting information (font, font size, and a bold header). In addition to the data, you can see all kinds of characters that explain to a word processor how to display the text, but don't represent data. This illustrates why you want to stick to a raw text editor like TextWrangler when handling data: the added formatting information can confuse the interpretation of the file by other programs. With a raw text editor, what you see is what you get—all the contents of the text file are shown and you know exactly what you are dealing with.

Now, in your computer's file browser (not the Open dialog box in TextWrangler), find the same ctd.rtf file and double-click it. This will open it in a word processor, probably the hybrid program TextEdit that we mentioned earlier. You will see that none of the formatting characters are shown when using a word processor, just the formatted data. It is difficult to know what the actual contents of the file are, and consequently how they would be interpreted by other programs. Though it is often tempting to open raw data files in a word processor and use different fonts, colors, and highlights to organize different types of information within the file, this peek inside should begin to give you a sense of the perils of following such an approach. It is far clearer, and ultimately more flexible, to use well-defined strategies for organizing and explicitly annotating data (e.g., with column headers) in a plain text file.

SUMMARY

You have learned:

- The nature of text files
- Broad considerations regarding the organization of data within text files
- What text editors are, and how to install the TextWrangler editor
- The different line endings that are used in text files
- How to install the example files used throughout the book

## Chapter 2

REGULAR EXPRESSIONS: POWERFUL SEARCH AND REPLACE

Many of the computing tasks in biology amount to a series of simple text manipulations. There is a wide variety of programs that facilitate these text modifications for specific types of data, but there is also a single set of general tools that can be used in many different contexts: regular expressions. These are flexible means of searching and replacing text. In this chapter you will learn the components of regular expressions, while later chapters will add further flexibility and control to these tools.

A widespread language for search and replace

One of the most common computer problems faced by a scientist is to reorganize a text file generated by one program so that its content can be understood by another program. In some cases only a few manual changes may be needed. At other times simple search and replace is sufficient, such as when a file contains data separated by commas and must be modified to suit a program that expects data separated by tabs. Often, though, the required manipulations can be too complicated for these approaches, for example when the order of elements in each line needs to be rearranged. All of us have at some point made these complex changes by hand, but this gets tedious if many files need to be fixed or the files are long. There is also the danger that you may be in for more than you anticipated: most of us have spent several hours manually reformatting a file, only to realize that we have to do it all over again because something about the original file wasn't quite right.

A surprising number of these seemingly complex problems can be addressed using a powerful language for search and replace known as **regular expressions**.[1]

---

[1] Regular expressions are also sometimes referred to as regexp, regex, or even **grep**. The latter is an acronym for "**g**lobal **r**egular **e**xpression **p**rint," a common tool for using regular expressions that is so frequently used that it is often used as a synonym of the language as well.

Regular expressions are widely used and are built into many environments, including text editors, programming languages, some Internet search engines, and many applications. Because regular expressions are so powerful and such a portable tool, they are the first skill you will learn here. As you proceed, they will allow you to perform complex text manipulations across a wide range of environments using one skill set.

Regular expressions can do anything that can be done by simpler search and replace tools, such as those found in a word processing program. This includes replacing one chunk of exact text (such as "jellyfish") with another ("scyphozoan"). You can also leave the replacement term empty, which leads to the deletion of the matched text from the file.

Like many other search and replace tools, regular expressions can employ **wildcards**. These are special characters in the query that can match more than one particular character in the text being searched. Wildcards greatly extend the utility of a search and replace tool when the text you want to match is variable. This would be the case, for instance, if you wanted to find all digits (i.e., number characters), but you don't know what the digits will actually be. Compared to most other search and replace tools, regular expressions allow more flexibility and precision in the way that wildcards are designed and used. There are many ready-to-use wildcards, and you can define your own wildcards that match any sets of characters you like.

The functionality of regular expressions extends well beyond customizable wildcards. Regular expressions also make it possible to capture all or part of the search term and use it in the replacement term. It is this capability that makes regular expressions so versatile for extracting data from complex text files and redisplaying them in a different sequence. You could, for example, find any sequence of digits followed by "cm" (not just a particular number), and then insert those digits (with or without the "cm") elsewhere in the text.

To start getting familiar with regular expressions, you will first use them through the dialog box of a text-editing program. In Appendix 2, you will find tables summarizing the permitted syntax. Once you get comfortable with the language, you will find yourself doing a lot of quick "one-off" file processing using a few well-conceived searches in a text editor. Regular expressions are so powerful that many of the programs you will write later will do little more than open a file, run a series of replacement operations on the contents (perhaps applying different transformations to different portions of the file), and save the results.

Understanding the components of this new toolbox

Setting up the text editor

 Open TextWrangler (the text editor introduced in Chapter 1) and create a new document. Windows and Linux users should consult Appendix 1 for comparable editors that support regular expressions.

Start just by doing a normal, old-fashioned replacement operation. In your document, type the text:

```
Agalma elegans
```

Now select Find from the Search menu (or hit ⌘ F).[2] In the dialog box, check the Grep box (Figure 2.1), and make sure that Case Sensitive is also checked. Unless you uncheck them, these options will remain selected each time you open the program.

Find:	galma				⊙ ▾	Next
					g ▾	Previous
						Find All
Replace:	.|					Replace
Matching:	☑ Case sensitive	☐ Entire word	☑ Grep			Replace All
Search in:	☐ Selected text only	☐ Wrap around				Replace & Find

FIGURE 2.1 The Find and Replace dialog box in TextWrangler

In the Find box type:

```
galma
```

and in the Replace box type a period by itself:

```
.
```

Then click the Replace All button. As expected, where it originally said:

```
Agalma elegans
```

it now says:

```
A. elegans
```

With the document window in the foreground, undo this last operation (⌘ Z) so that the text again says Agalma elegans. A literal search term works fine if you are dealing with only one species, but if your data file includes other species this one-size-fits-all approach breaks down:

Try ctrl Z.

```
Agalma elegans
Frillagalma vityazi
Cordagalma tottoni
```

---

[2] The symbol refers to the Command key on Apple keyboards, just to the left of the space button. It is not the ctrl key (Ctrl), which will be used later in terminal operations. In Windows programs, the find command will likely be triggered by ctrl F.

INVISIBLE CHARACTERS
By default, placeholder symbols for spaces are not shown in most editing programs, but it is sometimes helpful to see these and other invisible characters, such as tabs. A drop-down menu in TextWrangler allows you to turn on this option by selecting **Show Invisibles** and then **Show Spaces**. In older versions of the program, the menu is under a different small icon, and in other programs you might have to search the help to find the equivalent command. For the moment, turn on both invisibles and spaces, and you should see little place-holders in your text file.

Here, the same replacement will give you:

```
A. elegans
Frilla. vityazi
Corda. tottoni
```

Equally awkward will be the situation where the search term occurs outside the instance you want to replace:

```
Mus musculus
```

Replacing this text (using **Replace All** and the query us) will cause more problems than it solves:

```
M. m.cul.
```

Within the next few pages, you will be able to build regular expressions to handle all these cases and many more.

Your first wildcard: \w for letters and digits

Some programs do not have \w. See Appendix 1 or use jEdit.

To add more flexibility to your searches, you can use a **wildcard**. A wildcard is a special character that represents a specific variety of characters, such as any numeric digit. There are different wildcards that match different ranges or sets of characters, many of which are listed and explained in Appendix 2.

There are several different notations for using regular expression wildcards. Most commonly, wildcards are represented by a letter preceded by a backslash. The first wildcard introduced here is \w. It matches any letter (A–z) or digit (0–9), as well as the underscore character (_).

To test this wildcard, type the following latitude and longitude into a new blank document:

```
+40 46'N +014 15'E
+21 17'N –157 52'W
```

In this case, imagine that you want to get rid of all the trailing direction indicators, for example north (N) and east (E). You could run four different searches to get rid of N, E, S, and W in turn. Instead, to create one search term that covers all these cases, use the \w wildcard. Remember that \w does not match just alphabet characters, but also digits. If you search and delete \w, you will end up with:

```
+ ' + '
+ ' – '
```

(Try it, then undo to get the original text back.) To get around this problem, take advantage of the observation that the letters you want in this case always come after a single tick mark. So you can use this search query:

```
'\w
```

> **SPECIAL PUNCTUATION** The data tick marks (') in latitudes and longitudes, or in references to chemical structures (5'-AGCT-3'), are different from the "curly" quote mark (') used in contractions such as won't as well as in quoting text. Most text editors create tick marks when you use the double- and single-quote keys on the keyboard, but word processors often employ a "smart quotes" feature which automatically inserts curly quotes instead. On Mac OS X, you can insert curly quotes using the [option] and [shift] keys in combination with open and close brackets: [option]] makes opening single quotes and [option] [makes opening double quotes, while adding the [shift] key in either case makes the appropriate closing quotes.
>
> The degree (°) symbol, which you may encounter in latitudes, longitudes, angles, and temperatures, can be entered on Mac OS X using the key combination [option] [shift] 8. (Remember this as an alternative to *.)
>
> In general, if there is a special punctuation character in the text you are working with and you would like to include it in your search term, it is usually easiest to simply copy the character from the document and paste it into the search box. This will ensure that you are using the right character—far from a trivial concern, since many punctuation marks look quite similar. It also avoids the need to hunt down the particular keystrokes required to create obscure characters.

and replace with just a tick mark by itself, leaving:

```
+40 46' +014 15'
+21 17' -157 52'
```

As you can see, this searches for any letter, number, or underscore that comes immediately after a tick mark, and replaces it with the tick mark itself.

Note that the letters used to form wildcards are **case sensitive**, which is to say that \W does not mean the same as \w (in fact, it means the opposite!). On the other hand, letters in a search term which are *not* wildcards will only match the same letter of the same case, so searching for agalma will usually not find Agalma.

Capturing text with ( )

Perhaps the most important characteristic of regular expressions searches is the ability to use parentheses to capture portions of the original text and use them in

> **TROUBLESHOOTING** If you can't get your search and replace commands to work, there are a number of things to check. Your text editor must support grep searches, and you should make sure you have **Use Grep** selected. Capitalization and spaces matter, so check for unnoticed spaces at the beginning or end of your search term. (You can highlight all the text in the search term to reveal spaces.) While creating search terms, remember that just because you can't see spaces doesn't mean they can be ignored. They are treated just like any other character. **TextWrangler** will help clarify the structure of queries in the search dialog box by highlighting special characters in red and wildcards in blue, leaving normal characters in black. Other editors may have slightly different wildcards, as described in Appendix 1.

creating the replacement text. Datasets often have extraneous characters which make it difficult to import the files as-is into a graphing program or spreadsheet.

As a simple example, imagine you have a list of ordinal numbers:

```
5th
3rd
2nd
4th
```

You want to delete the letters after the numbers and have just the numbers remain in a column. It is often helpful to think in general terms about what you are hoping to achieve with a search and replace, before trying to translate that into a regular expressions term. Drawing from what you have learned so far, you could write a search term like:

```
\w\w\w
```

which would match the three characters in sequence. You want to keep the first character (that is, just the number) and remove the last two. If you try to just search for \w\w, it will match 5t, then 3r, etc. Regular expressions are non-overlapping. For instance, \w\w wouldn't match 5t and then th in the first line since the t would be in both matches.

The solution is to capture certain parts of the text that the search term finds and put these into the replacement term. To do this, place parentheses around the parts of the search term you want to save. In the example above, you can capture the first character (the number) of the three as follows:

```
(\w)\w\w
```

Other text editors and some programming languages use $1, $2, etc., instead of \1, \2. Confirm the behavior of the editor or language you are using with a simple test before spending time trying to figure out why things are not working.

To place that captured portion in the replacement term, use the backslash character, followed by a number indicating which set of parentheses you extracted the text from. In this case, you just have one set, so the replacement term would be:

```
\1
```

Searching with (\w)\w\w and replacing with \1 on the list of ordinal numbers above gives:

```
5
3
2
4
```

Try it out in TextWrangler and move the parentheses to capture different parts of the three characters.

To capture two parts of the text, you could use a search query like (\w)(\w\w). In this case, the number will be inserted where you place \1 in your replacement term and the two letters that follow will be inserted where you place \2 in the replacement term.

Remember that both search terms and replacement terms can contain a combination of normal text and regular expressions. For example, the replacement term `Position: \1` converts the list to:

```
Position: 5
Position: 3
Position: 2
Position: 4
```

This example query will not work properly for numbers of two or more digits, but later you will see how to accommodate many more general cases.

Quantifiers: Matching one or more entities using +

Wildcards such as \w provide the flexibility of changing different types of characters at once. To be really useful, however, a search term should also be able to handle different counts of characters as well, and then be able to modify the replacement text depending on what it finds. By default, each wildcard matches a single character. Methods of adjusting the number of times an element such as a wildcard matches are called **quantifiers**. Plus (+) immediately after a character indicates that the term should match one or more times in succession. For example, the term \w+ could represent a single character, such as the letter a, or many characters, such as the number 123.

To illustrate further, let's return to the first example of replacing genus and species names with their abbreviated forms. Here is our original list:

```
Agalma elegans
Frillagalma vitiazi
Cordagalma tottoni
Shortia galacifolia
Mus musculus
```

> Instead of retyping text snippets from scratch each time you modify them, simply bring the text document to the foreground and choose **Undo** (⌘Z) to return them to their original state. You can also test search terms with **Find** rather than with **Replace**. Type ⌘G, the shortcut for **Find Next**, to find each instance if there are multiple matches.

You can use \w+ by itself to generate a search term that will match an entire word anywhere in this list of names. Searching for one or more word characters will match with any sequence of such characters found together up until the next non-word character (such as a space, punctuation, or the end of the line). To preserve the first letter of each word, add another non-repeated \w before it, and capture that first letter with ( ).

```
(\w)\w+
```

To generate the replacement term, use the captured letter (represented by \1) followed by a period:

```
\1.
```

You could test this search and replace against the example list now, but you can probably imagine why it won't work correctly yet.

There are several ways to fix the search term so it *does* work; which way is best will depend on the rest of the data file you are modifying. In the present case, the search term will be modified to capture the second word independently—that is, just the species name—so that you can add it back into the replacement text.

Try doing Replace All using the following search and replace pair:

Find	Replace
`(\w)\w+ (\w+)`	`\1. \2`

This combination will generate an abbreviated genus and species list as below:

```
A. elegans
F. vitiazi
C. tottoni
S. galacifolia
M. musculus
```

Notice that the second captured term is \w+, with the quantifier placed inside the parentheses, so all of the characters up to the next non-word character (in this case, an end-of-line character) are preserved and can be recovered with \2. Any spaces occurring in the text have been typed directly into the search query.

For jEdit use $2.

Analysis programs often require a shortened version of the taxon name separated only by an underscore. It should be clear how to modify the replacement term above to create a name in the format A_elegans. You can even reuse the captured text, so as to preserve the original genus and species pair while at the same time generating a shortened version:

Find	Replace
`(\w)(\w+) (\w+)`	`\1\2 \3 \1_\3`

```
Agalma elegans A_elegans
Frillagalma vitiazi F_vitiazi
Cordagalma tottoni C_tottoni
Shortia galacifolia S_galacifolia
Mus musculus M_musculus
```

With the three building blocks of regular expressions you have learned—wildcard, quantifier, and capture terms—you can already do some very powerful manipulations.

*Escaping punctuation characters with *

With all these special uses for + and ( ), you might begin to wonder how you would search for these characters themselves in your text. Again, the backslash is used to modify how a character is interpreted. To remove the special meaning of punctuation in a search term, put \\ before the character. This is a general trick you will see in other contexts, and is referred to as **escaping** the character. This technique even applies to the \\ character itself: to search for \\, use \\\\. In fact, so many punctuation marks have special meanings that it is often a good idea to use a preceding \\ whenever you want such a mark to be taken literally. When escaping, the \\ has the opposite effect that it has with letters, where it *gives* them special meaning, as with \\w.

For example, to obtain the final element in `Physalia physalis (Linnaeus)`, start by generating a search to match each part of the text:

    \w+ \w+ \(\w+\)

In this formulation, none of the text will be captured to \\1 \\2, etc., because the parentheses are escaped by the preceding backslashes. Now just place parentheses around the word characters in the final element:

    \w+ \w+ \((\w+)\)

The variable \\1 now contains the text `Linnaeus`.

As you will have noticed, the appearance of regular expressions can get very confusing very quickly. It is almost always easiest to copy an actual example of the text you will be searching, and paste two copies of it into your text editor window. Then progressively edit one of the copies into the search term. Once you have a good search term typed into the text window, copy it and paste it into the search box of the Find dialog. Provided that Use Grep is checked *before* you paste, the search should be interpreted correctly. If it is not checked, on the other hand, the program may escape out all your punctuation for you, so that your wildcards and quantifiers are interpreted as normal characters instead.

Here is a similar step-by-step approach to generating a complex search term, using the same example text. Spaces have been inserted in the first steps to make the elements more distinct:

Original text	`Physalia physalis (Linnaeus)`
Search with matches	`\w+ \w+ \( \w+ \)`
Text with captures	`(\w+) (\w+) \( (\w+) \)`
No extra space	`(\w+) (\w+) \((\w+)\)`
Replacement text	`\1_\2_\3`

Applying the search leaves the text without parentheses:

    Physalia_physalis_Linnaeus

TABLE 2.1 Some of the most common search wildcards	

Search term	Meaning
\w	A **word** character, including letters, numbers, and the underscore
\t	A **tab** character (can also be used in replacements)
\s	A **white space** character, including spaces, tabs, and the end-of-line
\r \n	**End-of-line** markers (can also be used in replacements) TextWrangler uses \r, but jEdit and Python programs will use \n
\d	A **digit**, from 0 to 9
.	**Any** letter, number, or symbol, except end-of-line characters

More special search terms: \s \t \r . \d

In addition to \w, there are many other wildcards and special characters. A few of the most general of these are listed in Table 2.1, and a more complete list is provided in Appendix 2.

The tab character (\t) is widely used in regular expressions searches because tabs often separate columns of data from each other. By inserting tabs between captured text in the replacement term (for example \1\t\2\t\3 in the genus-species-author search), you can rapidly reformat a plain text file into spreadsheet-suitable format. This works with data copied from PDF files or Web documents—anywhere that information occurs in a predictable manner.

As an example, consider latitude and longitude data that might be generated by a GPS (note that there is a single space between the degrees and minutes):

```
-9 59.8'S -157 58.2'W
+21 17.4'N +157 51.6'W
+38 30.5'N +28 17.2'W
+40 46.1'N +14 15.8'E
+10 24.8'N +51 21.9'E
```

To get these into a spreadsheet with different fields in different columns, the goal is to capture each degree–minute pair and separate them using the replacement \1\t\2\t\3\t\4. For the moment, it is okay to preserve the positive and negative symbols at the beginning of each value.

The latitude and longitude formats are the same, so one general search term can be duplicated to handle both. The job of this search term is to capture all the values before the space, then the values between the space and the tick mark (').

First look at some of the challenges to generating a universal query for these data:

- The symbol right before the number can be either a plus or a minus (+ or –).

- The number of digits used to specify degrees can have one, two, or three digits, so this calls for a flexible quantifier (+).

- There is a decimal place in the middle of the minutes value, which must be accounted for (and escaped so it is not read as a wildcard).

- The character at the end of the field can be either N, E, S, or W.

This is a lot to think about, but well within the abilities you already have. To begin, copy one of the lines into a new document so you can begin substituting regular expressions text for the original text. Then put parentheses around the items to capture:

Original text	`+38 30.5'N`
Mark captures	`(+38) (30.5)'N`
Regular expression with wildcards	`(.\d+) (\d+\.\d)\'\w`
Replacement text	`\1\t\2\t`

In order to read regular expressions, look at the backslash and the character that follows it as a single entity. Otherwise, it becomes too confusing, especially with parentheses and periods. Reading the search expression in the third line from left to right would translate to something like:

"Any character followed by one or more digits, to be saved as \1. Next, a space symbol. Then save as \2 one or more digits followed by a decimal point, followed by a single digit. Finally, a tick mark and a single word character; these go uncaptured and are therefore deleted."

To search for the longitude values that follow, duplicate this general term, separating the fields with one or more white space characters:

`(.\d+) (\d+\.\d)\'\w\s+(.\d+) (\d+\.\d)\'\w`
↳first values   spaces↵   ↳second values

Account for all the captured fields with an expanded replacement term:

`\1\t\2\t\3\t\4`

When generating the search term, remember that a literal period character needs to be preceded by a backslash, and that + applies to the character immediately before it, either by itself or in a sequence of like characters.

By default, TextWrangler and other editors apply the search term line-by-line to a file. If you want to search across lines, you can end the search term with the line ending character (\r, or sometimes \n). This is a way of joining lines, and if you want to preserve line endings, you will have to add that \r to the end of the replacement term.

Example: Reformatting molecular data files

Now that you understand the key elements of regular expressions, you are ready for a larger reformatting job. Regular expressions are very useful when data are provided in one format but are needed in another format. The following protein sequences (available in the examples folder as FPexamples.fta) have headers with the accession number, a description of the protein, and the genus and species of origin in brackets:

```
>CAA58790.1= GFP [Aequorea victoria]
MSKGEELFTGVVPILVELDGDVNGQKFSVRGEGEGDATYGKLTLKFICTTGKLPVPWPTL...
>AAZ67342.1= GFP-like red fluorescent protein [Corynactis californica]
MSLSKQVLPRDVKMRYHMDGCVNGHQFIIEGEGTGKPYEGKKILELRVTKGGPLPFAFDI...
>ACX47247.1= green fluorescent protein [Haeckelia beehleri]
MEFEPEFFNKPVPLEMTLRGCVNGKEFMIFGKGEGDASKGNIKGKWILSHSEDGKCPMSW...
>ABC68474.1= red fluorescent protein [Discosoma sp. RC-2004]
MRSSKNVIKEFMRFKVRMEGTVNGHEFEIEGEGEGRPYEGHNTVKLKVTKGGPLPFAWDI...
```

Instead of this format, it might be preferable to work with these sequences if the names were shortened to preserve just the initial identifier and the genus name, while removing spaces and leaving the sequence information untouched:

```
>CAA58790_Aequorea
MSKGEELFTGVVPILVELDGDVNGQKFSVRGEGEGDATYGKLTLKFICTTGKLPVPWPTLV...
>AAZ67342_Corynactis
MSLSKQVLPRDVKMRYHMDGCVNGHQFIIEGEGTGKPYEGKKILELRVTKGGPLPFAFDIL...
```

You can use the fact that each name begins with > and the species names are in [ ] to construct a query that only modifies the headers and not the sequences.

Start by copying a line of data into a new file and identifying the portions that you would like to capture:

Original text	>CAA58790.1= GFP [Aequorea victoria]
Mark the captures with ( )	(>CAA58790).1= GFP [(Aequorea) victoria]
Add wildcards (extra space for clarity)	(>\w+) .+ \[(\w+) .+
Final query (no spaces)	(>\w+).+\[(\w+).+
Replacement	\1_\2
Result	>CAA58790_Aequorea

This query finds and saves the first word after the > symbol, up to the period, which is not in the \w character set. The period is used as a wildcard to match everything up to the next square bracket, indicated by \[. The first word in the brackets is captured as \2 and the rest of the line after the space is discarded. It is a little confusing that the period is acting as a wildcard, yet in this situation matches an actual period. Also, in the search query, the > does not need to be escaped with \, but it wouldn't hurt to do so.

For a replacement string, you can join the captured identifier (which includes the > since it is inside the parentheses) and the genus using an underscore.

Comments about generating regular expressions

Regular expressions searches are picky. Although you can write an expression to capture almost any type of text, if you specify something and it is not there, the search will fail. Getting one part of the query wrong often leads to the entire query failing to match. It is therefore important to anticipate the full range of variability in your data.

Another problem can occur if hidden characters find their way into a regular expressions Find box; typically this happens when pasting text from the document into the box. Check for spaces and extra return characters both in your document and in the query itself—they won't be visible in the Find box, even if Show Invisibles is checked in the document options.

A table of regular expression terms and their usage can be found in Appendix 2.

> A general approach to debugging your expression is to cut out various portions of the search, and test different subsets of the search to see where it fails when you add something back in. It can be informative to do **Find** without **Replace** to highlight the matching text. Proofreading a manuscript requires close attention, but proofing regular expressions often requires even more careful scrutiny of every character.

SUMMARY

You have learned how to:

- Create search and replace queries and apply them in a text editor
- Use the following wildcards and special characters:

  \w for letters, numbers and the underscore

  . for any character except line breaks

  \d for numbers

  \r (or \n) for line breaks

  \s for spaces, tabs, and end-of-line characters

  \t for tabs

- Capture portions of the search with ( )
- Reuse captured text with \1
- Escape punctuation and special characters with \
- Use the plus + quantifier to repeat a character or wildcard

Chapter 3

EXPLORING THE FLEXIBILITY OF REGULAR EXPRESSIONS

Now that you understand some of the different categories of tools and approaches to regular expressions, you will explore the flexibility of the language in greater detail. We will show you how to tailor queries to accomplish specialized tasks and address many text manipulation challenges you are likely to face.

Character sets: Making your own wildcards

Defining custom character sets with [ ]

Although the wildcards you now know are sufficient for many common searches, it is often necessary to create your own special wildcards. For example, there is no built-in regular expressions wildcard which represents only the letters of the alphabet. You can create wildcards by designating a set of characters inside square brackets [ ]. The characters within the brackets are treated as a single character in the search. For example [AGCT] matches just one A or G or C or T. Just as with \w, if you want to match one or more characters in a row, you would put a plus sign after the closing bracket:

[AGCT]+

You can also indicate a range of numbers or letters so you don't have to specify each one, or you can even mix individual characters with ranges. Any uppercase letter is matched by [A-Z], while [0-9\.] matches any digit or a decimal point. The expression in the brackets is case-sensitive, so to match any letter (lowercase or uppercase) specify [A-Za-z]. Note that there is no separator between the two ranges; each dash is defining a range between the two characters adjacent to it. You can of course include a dash in the character set by escaping it, so [A-Z\-] matches any uppercase letter or a dash.

> **CHARACTER RANGES** These are based on the order in the ASCII character set
> (see Chapter 1 and Appendix 6). This makes it possible to specify a single range that
> spans both upper and lowercase characters, in the form [A-z]. However, this range
> will also include the six punctuation marks [\ ] ^ _ ` which fall between z and a in
> the ASCII character set. Consult the table in Appendix 6 to see what ranges are both
> permissible and useful.

Applying custom character sets

It is often possible to perform what would at first seem to be a complex reformatting task by breaking it into a sequence of a few simple search and replace operations, one after the other. In the next example, you'll convert latitude and longitude values indicated by N, E, S, W into positive and negative values. You'll also join two related lines of data into one line.

Our earlier latitude and longitude example was a bit unusual in that the starting data had both +/– at the beginning and N, E, S, or W at the end. Usually you would only have one or the other. The following data set (available as LatLon.txt in the examples folder) has five locations, with latitudes and longitudes on consecutive lines:

```
 21 17'24.68"N
157 51'41.50"W
 38 30'36.62"N
 28 17'16.87"W
  8 59'53.30"S
157 58'13.70"W
 10 24'47.84"N
 51 21'54.61"E
 22 52'41.65"S
 48 9'46.62"E
```

Start writing an expression to combine each pair of latitude and longitude lines into a single tab-delimited line using positive and negative values. For example, the record specified by the first two lines will become:

```
 21 17'24.68"    -157 51'41.50"
```

Notice that you want to remove the invisible end-of-line character \r from lines that end with "N or "S, but not from lines that end with "E or "W. This can be accomplished by searching for:

```
(\"[NS])\r
```

and replacing it with:

```
\1\t
```

[NS] means "either N or S," and because \r is outside the ( ), it is not part of the replacement \1, so the line breaks will be replaced by tabs.

Try \n for \r.

   In this example, you could search without the preceding \", and you could also write a query for values that end with single quotes. By using the quote marks instead of the letters alone, it makes our query more robust if the data file includes other names, but less flexible in situations where the seconds (") might not be specified.

   After this initial search and replace, the data appear as follows, with the value pairs joined to form single lines:

```
21 17'24.68"N     157 51'41.50"W
38 30'36.62"N      28 17'16.87"W
 8 59'53.30"S     157 58'13.70"W
10 24'47.84"N      51 21'54.61"E
22 52'41.65"S      48 9'46.62"E
```

Next, use the compass points (characters N, S, E, or W) to decide whether to add a minus sign (–) before removing the compass points from the file. Latitudes to the south of the equator and longitudes west of the prime meridian are by convention negative, so you can find coordinates followed by W or S and replace them with the same coordinates preceded by a minus sign.

   The first part of the query is any combination of the digits and symbols used to indicate latitude or longitude values, but it matches only if they are followed by a W or S:

```
([0-9]+ [0-9 \'\"\.]+)[WS]   ← This matches either W or S, not both
                                together
```

Replace with –\1:

```
21 17'24.68"N     –157 51'41.50"
38 30'36.62"N      –28 17'16.87"
–8 59'53.30"      –157 58'13.70"
10 24'47.84"N       51 21'54.61"E
–22 52'41.65"       48 9'46.62"E
```

To remove north and east, search for [NE] and replace with nothing. You do not need to capture any text.

   The time it takes to complete these three commands in succession on a single file with many entries will be much faster than doing it by hand. Of course, if you have to do such conversions on many different files, or repeatedly over a long time, you will want to write a program or script that automates the full sequence of modifications. In later chapters, we will show you how to do exactly that.

Negations: Defining custom character sets with [^]

You can place as many characters as you like within [ ], but this can quickly get out of hand. If the character set is large you may forget something. Thus it is sometimes easier to specify the characters you *don't* want to match. This can be accomplished using the caret (^, typed by [shift] 6 on U.S. keyboards) as the first character inside the square brackets. In these cases, the bracketed term matches any character (letters, numbers, punctuation, and white space including the end-of-line \r) *except* those that follow the caret. For example, searching for [^A-Z\r] and replacing it with x will x-out anything in the document except capital letters and the end of the line.

This may seem like a relatively esoteric trick, but it has at least one very important use: it is a good shortcut for recognizing the columns of a character-delimited file. The delimiter is just a character that is reserved to separate the values of each column, most commonly a tab or comma. For instance, because \t specifies a single tab character, each row in a three column tab-delimited file has this format, where X, Y, and Z are the data values:

```
X\tY\tZ\r
```

The act of pulling individual pieces of data from a line like this is called **parsing**, and this is a critical skill for analyzing data. To separate tab-delimited data quickly, you can use the negation character to build a wildcard [^\t]+ that matches "anything except tab."

Using this approach, a general search term for capturing a column of tab-delimited data would be:

```
([^\t]+)\t
```

In other words, "anything except a tab, followed by a tab." There are tabs between each field, not after each field. In a line with three columns of data, for instance, there are only two tabs since no tab follows the final column.

To pull values out of tab-delimited files, copy and paste this once for each column of data in your data set, which will load your data into \1, \2, \3, etc. This provides a very simple way to change the order of columns in a text file, or to modify certain columns. If one program outputs data in a tab-delimited text file in the order X\tY\tZ and another program expects these data in the order Z\tX\tY, you can search for ([^\t]+)\t([^\t]+)\t([^\t]+) to capture the values and then write them out in the new sequence with \3\t\1\t\2.

Note that since you are using uncaptured tabs in the search term to demarcate the text you want to capture, it is necessary to include tabs in the replacement term or they will be removed. Of course, you can use another delimiter in the replacement term if you wish. For instance, to switch the column order and change the delimiter to a comma in one step, you would use a replacement term of \3,\1,\2. You would also want to use \r either at the end or within the square brackets, because the end-of-line character is also included in [^\t].

As written, ([^\t]+)\t([^\t]+)\t([^\t]+) will match only the first, second, and third columns of each line. This is because regular expressions match the first instance of the specified pattern that they encounter, and usually each non-overlapping instance of the pattern after that (i.e., the last part of one match can't be the first part of the next match). If there are only three columns, this regular expression will match all three. If there are more than three columns but less than six, it will match the first three and leave the others untouched. If there are more than six columns, it will match two or more times, depending on the actual number of columns. If there are less than three columns, it won't match at all since no line would contain the full search term. Sometimes you may want to explicitly match only lines that contain exactly three columns, match only the first three columns no matter how many columns there are after that, or even anchor the search to the end of the line to modify only the last columns. The next section will give you the tools required for all these tasks.

Boundaries: ^beginnings and endings$

Some regular expressions symbols don't correspond to characters, but instead match the boundaries next to characters at certain places in a word or line. The most useful of these are shown in Table 3.1.

The ^ can be a bit confusing since it has different meanings in different contexts. We've just told you that it is used to negate character sets, but that was within square brackets. Outside of brackets it has this different meaning of matching the beginning of a line, right before the first character. The $ matches the end of the line, just before the special line-ending character. If you "replace" the beginning or ending anchor (you aren't really replacing it, because the beginning of the line is still there), it has the effect of inserting a character before or after the rest of the line. In contrast, if you replace the line-ending \r you'll end up combining multiple lines into one.

An example of using ^ would be to add a leading column that says `Sample` to each line of a tab-delimited text file. To do this, search for ^ and replace it with `Sample\t`. You can also be more specific than this. For instance, you could append a > to the beginning of lines that start with a digit but not other lines by searching for ^(\d) and replacing it with >\1.

Boundary matching has another important function beyond tacking text onto the beginning or ending of a line: it is a convenient way to anchor searches. For example, your tab-delimited file may identify illegal or missing values using the placeholder –1 in the first column, but you want to read it with another program that expects `NaN` ("Not a Number," a standard symbol for illegal values, or sometimes inapplicable or missing data, used by many programs). In other columns, a –1 could be part of a legitimate value that you don't want to replace with `NaN`. Here you would use the anchored search term ^–1\t and replace with `NaN\t`. This will only find that combination when it occurs at the beginning of a line. You don't need to include ^ in the replace term since the process doesn't eliminate the boundary.

In the earlier example of replacing genus names with the first initial and a period, you had to capture the species name because you didn't want it replaced with a period. Now, with anchoring, it is easy to write a robust query which will only replace the first word

TABLE 3.1 Boundary terms for regular expression queries	
Term	**Matches**
^	The boundary at the beginning of a line
$	The boundary at the end of a line

with its initial, even if other words or fields follow after. Here is the example text we worked with before:

```
Agalma elegans
Frillagalma vitiazi
Cordagalma tottoni
Shortia galacifolia
Mus musculus
```

This time, use this search term:

```
^(\w)\w+
```

and this replacement term: `\1\.`   ← Note the escaped dot which replaces the characters after the first

to obtain this result:

```
A. elegans
F. vitiazi
C. tottoni
S. galacifolia
M. musculus
```

Adding more precision to quantifiers

Another quantifier: * for zero or more

Whereas the + quantifier indicates that the preceding element can match one or more times, the * quantifier indicates that it can match zero or more times. This is useful, indeed critical, if you are unsure whether the element will be present in all cases.

At times, you may want to grab part of a line and not spend time parsing the rest. Here, the all-inclusive wildcard .* can be very useful. If you put this dot-asterisk after a more specific search expression, it will match everything left over by the query, up to the end of the line. (Remember that . matches everything except \r or \n.) You can either capture this information and re-use it in the replacement query, or delete it if the data are unwanted.

Modifying greediness with ?

One important aspect of the + and * quantifiers is that they are "greedy," meaning they match the maximum number of characters that they can. This can sometimes cause surprising and difficult to understand results. For example, you might want to search through a sequence and find all the characters up until a series of repeated As at the end, and then delete that tail of As:

CCAAGAGGACAACAAGACATTTAACAAATCACATCTTTGTATTTTTGGTTAGAGTTGAAAAAAA

Your first inclination might be to search for (\w+)A*, or ([ACGT]+)A* and replace with \1. If you try this, though, you'll notice that your A* term (zero or more

As) doesn't find anything! Even if you use A+ (which runs the risk of failing if the sequence doesn't end in A at all) you will only trim off the last A of the sequence. Remember, regular expressions are eager to please, and they try to match as much of a line as possible. Quantifiers likewise also try to match, left to right, as many characters as they possibly can. In this case, \w+ is able to match the final As, while still remaining consistent with a match by A* (zero or more). To get around this, you could specify that there must be a "non-A" character before the final A*, which leaves A* with something to match:

(\w+[CGT])A*   ← This works as long as there are no gaps or other unanticipated characters

Another way to do it, though, is to force the first quantifier to produce a *minimum* match by adding a ? after the + sign:

(\w+?)A*   ← This won't work quite yet

This search term represents the shortest series of letters which match a word followed by zero or more As. It still won't work because a single C by itself would match that term minimally. To make it extend to the end of the sequence, you have to anchor the final string of As using the $:

(\w+?)A*$

How you solve these problems often comes down to a matter of personal preference, but you should be aware of potentially unanticipated consequences of both greedy and non-greedy quantifiers.

Controlling the number of matches with {}

We've shown you how to match an entity once (with just a single character, character set, or wildcard), one or more times (by adding + after an entity), or zero or more times (by adding * after an entity). It is often necessary, though, to have finer-scale control than this, such as defining the exact number of times an entity should match in a sequence, or the minimum and maximum number of times an entity can match.

You can do this with curly brackets {}. The curly bracket term is placed after a character or character set. For example, {3} matches the preceding entity exactly 3 times, while {2,5} matches the preceding entity at least 2 times but no more than 5 times. You can also leave out the maximum value: {3,} matches at least 3 times, but more if it can. You can see additional examples in Table 3.2.

TABLE 3.2 Some examples of using curly brackets to control the number of matches

Start	Search	Replace with	Result
CTAAAAGCATAAAAAAAAAAAA	A{8,}		CTAAAAGCAT
34.2348753443	(\d+\.)(\d{3})\d+	\1\2	34.234

Putting it all together

Imagine that your colleague has sent you a data file (available as examples /Ch3observations.txt) with dates and variables in the following format:

```
13 January, 1752 at 13:53△ -1.414△ 5.781△ Found in tide pools
17 March, 1961 at 03:46△ 14△ 3.6△ Thirty specimens observed
1 Oct., 2002 at 18:22△ 36.51△ -3.4221△ Genome sequenced to confirm
    ...800 more lines like that...
20 July, 1863 at 12:02△ 1.74△ 133△ Article in Harper's
```

Data are separated by a combination of tabs (represented by the little grey triangles) and spaces. The month names may or may not be abbreviated, and the X and Y values can be positive or negative. The last column contains comments that can be discarded. Assume that our goal is to rearrange the data into columns containing the following items, separated by tabs. As an added challenge, you will switch the position of the year and day values from how they are presented in the file. Our desired output fields will look like this:

Year	Mon.	Day	Hour	Minute	X data	Y data

This operation requires separating values from each other based on several different properties. The first step is to look at the lines of data and think what you can use to identify and separate the values. At this point, don't even think about how you will phrase the actual search term, just describe the characteristics in words.

If you separate the fields just by spaces and tabs, you will end up with the following elements:

13	January,	1752	at	13:53	-1.414	5.781	Found in tidepool

This is a promising start. Now put parentheses around the parts you want to keep, and then put numbers to go with the parentheses. This is still being done on the big-picture scale of just thinking about what you have relative to what you want:

(13)	(Jan)uary,	(1752)	at	(13):(53)	(-1.414)	(5.781)	etc.
1	2	3		4 5	6	7	

Some things to notice: In group 2, you only want to save the first three characters, which constitute the month abbreviation. There are no group numbers for the "at" or the comments at the end of the line because you don't want to save that in the final file and therefore don't capture them. In the time field, you want to save hours and minutes separately. Finally, you want to save the X and Y data fields.

This is a lot to think about at once, but if you go through piece by piece like this, it becomes a more tractable problem.

Now that you have a clear view of the "keepers" from the original file, you need to figure out how to rephrase them as a query. Take the specific characters from the example line and replace them with wildcards as needed. First take care of the characters inside the parentheses and then move on to everything between:

(13)	(Jan)uary,	(1752)	at	(13): (53)	(−1.414)	(5.781)	etc.
(\d+)	(\w\w\w)[\w\,\.]*	(\d+)	at	(\d+):(\d+)	([−\d\.]+)	([−\d\.]+)	.*
1	2	3	4 5	6	7		

Taken together, the second line of this table—our emerging query string—looks quite complicated. But each element on its own is not that bad. Group 1 is pretty straightforward: the day should always be one or more digits.

Group 2 will be three or more letters, with an optional period or comma (or both) after it. You only want to capture the first three characters. Because the period is treated as a wildcard by default, you need to escape it with a preceding backslash, and for good measure you also escape the comma.

Skip the at and move on to the time in groups 4 and 5, which should always be some digits separated by a colon. The data values that follow can be positive or negative, so our set in the brackets includes digits, a dash, and a period (since the number may have a decimal point).

Before forming the replace string, take another pass through the query and deal with the spaces and intervening characters which you don't want to keep:

(13)	(Jan)uary,	(1752)	at	(13):(53)	(−1.414)	(5.781)
(\d+)\s+(\w{3})[\w\,\.]*\s+(\d+)\sat\s(\d+):(\d+)\s+([−\d\.]+)\s+([−\d\.]+).*						
1	2	3	4 5	6	7	

The spaces require inserting \s+ (meaning one or more white space characters) between the other elements. You could also use actual spaces.

At this point in your query building, or even before, it is good to run a test to make sure you haven't missed something. You can open your data file in TextWrangler and type the query into the Find field (check again that Use Grep is turned on). Then click the find button and make sure that the whole line gets highlighted. This means that the text has been found. From there you can hit ⌘G to repeat the last find, and each line of the data file should highlight in succession.

Take a minute to look back at the search query, and appreciate how utterly inscrutable this pattern appears. Building it up part-by-part made it easier for you to keep track of things as you went. This divide-and-conquer approach is one of the most important work practices you can take away from this book.

Generating the replacement query

Despite all this work, you aren't yet done. You still need to carry out the replacement, so that you can generate the desired output. Fortunately, generating a replacement query is usually easier than making the search query.

Recall that the parts of your query between parentheses get saved in memory for use in generating the replacement text. You can access these elements in your replacement text using a backslash followed by the number of the group.

To just replace the original text with the fields in the order they were found, but separated by tabs, the following would do:

```
\1\t\2\t\3\t\4\t\5\t\6\t\7
```

The required replacement is a bit more complicated, because you need to rearrange some of the variables in the output.

Here are the search groups in the order that they appear in the original dataset:

Field:	Day	Mon	Year	Hour	Min	X	Y
Group Number:	1	2	3	4	5	6	7

The next step toward generating the replacement string is to rearrange the fields into the right order. After that, insert any desired punctuation (in this case a period after field 2), and finally insert the tabs which will let the fields be imported into separate columns:

Year	Mon.	Day	Hour	Min	X data	Y data
\3	\2\.	\1	\4	\5	\6	\7
		\3\t\2\.\t\1\t\4\t\5\t\6\t\7				

This replacement string will let you use a single search and replace command (albeit complex) to reformat, edit, and reorganize your data files. Here we see the result:

```
1752△ Jan.△ 13△ 13△ 53△ −1.414△ 5.781
1961△ Mar.△ 17△ 03△ 46△ 14△ 3.6
2002△ Oct.△ 1△ 18△ 22△ 36.51△ −3.4221
```

 When working with data, you don't have to accomplish all the substitutions in one step. You can often do something with multiple steps—although if it is a series of operations you do repeatedly, you might want to consider revisiting the process after reading the Python programming chapters. You can also sometimes use a handy trick of creating an intermediate replacement with a unique placeholder (for example, ZZZ or ### to represent a line break), which you then replace in a later search operation.

Constructing robust searches

If you have been following along and typing the examples, or trying some of these tools out on your own data, at least a few regular expressions will probably have failed due to typos or other problems. Failure can result from mistakes in designing the query, errors in the data file, or not anticipating the full range of potential variation within the data file. Since at least some regular expressions you try will fail some of the time, it is important to think not only about how to make your queries work, but how to make them fail. This isn't to say you don't want to design them to work, but that you want to design them to fail in particular ways when they don't work. It is preferable for a search to fail entirely than for it to seem to work, but fail in subtle and imperceptible ways.

When a query fails, there are two possible results: First, it can choke entirely, failing to provide any output. Second, it can make unanticipated and incorrect replacements. In the latter case, you may not even notice, and this will jeopardize your analyses. Because this can be disastrous, it is often best to overspecify your searches beyond what is required to accomplish a task. Overspecification involves using as many elements in your search as there are in the search line, and making them as specific as possible. For example, instead of capturing (.+) at the end of the line, spell out the fields that you expect to see there. This is most important for complex queries that are difficult to understand at one glance, as well as for queries that will be applied to large datasets, meaning that only a small fraction of the results can be verified by eye; it is also important in cases where the consistency of the input file is dubious, such as when the file format is unclear and you are limited to making a best guess, or when the data were entered by hand and there is a good chance of typos.

To make your assumptions explicit, add in some redundancies. For instance, if you think that headers are designated with the character > at the start of a line, look for ^> instead of just > or just the start of the line (^). This is because if you only use >, you could be misled if > also occurs within the header or in some other part of the file.

Another easy way to make your query more robust is to force the search to match the entire line. You can do this by including terms to match even the parts of the text you aren't processing, or by anchoring queries with ^ and $ at each end. This is a good way to increase the chances your queries are behaving as you expect.

You should look for other ways to add redundancy once your basic query is working. For example, most programs will report back the number of times a substitution has been made, so be sure to check this against the anticipated number of changes (or the number of lines in the file) to see whether there are some improperly formatted lines buried within the file that the query did not match.

The importance of thinking about the way your custom-built computational solutions will fail, and engineering them to fail in a defined way, extends well beyond regular expressions. Most commercial software goes through extensive testing before it is released, yet every major package still has bugs. Chances are that you will be using the computational tools you make for important analyses after far less testing. In addition, since many types of biological data don't have explicitly defined data formats, you are likely to encounter bad data files on a regular basis. You need to build in safety nets for yourself—putting data in and getting an answer out is not enough to verify that you are on the right track.

SUMMARY

You have learned how to:

- Define your own character sets with [ ]

- Isolate fields of tab-delimited text with ([^\t]+)\t

- Match the boundaries at the beginning (^) and endings ($) of lines

- Define a wider array of quantifiers with * and {}

- Control the "greediness" of your quantifiers with ? to produce a minimum match

- Build up complex queries step by step

Moving forward

Regular expressions, applied in the context of a text editor, can be one of your most versatile and accessible tools (Table 3.3). We use them nearly every day in our own work. You can use regexp to insert, delete, and simplify text as shown in this chapter, and you can also perform other general manipulations. The table found in Appendix 2 can serve as a quick reference guide for your query constructions.

Try to subdivide your task into combinations of these operations. Using them together with some of the other features of TextWrangler, such as Sort and Process Lines Containing... can support advanced file manipulations without writing a script. You can also use the editor's ability to search across multiple files (Search▸MultiFile Search...) as a means of extending your capabilities.

TABLE 3.3 Some common operations that can be performed with regular expression

Operation	Find	Replace
Split elements like `nano_128.dat`	`(\w+)_(\d+)\.(\w+)`	`\1\t\2\t\3`
Merge or rearrange columns of values	`(\w+)\t(\w+)`	`\2 \1`
Join multiple lines into one (\n or \r, depending on the program)	`\r` `\n`	`,`
Split a single line into multiple lines	`,`	`\r`
Transform a list of unique elements into longer items (e.g., URL, table, line of script)	See examples in Chapter 6	
Convert a list of names to move the first name and initials to the end, separated by a comma (won't work with "Jr.," etc.)	`(.*) (\w+)$`	`\2, \1`
Delete relative to the occurrence of X Beginning to last occurrence Beginning to first occurrence Same as above After the first occurrence to the end Up to the last occurrence Caution: Finding no match for x will result in deleting the whole line	`^.*X(.*)` `^[^X\r]*(X)` `^.*?(X)` `(X).*?$` `^.*(X)`	`\1`

PART II

THE SHELL

COMMAND-LINE OPERATIONS: THE SHELL

In this chapter you will learn the rudiments of the shell, a tool for interacting with your computer through typed instructions at the command line. For now the focus will be on navigating your computer from the command line, with later chapters focusing on handling text, using programs, and automating tasks with this powerful interface.

Getting started: Don't fear the command line

In order to appreciate what the shell is, it is useful to contrast it with something more familiar to you: a **graphical user interface**. Graphical user interfaces, or GUIs, are the displays through which most users interact with their computer. Some examples of GUIs include the interfaces for the Windows and Mac OS X operating systems, the KDE and GNOME interfaces for Linux, and X Window. With each of these, you can open and manage files and folders, as well as interact with programs in an intuitive way, using a mouse to move a cursor among icons and menus on the screen.

GUIs have dramatically improved the friendliness and usability of computers for many tasks. These advantages do come with trade-offs, though, some of which are felt far more acutely by scientists than by other users. Potential disadvantages of GUIs include the following:

- Many of the analyses that scientists need to perform consist of a long sequence of operations which may need to be repeated on different datasets, or with slight modifications on the same dataset. This does not lend itself well to GUIs. Everyone has had the frustrating experience of clicking through the same sequence of menus and dialog boxes over and over again.

- Most programs do not keep a log of all the user commands issued through the GUI, and if you want to recreate the steps it took you to analyze your data, they must be documented separately.

- GUIs are not conducive to controlling analyses on a cluster of computers or on a remote machine. This will be a problem when you need to access increased computing power beyond your personal computer for complex or large-scale tasks.

- GUIs are labor-intensive to make and typically work on only the particular operating system for which they were developed.

Fortunately, there is another way of interacting with your computer that avoids many of these problems. This is the oft-dreaded **command line**, an entirely text-based interface. Using the command line requires an up-front investment in a new skill set, but provides a huge net gain in the long run for many data management and analysis tasks. To the uninitiated, the command line can seem like a primitive throwback to the days before the mouse. However, the command line is alive and well, and is in fact the interface of choice for a wide range of computational tools. In many cases it is a more convenient environment for data manipulation and analysis than a GUI could ever be.

We will start by explaining what the command line is and how to navigate your computer from this powerful interface. Later chapters will illustrate what you can do with further commands, particularly for text manipulation. This book is intended to get you comfortable with the command-line environment, but it is not a comprehensive guide. To go further, you can consult Appendix 3 on shell commands, as well as online references and any of the many books devoted to the topic. (We recommend a few in particular at the end of this chapter.)

Starting the shell and getting oriented

Starting the shell

Shells run in what are called terminal emulators, or more simply, terminals. In Mac OS X, the default terminal program is in fact called Terminal, and it is located in the Utilities folder within the Applications folder.[1] Launch the program by double-clicking its icon. The terminal window opens with a greeting, a prompt, and a cursor (Figure 4.1). The greeting is usually brief, indicating information such as when you last logged in. The exact text of the prompt will vary, and can depend upon your computer's name, your user name, your current network connection, and other factors. Throughout this book we are going to assume you are a computer guru with the user name lucy, so the text of the demonstration prompts will include the word "lucy."

Windows users, see Appendix 1; Linux users can use their Terminal program.

---

[1] Another terminal program sometimes installed on OS X is xterm, available through a menu for the X11 application. Since xterm operates differently than Terminal, avoid using it for these sections of the book.

```
Terminal — bash — 80×28
Last login: Mon Dec  8 21:45:23 on ttys003
haeckelia-2:~ lucy$ ▮
```

FIGURE 4.1 A portion of the Terminal window in OS X, showing the greeting, prompt, and cursor

The command line, as the prompt and cursor are known, is displayed within the terminal window by a program called a **shell**. Shells are customizable and powerful—and at times they can seem frustratingly simple-minded. You can issue short commands or write simple scripts that do amazing things, but you can also wipe out a chunk of your hard disk with a few ill-chosen punctuation marks. The shell assumes that you mean what you say, and rarely gives you the chance to opt out of a command you issue or to undo the results.

There are many different shell programs, and each has its own camp of loyal advocates. Shells mostly have the same capabilities, but employ subtly different ways of saying things. OS X has several shells installed that can be used in the Terminal window, but since version 10.3 the default has been the bash shell. We will use bash throughout this book.

It's time to try your first command. To check that you are set up to use the bash shell, at the prompt in the open terminal window type the following:

echo $SHELL ← This name must be in uppercase

The terminal will respond by printing out the name of the active shell, which should be /bin/bash. You can also look at the top of the Terminal window to see if it says bash. If the active shell is not bash, open Terminal preferences, change

COMMAND-LINE INTERFACES ON DIFFERENT OPERATING SYSTEMS The examples in this book assume you are using the OS X operating system provided with Apple computers, which is a Unix operating system. If you are using an operating system other than OS X, see Appendix 1 for specifics about your system. Many other operating systems that are also based on Unix, such as Linux, are similar enough that most of the examples will work as written or with very minor modifications. In fact, the exact same command-line programs are widely distributed on many Unix systems. However, if you are on a Windows system, *you cannot follow along at the DOS or PowerShell prompt*. The Microsoft command-line interface provided with Windows operating systems is fundamentally different than, and not equivalent to, the **Unix** command line. As described in Appendix 1, you can install a Unix-like command-line tool called Cygwin in Windows which provides partial functionality. You can also install Linux alongside Windows to take full advantage of these command-line skills.

the default startup shell to /bin/bash, and relaunch Terminal by quitting and reopening it.[2]

A command-line view of the filesystem

The nested hierarchy of folders and files on your computer is called the **filesystem**. You should already be familiar with the filesystem through the variety of windows you see when using the graphical user interface of your computer. For example, you have browsed to folders and files within the Save and Open dialog boxes of various programs, and on OS X, you have used the Finder to create new folders and drag files between folders. These graphical dialog boxes and tools give you a bird's-eye view of how files are organized on your computer.

Your perspective of the filesystem from the command line is different than the detached view provided by the GUI. It is more like a first-person perspective—as if at any given time, you are viewing the filesystem from where you are standing within it. You get a different view depending on where you stand. Sometimes it is easier to stay right where you are and reach out to interact with a file from afar; other times, it is more convenient to first move to where the file is, and then work with it at close range.

When working from a graphical user interface such as the Finder, it is often unclear what is meant by the **root directory**—that is, the most inclusive folder on the system, serving as the container for all other files and folders. (Directory and folder mean the same thing; however, it is more common when working with the command line to speak of directories than of folders.) As you will see, the meaning of the root directory is much more obvious, as well as more important to know, when working at the command line. A Unix-based system such as OS X has a single root directory. You can think of it as the base of the tree of files on your computer, or as the most inclusive container that contains all other folders and files (Figure 4.2). Even the hard drive is located within this single filesystem sprouting from root. If you insert a disk into your DVD drive or plug in a thumb drive, these also get their own folders within the tree.

Misunderstandings of just what constitutes the root directory in Unix are common. Windows users are more used to each disk drive having its own root, in effect creating multiple roots (C:, D:, etc.) depending on the hardware configuration of the computer. On Mac OS X, some users think of their Home folder as the root directory, but in fact each user of that computer has their own Home folder; it is merely the default location for personal account settings and user-generated files. Other users get the impression that the root of the filesystem is their Desktop, but in fact the Desktop is just a folder nested within the Home folder.

When working with the command line, your view of the filesystem is always based on the directory you are currently in. So the first thing you need to learn

---

[2] If you cannot find an appropriate setting, join us in the PCfB support forum at practicalcomputing.org.

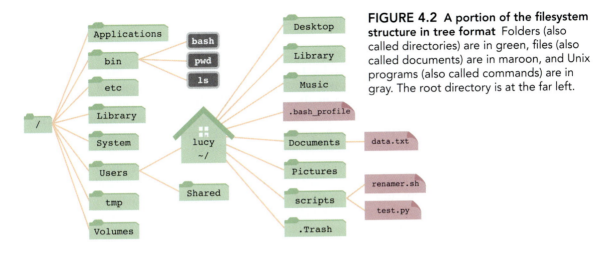

FIGURE 4.2 A portion of the filesystem structure in tree format Folders (also called directories) are in green, files (also called documents) are in maroon, and Unix programs (also called commands) are in gray. The root directory is at the far left.

is how to move to different locations, figure out where you are, and look around from there to see what other files and directories are close by.

The path

The written description of where something is located in the filesystem is called the **path**. The path is a list of directory names separated by slashes. If the path is for a file, then the last element of the path is the file's name. In Unix-based operating systems, the separator is a normal forward slash (/), *not* a backslash (\) as in Microsoft DOS. An example of a path is /Users/lucy/Documents/data.txt, which points to the file called data.txt in the Documents folder in the lucy folder in the Users folder in the root directory. Paths can refer to files or to folders. For instance, in this example, /Users/lucy/Documents/ indicates the folder that contains data.txt, or in other words, the Documents folder for the user lucy.

A path can be absolute or relative. The **absolute path** is a complete and unambiguous description of where something is in relation to the root, the very base of the filesystem. Since there is only one root in the filesystem, you don't need any other information to know exactly which file or folder is designated by an absolute path. In some ways this is like knowing the latitude and longitude of a file or folder, in that it describes an unambiguous location. In contrast, a **relative path** describes where a folder or file is in relation to another folder.

In order to interpret a relative path you need to know the reference point as well. A relative path is essentially an offset. In this way it is like saying "the kitchen in this house." In order to know where the kitchen is, you need to know where the house is. Why even bother with relative paths when absolute paths can always be used unambiguously? In large part this is a matter of convenience. It just gets too cumbersome to fully specify the location of each object, especially as you move further and further away from the root. For example, why say

/mycity/mystreet/myhouse/kitchen if you already know you are in your house?

The slash has two different meanings depending on its placement. It is most commonly used to separate the name of a directory from the name of a directory or file which follows. If it is the *first* character in a path, however, it denotes the root directory. In fact you can always tell an absolute path because it starts with /. It is clear at a glance, for instance, that the path we mentioned earlier, /Users/lucy/Documents/data.txt, is an absolute path because of the leading slash that designates the root, and since there is only one root, you have all the information you need to know exactly where this particular data.txt file is.

Your home path will differ.

Remember that when you are interacting with the shell, you are always seeing your filesystem from within a particular directory. This vantage point is called the **working directory**, and by changing the working directory you can change your perspective on the filesystem. Related to this idea, relative paths have no leading slash and start with the first letter of the file or directory name. They describe the path to a file or folder in relation to the current working directory.

If the working directory were /Users/lucy/, for instance, then the relative path Documents/data.txt would specify the same file as the absolute path /Users/lucy/Documents/data.txt. Notice there is no leading slash in Documents/data.txt, so you know that this isn't an absolute path, and it is interpreted as relative to where you are. (In practice, it usually doesn't matter whether or not there is a trailing slash at the end of a path to a folder.) If you refer to /Documents/data.txt, you will get an error because of the leading slash; there probably is no folder and file by that name relative to the root of your system.

The relative path to a file in your working directory is just the name of the file itself, since it is in the same folder as you are. For example, within the folder /Users/lucy/Documents/, the relative path data.txt by itself is sufficient to specify /Users/lucy/Documents/data.txt. In practice, this is how you'll usually end up interacting with the filesystem from the command line. That is, if you want to perform a set of operations on a file, you'll first change your working directory to the folder that contains the file and then do everything from close range.

Navigating your computer from the shell

Listing files with ls *and figuring out where you are with* pwd

Each time you open a new terminal window, you are logging onto the Unix system of your computer anew. You will be located in what is known in Unix as your home directory, just as if you had clicked on the house icon representing your home folder in a Finder window. Each user with an account on your computer has their own home directory, in the form of their username, and all of these directories are located within the directory/Users. Since we assume the username lucy throughout the book (your username may of course be different), the home folder we'll use in all examples is /Users/lucy.

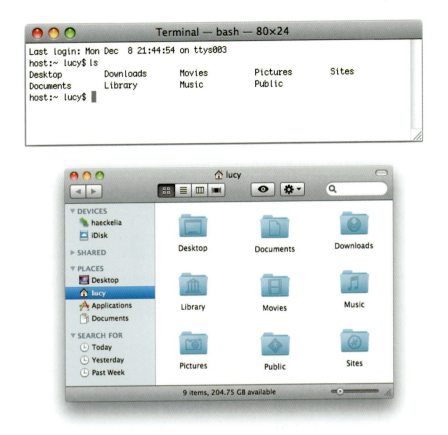

FIGURE 4.3 Viewing the interior of the home folder from two different interfaces, the Terminal window (top) and the familiar Finder window (bottom) Notice that the same folders are listed by both interfaces.

The first command you will learn, and one you'll use frequently, is `ls` (short for **list**). It prints out a list of files and directories contained within a specified directory. If you don't specify a directory, it shows you what is in the working directory—that is, the directory where you are currently located. Since all you have done so far is open a terminal window and start a new shell, by default the working directory is your home directory (`/Users/lucy`). Thus, `ls` shows you its contents (Figure 4.3).

In these examples, the bold characters are what you type, and the regular characters are what the system prints, and we may or may not show the prompt that begins each line. In place of `lucy` you'll see your own user name, and you will also see a different host name at the beginning of the line. This is your network identity, created when you first set up an account on your computer.

Type `ls` by itself, then use another folder name from that list.

To see the contents of the `Desktop` folder while sitting in your home directory, type:

```
host:~ lucy$ ls Desktop
```

On other systems you may not have a Desktop folder.

Without anything before `Desktop`, `ls` expects that `Desktop` is a directory within the working directory. This is a relative path to `Desktop`, because it is in relation to where you are now. Upon executing the command, you should see a list of whatever files are stored on your `Desktop` at the moment.

To see the absolute path of the working directory, use the command `pwd` (**p**rint **w**orking **d**irectory):

```
host:~ lucy$ pwd
/Users/lucy
```

Since you haven't moved anywhere since starting the shell, you are seeing the absolute path of your home directory. Because `Desktop` is a folder within your home directory, and the absolute path to your home directory is `/Users/lucy/Desktop`, you could also list its contents with:

```
host:~ lucy$ ls /Users/lucy/Desktop
```

This command has the exact same results as the previous `ls Desktop` because it is showing the contents of the same folder, just specified in two different ways.

How to move around with `cd`

To move to another directory, use the command `cd` (**c**hange **d**irectory). Move inside your `Desktop` directory by typing

```
host:~ lucy$ cd Desktop
host:Desktop lucy$
```

Notice that the prompt changes from `host:~ lucy$` to `host:Desktop lucy$`, showing the name of the new working directory.

You will need to keep in mind that capitalization generally counts in all types of Unix, including OS X. Getting the capitalization of a command wrong can be as bad as misspelling it. This capitalization behavior is hard to predict, though, so you should not rely on it to distinguish similarly named files.

To see the contents of the `Desktop` directory, but this time from within the directory, you can now type `ls` by itself:

```
host:Desktop lucy$ ls
```

From here, if you type ls Desktop, you will generate an error, because there is no Desktop folder within your Desktop folder (unless you have created one).

```
host:Desktop lucy$ ls Desktop
ls: Desktop: No such file or directory
        ↳Gives an error because you already moved to the Desktop folder
```

The same error happens when you try this, but for a different reason:

```
host:Desktop lucy$ ls /Desktop
ls: /Desktop: No such file or directory
        ↳Gives an error because the root folder doesn't have a Desktop folder
```

To move back towards the root in the directory structure (that is, to the left in Figure 4.2), use the command:

```
host:Desktop lucy$ cd ..
```

Make sure to include a space between the cd and the two dots. This command won't return any feedback after the prompt. However, since the prompt by default shows a shortened version of your current working directory, it will again change to reflect your new location.

The two dots in the command indicate the folder that contains the current folder. This is just another type of relative path. This will always move you from the current working directory to the directory which contains it—unless, of course, you are in the root directory, which is absolute and contained by nothing.

The .. symbol can be used in conjunction with other directory names. This allows you to move with a single command from one directory to another that is sister to it in the hierarchy. For example, if you want to go from your Desktop directory to your Documents folder, this is equivalent to backing out one level (left in Figure 4.2) and then moving one level into a different but parallel directory relative to where you started. From the Desktop folder you would type:

> A convenient way to enter long paths in the terminal window is to just drag and drop the file or folder from the **Finder** into a terminal window. So if you are working in a folder in the GUI or editing a file in another program, you can type **cd** and a space, then drop the folder icon to fill in the rest.

```
host:Desktop lucy$ cd ../Documents
```

From Documents you can type cd .. to go back to the home directory again. Note that you can use .. more than once in a command. For example, cd ../.. will send you back two directories in one step.

Signifying the home directory with ~

A shortcut for referring to your home directory is the tilde symbol (~, the wavy line below the $\boxed{esc}$ key on most keyboards). So, for instance, no matter where you are on the system, you can type:

```
host:Desktop lucy$ cd ~/Documents
```

to jump straight to the Documents folder in your home directory. This is equivalent to typing the absolute path, specified all the way from root:

```
host:Desktop lucy$ cd /Users/lucy/Documents
```

Think of the text that follows ~/ as another relative path. However, instead of being relative to the working directory where you are now, it is relative to the home directory of whoever you are logged in as. If you are lost in your filesystem, you can still refer easily to files in your home directory using ~/ at the beginning of a path, just as you can take yourself directly home with cd ~ . (The cd command by itself will also take you to your home directory.) Don't hesitate to use pwd to figure out what directory you are in at any given point.

    If you are in the middle of a command and wish to start over, typing $\boxed{ctrl}$ C will clear the line. (Hold down the key labeled $\boxed{ctrl}$ or $\boxed{control}$ while pressing C.) This key sequence is a universal way to interrupt a program or process.[3]

Adding and removing directories with mkdir and rmdir

So far we've discussed moving around the filesystem, but you have been a fly on the wall and have not actually done anything to affect the filesystem in your travels. One basic modification you can make to your filesystem is to add a directory—a task accomplished with the command mkdir (abbreviated from **m**ake **dir**ectory).

    The following sequence of commands will make a folder called latlon in the sandbox folder you installed along with the example folder in Chapter 1. While you do these operations, it is illustrative to have a graphical browser window open in the Finder with the contents of the pcfb/sandbox folder showing. The ls commands give a before and after view of the changes resulting from mkdir:

```
host:~ lucy$ cd ~/pcfb/sandbox   ←Only there if you installed examples
host:sandbox lucy$ ls
host:sandbox lucy$ mkdir latlon
host:sandbox lucy$ ls
latlon
```

---

[3] Unfortunately, Windows uses this command for copying, so it is not quite as universal as it might be. If you are working with Cygwin, as described in Appendix 1, you might have to re-learn some keyboard shortcuts.

```
host:sandbox lucy$ cd latlon
host:latlon lucy$ ls
host:latlon lucy$ cd ..
host:sandbox lucy$ ls
latlon
```

After the `mkdir` command is issued, you find a new directory has been created called `latlon`. If you look at your `sandbox` folder using the Finder (remember that GUI you used to use?) you should also see it there, represented by a new folder of the same name. In the last few commands you simply moved into the new directory, looked around (there was nothing there to see since you hadn't put any files in it), and moved back out. Next you get rid of the empty directory with the command `rmdir` (**rem**ove **dir**ectory):

```
host:sandbox lucy$ rmdir latlon
host:sandbox lucy$ ls
```

You can see that `latlon` is now gone. By default, `rmdir` is conservative in that it only removes empty directories. The equivalent to `rmdir` for deleting files is `rm`. Use caution with `rmdir` and even more so with `rm`. These commands are not like putting something in the Trash, where you can pull it out later. Files deleted this way are gone and cannot be recovered.

Copying files

The command to copy a file is `cp` (**c**o**p**y). It is followed by the file's present name and location (known as the source) and the location where you would like the copy to end up (known as the destination). In the following hypothetical example, the source is a file named `original.txt` and the destination is an identical file but with a different name, `duplicate.txt`:

```
host:~ lucy$ cp original.txt duplicate.txt
```

You can use `cp` to copy a file within the same directory, giving the new file a new name, as we just did. You can also copy a file to another directory, either giving it a new name or retaining the old name. These behaviors may seem different at first glance, but are actually quite similar and follow naturally from the ways that the source and destination are specified. If only the filenames are specified, everything takes place in the working directory. This is because a naked filename is just a relative path to a file in the working directory. If the source or destination is a path to a file in a different directory, then files can be moved between directories. The working directory can even be different than either the source or destination directory, provided that paths are specified for both source and destination. If the

⚡ When moving or copying a file, by default the shell will not warn you if a file or folder of the same name exists in the destination directory. If a file of the same name does exist, it will get "clobbered"—that is, deleted and replaced without warning or notification. You may not realize this until much later when you go looking for the original file. In Chapter 6 you will see how to modify this behavior.

source and destination directories are different, but the filenames are the same, then the file will be copied to the destination directory with the same name.

There is one other thing that may seem a little odd at first about cp, but saves quite a bit of typing in the long run: if you only specify a path to a directory for the destination, without an actual filename, then the file will be copied to that directory with its original name.

Remember that the command option consisting of two periods (..) refers to the directory that contains the current working directory. Similarly, a single period (.) represents the current working directory itself, which is useful if you want to copy a file from elsewhere to the working directory. In the following example, you will move to the sandbox folder and copy a file from the examples folder to it:

```
host:~ lucy$ cd ~/pcfb/sandbox
host:sandbox lucy$ ls    ← Probably nothing in this directory yet
host:sandbox lucy$ cp ../examples/reflist.txt ./
host:sandbox lucy$ ls
reflist.txt
```

The source, ../examples/reflist.txt, in essence says "Go into the directory containing the working directory, then from there go into the directory examples and get the file reflist.txt within it." The dot followed by a slash, ./, says "Place a copy of that file, with the same name, in the present working directory." Since only a directory is specified for the destination, with no accompanying filename, the copy of the file has the same name as the original, reflist.txt.

Now copy reflist.txt again, but to a new file with a different name in the same directory:

```
host:sandbox lucy$ ls
reflist.txt
host:sandbox lucy$ cp reflist.txt reflist2.txt
host:sandbox lucy$ ls
reflist.txt        reflist2.txt
```

Where there was once only one file, reflist.txt, there are now two files, reflist.txt and reflist2.txt. Their contents are exactly the same.

Moving files

The command for moving files is mv (**move**). You move files in the exact same way that you copy them, except that the original file disappears in the process. In the following example, you move the reflist2.txt file you just made to Desktop:

```
host:~ lucy$ cd ~/pcfb/sandbox
host:sandbox lucy$ ls
reflist.txt       reflist2.txt
host:sandbox lucy$ mv reflist2.txt ~/Desktop
host:sandbox lucy$ ls
reflist.txt
```

Remember that when you leave off the destination filename, it defaults to the same name as the original. If you now look on the Desktop you should see the `reflist2.txt` file there. Feel free to delete it if you like. You could do this with the command `rm ~/Desktop/reflist2.txt`.

The command mv is also used for renaming files, since renaming a file is equivalent to moving it to a new file with a different name in the same directory. The following example illustrates this:

```
host:~ lucy$ cd ~/pcfb/sandbox
host:sandbox lucy$ ls
reflist.txt
host:sandbox lucy$ mv reflist.txt reflist_renamed.txt
host:sandbox lucy$ ls
reflist_renamed.txt
```

In a later section of this chapter, you will see how to quickly move or copy multiple files at once.

Command line shortcuts

Up arrow

It is not too soon to introduce two shell shortcuts that will save you a lot of typing over the years. After all, the point of this book is to reduce the amount of extra work you have to do. Try to get into the habit of using these, because they will not only save you time, but will also reduce the number of typographical errors you have to correct.

The first shortcut is the ⬆ key. In most shells, the ⬆ moves back through your previous command history. For example, if you type in a long command and hit return , only to realize that you were in the wrong directory for the command to do what you wish, you can easily move to the correct directory, press ⬆ one or more times until you see the correct command, and then hit return again to execute the command. Try pressing the ⬆ key now to step back through and see all the cd'ing and ls'ing that you have done so far.[4]

---

[4] Pressing the ⬇ will show future commands that you haven't typed yet.

Previous commands can also be edited before you re-execute them. If you enter a command but it has a typo somewhere in it, press ⬆ to show it again, then use the ⬅ and ➡ keys to move to the place where the error occurred and fix it. Once you have corrected the typo and the command looks good, hit ⟨return⟩. You don't need to use the ➡ key to move the cursor back to the end of the line. When editing commands, depending on the terminal program and operating system, you can also change the position of the cursor within the terminal window using the mouse rather than the ⬅ and ➡ keys. Try it out now: press the ⬆ key to bring up a command, and point with the mouse to where you would like the cursor to be inserted. Hold down the ⟨option⟩ key and click, and the cursor should be inserted at that point for quick editing.

Tab

Another big time-saver is the ⟨tab⟩ key. This is in effect the auto-completion button for the command line. For example, at the prompt, start typing:

```
host:~ lucy$ cd ~/Doc
```

Then, before pressing ⟨return⟩, press ⟨tab⟩. The command line should fill in the remainder of the word Documents and append a trailing slash. Now try this command:

```
host:~ lucy$ cd ~/D⟨tab⟩
```

You should get a beep or a blank stare from your terminal. This is because there is more than one folder starting with D in your home directory. The shell can't tell if you are referring to the Documents folder or the Desktop folder. To see what the choices are, press ⟨tab⟩ again. The terminal will return a list of the matches to what you have typed so far. If nothing shows up, check for incorrect slashes at the beginning, or else a missing tilde, to make sure you haven't entered the start of a path that doesn't exist.

Completion of directory and filenames is especially convenient when the name has a space in it. If you try to type a command with a space in it, such as cd My Project, the shell thinks that there are two different pieces of information (arguments) being sent to the cd command. Since you probably don't have a directory called My, it returns an error.

When you use ⟨tab⟩ auto-completion in this context, the shell types the command with extra backslashes inserted before the spaces:

```
host:~ lucy$ cd ~/My\ Project/
```
If you type ⟨tab⟩ here...↵ ↳...the computer will complete the rest

Remember from Chapters 2 and 3 on searching and replacing with regular expressions that a backslash (\) modifies the interpretation of the character that follows. In this case, it causes the shell to interpret the space as part of the filename rather than as the separation between two filenames. You don't need auto-completion to take advantage of \. You can also type it in yourself as part of the path.

Spaces can be accomodated in file or directory names by using either single or double quotes to enclose the full path, including spaces. However, you cannot do so if you are trying to use the tilde shortcut or wildcards as part of the path, since they won't be expanded.

It is generally best to just AvoidUsingSpacesAltogether when naming data files and scripts. This will make things easier later on. Other punctuation marks in names (?, ;, /) have special meanings at the command line, and should therefore be avoided as well. Underscores (_) are acceptable to use.

Modifying command behavior with arguments

All of the commands we've discussed so far, such as `ls`, `cd`, and `pwd`, are actually little programs that are run by the shell. The programs each read in bits of information and do something as a result. Pieces of information that are passed to a program at the command line are called **arguments**. Each argument is generally separated from the program name and any other arguments by a space, but beyond that there aren't any global rules for how programs receive and interpret arguments. You've already used arguments, such as when you told `cd` which directory to change to, or when you told the `mkdir` program what directory to create. Some programs interpret arguments in a particular order, while others allow arguments to be in any sequence.

Besides a path name, the `ls` command has many optional arguments; in format, these consist of a hyphen followed by a letter. Two of the most commonly used arguments or options are `-a` and `-l`. Consider `-a` first. Some files and folders on the system, those whose names begin with a single period, are normally hidden from view. To tell the `ls` command to show all files, including those that are hidden, type `ls -a` (meaning list **a**ll). Try `ls -a` from within your home directory (or `ls -a ~/` from any folder), and you will notice some of the hidden files that show up. For example, the directory `.Trash` is normally a hidden folder into which your not-yet-deleted files are placed when you drag them to the Trash icon from within the graphical user interface. Other hidden files you see may be settings files used by the system. In a later chapter, you will edit an important hidden settings file called `.bash_profile`.[5]

You can also tell `ls` to return a more detailed list of directory contents with the `-l` argument. (This is a lowercase L, not the number 1.) This prints a list that includes, in columns from left to right, the file or directory permissions; the number of items in the folder, including hidden items; the owner; the group; the file

---

[5] These hidden files are not normally visible when working in the Finder or in other parts of the Mac GUI. To make them visible in the Finder, type the following in a terminal window: `defaults write com.apple.finder AppleShowAllFiles TRUE`. Then restart the Finder by logging out or typing: `killall Finder`. You can also go to hidden folders using the ⌘ shift G shortcut.

size; the modification date; and the file or directory name. We won't talk about permissions until later, but much of this information will eventually prove useful or necessary as you progress:

```
host:~ lucy$ ls -l
total 0
drwx------+  3 lucy   staff    102 Mar  9  2008 Desktop
drwx------+  4 lucy   staff    136 Mar  9  2008 Documents
drwx------+  4 lucy   staff    136 Mar  9  2008 Downloads
drwx------+ 30 lucy   staff   1020 Sep 28 13:07 Library
drwx------+  3 lucy   staff    102 Mar  9  2008 Movies
drwx------+  3 lucy   staff    102 Mar  9  2008 Music
drwx------+  4 lucy   staff    136 Mar  9  2008 Pictures
drwxr-xr-x+  5 lucy   staff    170 Mar  9  2008 Public
drwxr-xr-x+  5 lucy   staff    170 Mar  9  2008 Sites
drwxr-xr-x   3 lucy   staff    102 Apr 19 22:48 scripts
```

If you would like to both see a detailed view *and* all the hidden files, at the same time, use both arguments in the same command:

```
host:~ lucy$ ls -l -a ~/
```

This shows a detailed view that includes all the hidden files in your home folder. With many programs, you can combine arguments using one dash followed by the arguments together with no spaces; thus ls -la ~/ is equivalent to the command above. The order of the arguments for ls isn't critical either, so ls -al ~/ is also equivalent. See Appendix 3 for a more complete list of options.

Viewing file contents with less

There are several commands that display the contents of text files. This is useful when you have a huge file and want to take a quick peek inside, or just view a certain part of a file within the terminal without launching another program. The most commonly used file viewer is less.[6] Typing less followed by the path of a file presents the contents of that file on the screen one page at a time. Table 4.1 shows the keyboard shortcuts you can use to navigate in less. To move to the next page, type a space (that is, press the [space] key); to move back to the previous page, type b. Scroll up and down a line at a time with the arrow keys. To go to a

---

[6] If this seems hard to remember, know that Unix developers like to choose quirky names for their programs. The original file viewer was named more, and so when it was re-written to add functionality the name was changed to less because less is not more. If your system does not seem to have the program less installed, try more instead.

TABLE 4.1 Keyboard commands when viewing a file with `less`			
q	Quit viewing	⬆ or ⬇	Move up or down a line
space	Next page	/abc	Search for text abc
b	Back a page	n	Find next occurrence of abc
## g	Go to line ##	?	Find previous occurrence of abc
G	Go to the end	h	Show help for `less`

specific line, type the line number followed by g. (Typing a capital G instead will take you to the bottom of the file.)

Try viewing some of your documents or data files with the `less` command, either by typing the full path of the file (remember to use tab auto-completion to make this easier), or by moving to the directory with cd and running `less` followed by the file name.

To find a word in the file you are viewing, type / followed by the desired word or character and hit return. If found, the line containing the word will be placed at the top of the screen. To find the next occurrence, type n.

When you have seen enough, type q to quit. These navigation shortcuts show up repeatedly in shell programs (though they are not universal), so it is worth remembering them as you work. They are summarized in Appendix 3 for quick reference. If you try viewing a file in the terminal and it appears on the screen as strange characters or gibberish, it is probably not a plain text file and has to be opened with a special program which understands how the binary data are laid out.

The program `less` does not load the entire file into memory before showing it to you; it loads only the part it is displaying. So if you have a 500 megabyte dataset and just want to see the very first part of it (perhaps to check that it contains what you think it does, or to figure out the column headers), you are much better off taking a look at it with `less` than you would be opening the entire file in a text editor (which could take some time and bog the entire system down). Later you will see ways to search inside a file without having to open it, using `grep`.

Viewing help files at the command line with `man`

Most command-line programs have user manuals already installed on your computer. These explain what the programs do, which arguments they understand, and how these arguments should be entered. These manuals are viewed with the program `man`, which takes an argument consisting of the name of the program you want to see the manual for. So, for instance, `man ls` will display lots of practical information about the `ls` program and the optional arguments it understands.

The `man` program uses the `less` program described above to display the help files, so you navigate within it using the same keys: space advances the file page by page, a slash lets you find a certain word, and pressing q lets you quit the file and return to the command line when you are done reading.

Try using man to investigate some of the options for other programs you've already used, for example man pwd or man mkdir. Using man, you will be able to investigate any of the other programs which you encounter in coming chapters. Strangely, some of the most basic commands, including cd, alias, and exit, don't have their own man entries.

Many times you might not know the exact command that will help you do something, only the general topic. In this case you can search through the contents of all the manuals on your system using man -k followed by a keyword. For example, to find out commands dealing with the date, type man -k date and look through the brief descriptions of the commands that match. There are many utilities built in to Unix installations: calendar programs, calculators, unit converters, and even dictionaries.

The command line finally makes your life easier
Wildcards in path descriptions

So far you have not done much that you could not have accomplished just as easily (or maybe more easily) with a graphical user interface. It is the ability to use **wildcards** that suddenly makes the command line more powerful than a graphical user interface for processing many files in one operation.

While generating search queries back in the chapters on regular expressions, you encountered several types of wildcards—that is, shorthand ways to represent multiple characters. The shell has its own wildcards, but it can be confusing because it uses many of the same wildcard characters we have already encountered, but with significantly different meanings. The wildcard you will use most frequently in the shell is the asterisk (*). In the context of the shell it is the wildest of wildcards, representing any number of any kind of characters (except a slash). In regular expression language, this would be the equivalent of .* (the dot meaning any character, the asterisk meaning zero or more).

For example, to list all the files and folders that begin with D in your current directory, as well as the contents of those directories, you could type:

```
host:~ lucy$ ls D*
```

To list just files ending in .txt, you could type:

```
host:~ lucy$ ls *.txt
```

You can use several wildcards in the path. So to list all text files within all directories that start with D, you can type:

```
host:~ lucy$ ls D*/*.txt
```

Notice that the asterisk doesn't include text beyond the path divider (/). Therefore, `ls D*.txt` is not equivalent to the command above, because it would only show files in the current directory that start with `D` and end with `.txt`.

The construction of `*/*.txt` is a convenient way to indicate all the text files in all immediate subfolders of your current directory. It is even possible to search deeper folders, for instance with `*/*/*.txt`. These commands only show files at the exact specified number of levels, so they would not show a `.txt` file in the current directory.

When using wildcards, if `ls` finds a file that matches, it lists the filename; if it finds a directory that matches, it lists the directory along with its contents.

> ⚠ Wildcards carry the risk of causing problems through unanticipated matches. You especially want to be careful before using wildcards to remove, copy, or move important files; this will help keep you from accidentally modifying files you didn't realize were there. It is often wise to test wildcards with a harmless `ls`. This will give you a list of files recognized by the command so you can check to see if it includes any that you didn't anticipate.

Copying and moving multiple files

Using wildcards, you can quickly copy or move multiple files with names that match a certain pattern. In the following example, all of the files that end with `.txt` in the examples directory are copied to the sandbox folder:

```
host:~ lucy$ cd ~/pcfb/sandbox
host:sandbox lucy$ ls
host:sandbox lucy$ cp ../examples/*.txt ./    ← Very different from /  by itself
host:sandbox lucy$ ls
reflist.txt    ← You will see others too...
```

This should begin to give you a sense of some of the tasks that can be easier working at the command line rather than in a graphical user interface. Both the copy and move commands, used in conjunction with wildcards, are significant time-savers when dealing with large numbers of files. You can imagine, for instance, how you could gather all your physiology data from a particular species by using the taxonomic name at the beginning and the data format at the end of a path (`Nanomia*.dat`) while ignoring any similarly named images or documents that might be in that folder.

Ending your terminal session

To end your session, type `exit`. You can just close the window, but this is a bit like unplugging your computer without turning it off first: if there are any programs still running in the terminal window, they will come unceremoniously to a halt. Closing the window is also not an option when you log in to a different computer from within your terminal. If `exit` does not work, some shells use `logout` or `quit`.[7]

---

[7] If none of those work, *then* pull the plug.

SUMMARY

You have learned:

- The difference between a GUI and a command line
- The difference between absolute and relative paths
- Ways of indicating relative paths, including:

    ~ for home

    .. for the enclosing directory

    . for the current directory

- How to move around in the terminal using cd
- Common commands, including:

    ls to list the contents of a directory

    pwd to print the name of the current working directory

    mkdir, rmdir, rm to make and remove directories and files

    less to view files

    cp, mv to copy and move files

    man, man -k to get help on a command or search for help

    exit

- How to use the wildcard * in shell commands
- Command line shortcuts, including:

    tab to complete program and path names

    ↑ to recall prior commands in your history

Recommended reading

Kiddle, Oliver, Jerry D. Peck, and Peter Stephenson. *From Bash to Z Shell: Conquering the Command Line*. Berkeley, Calif: Apress, 2005.

Newham, Cameron. *Learning the Bash Shell*, 3rd Edition. Sebastopol, Calif: O'Reilly Media, 2005.

Chapter

HANDLING TEXT IN THE SHELL

Now that you are familiar with moving around in the shell, you can start to use it to do work. In this chapter you will begin to process large data files, viewing them, combining them, and extracting information from them. You will also see how to retrieve data from the Web at the command line.

Editing text files at the command line with nano

You are already familiar with editing text files through the graphical user interface of TextWrangler. It is sometimes more convenient, though, to create or modify a file right at the command line. Later, when you work on a remote machine, the only way to directly modify a file may be through a command-line editor.

Although less, which you learned about in the previous chapter, is a convenient command-line viewer for text files, it does not allow you to edit the contents of a file. For that, you will need one of the many command-line text editors available, such as vi or emacs. Each of these programs is optimized for different tasks and has its own devoted adherents. Each also has its own set of unique keystrokes for performing similar functions, so learning one of these editors doesn't necessarily enable you to easily use others. For starting out, we recommend nano, a widely available, general-purpose command-line text editor.[1] It displays the options in a menu-like array at the bottom of the screen.

If you call nano without any arguments it will create a new blank file. Move to your work folder and start a blank document:

```
host:~ lucy$ cd ~/pcfb/sandbox
host:pcfb lucy$ nano
```

---

[1] If your shell environment does not seem to have nano, try using pico.

Enter some text into the blank document:

> `Helpful shell commands`

As you can see, most characters you type show up just as you would expect in the document. Along the bottom of the window, though, you will see a series of characters, each preceded by a caret (^), and each followed by a short description of a corresponding function (Figure 5.1). The caret signifies that to execute these functions, you hold down the `ctrl` key while pressing the relevant character.

Go ahead and try `ctrl` X, which exits nano. In this case, however, you won't actually exit. Instead, you will see the options at the bottom of the screen change. This is because the document has not been saved, and nano is wondering what you want to do with the contents. If the document had been saved (or had not been modified since it was opened), you would have returned immediately to the command line. As it is, above the new commands, you will see that nano is asking if you want to `Save modified buffer`. The options are Y for Yes, N for No, or ^C (`ctrl` C) for Cancel. Just as you would expect from similar questions asked by programs with a graphical user interface, Y will save the file, N will discard any changes you have made, and ^C will return you to editing the document.

Press Y to save the document, and nano will now give you a prompt that says `File Name to Write:` with a new set of options below. Type in `shelltips.txt`, and press `return`. The nano program will quit, and you will be back at the command line. Listing the directory contents with `ls` will show the `shelltips.txt` file you just created, and `less shelltips.txt` will let you view the file's contents. Open it again with nano, this time specifying the filename:

> `host:pcfb lucy$ `**`nano shelltips.txt`**

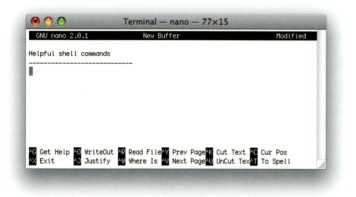

FIGURE 5.1 A view of the popular command-line text editor nano with the text you just entered   Some of the available commands are displayed at the bottom of the window.

You can move the cursor with the arrow keys. In documents that don't fit in one screen, you can scroll up a page at a time with `ctrl` Y and down with `ctrl` V (again, you can see these listed among the key combinations at the bottom of the screen). Another useful key combination is `ctrl` O, which saves the file as you work without closing it. Once you are done exploring, exit nano with `ctrl` X.

You can delete the test file either the old-fashioned way in the GUI, by dragging it from the Finder window to the Trash, or from the command line with `rm shelltips.txt`. Remember that `rm`, like the `rmdir` command for removing empty directories, deletes the file immediately after confirmation (depending on your configuration), without moving it to the Trash folder, so be careful when using it.

Although `nano` is available on all computers with OS X, you may find yourself working on a computer with a different type of Unix installation that doesn't include nano. In that case, see if `pico` is available. This is very similar to `nano`, and in fact `nano` is actually based on it. If `pico` isn't present on your computer either, you will probably need to give yourself a crash course on one of the other command-line text editors. You can also open a file in a separate TextWrangler window from the command line with the `edit` command, provided that you installed the TextWrangler command-line tools when prompted during installation. If you would like to install these tools after the fact, you can find the command to do so under the TextWrangler menu.

Controlling the flow of data in the shell

Redirecting output to a file with >

There are many times when it is useful to send the output of a program to a file rather than to the screen. Though this might at first sound like a nice but esoteric trick, it adds tremendous flexibility to the way that you can extract and combine data. It also creates the possibility of hooking software together in new ways. This is helpful when an analysis program normally sends its results to the screen, but you want to save the results instead. You could use the GUI's copy and paste functions to copy the results from the screen, but this can't be automated and gets cumbersome for big files or lots of files.

To redirect the output of a program to a file instead of the screen use >, the **right angle bracket** or greater-than sign found above the period on a U.S. keyboard. You type your command as you normally would, then on the same line add a > followed by the name of the file you want to send the output to. Think of > as an arrow that is pointing to where the output should go.

If the file doesn't already exist, the redirect will create it. Be careful, however: if a file with the same name does exist, it will be erased and replaced with the program output. It is therefore very easy to accidentally destroy an important document. (Later we will describe a way to avoid this behavior by redirecting with two right angle brackets together, that is, >>.)

To try out redirection, `cd` to the `sandbox` folder and use `ls` to list those files in the `examples` folder which end in `.seq`:

```
host:~/Desktop lucy$ cd ~/pcfb/sandbox/
host:sandbox lucy$ ls -l ../examples/*.seq
-rw-r--r--  1 lucy staff  524 Nov 1 2005 ../examples/FEC00001_1.seq
-rw-r--r--  1 lucy staff  600 Nov 1 2005 ../examples/FEC00002_1.seq
-rw-r--r--  1 lucy staff  538 Nov 1 2005 ../examples/FEC00003_1.seq
```

```
-rw-r--r--  1 lucy staff  622 Nov 1 2005 ../examples/FEC00004_1.seq
-rw-r--r--  1 lucy staff  490 Nov 1 2005 ../examples/FEC00005_1.seq
-rw-r--r--  1 lucy staff  548 Nov 1 2005 ../examples/FEC00005_2.seq
-rw-r--r--  1 lucy staff  495 Nov 1 2005 ../examples/FEC00006_1.seq
-rw-r--r--  1 lucy staff  455 Nov 1 2005 ../examples/FEC00007_1.seq
-rw-r--r--  1 lucy staff  501 Nov 1 2005 ../examples/FEC00007_2.seq
-rw-r--r--  1 lucy staff  569 Nov 1 2005 ../examples/FEC00007_3.seq
```

Now, try the same command, but redirect the output to a file called `files.txt` (you can just press ⬆ and type the new text from > onward):

```
host:sandbox lucy$ ls -l ../examples/*.seq > files.txt
```

This time, there won't be any output to the screen and you'll get the command line right back. Check the contents of `sandbox` with `ls`, and you will see that the file `files.txt` has been created. Take a look at this new file with `less` or `nano`. It contains the exact text that `ls` would have sent to the screen if you had not redirected the output to a file. Also note that `files.txt` lists the contents of the `examples/FEC` folder, even though `files.txt` itself is in the `sandbox` directory. This is because the current working directory is `sandbox`, but by using `../` we told `ls` to look back starting in the `examples` folder instead.

It is not uncommon to want to create a file listing the contents of a directory, perhaps to send to a colleague, or to use as a starting point in automating a task involving those files. In these cases and others, a simple `ls` with a redirect can be very helpful.

Displaying and joining files with `cat`

Another useful command is `cat`. It is very simple, taking a list of one or more file names separated by spaces and outputting the contents of these files to the screen. Instead of displaying the contents in a special viewer or editor, as `less` and `nano` do, `cat` dumps them right into the display without a break, in the same manner as the results of many other commands, such as `ls`. Though this may not seem very useful at first, there is a good chance `cat` will become one of your most frequently used commands.

To begin with, `cat` is a convenient tool to view the contents of a short file. For example, `cat files.txt` will output the contents of the `files.txt` file from the previous example. You don't want to view large files with `cat`, as it will take some time for the file contents to scroll by. (Remember, `less` is very good at viewing large files because it only loads a small chunk of them into memory at a time.) If your file seems to be scrolling past for too long, press ⌈ctrl⌉ C to kill `cat` and get the command line back. This is a useful trick whenever a program stops responding in the terminal.

Take a look at the contents of another example file. From within sandbox, type the following command (or press ⎡tab⎤ after the F to partially complete the file name):

```
host:sandbox lucy$ cat ../examples/FEC00001_1.seq
```

As you might expect, this will dump the contents of the file FEC00001_1.seq in the examples directory to the screen.

Now, use cat to view two files. Notice that there is a space between the paths to the two files:

```
cat ../examples/FEC00001_1.seq ../examples/FEC00002_1.seq
```

You can see that cat just dumps the contents of both files to the screen, one after the other, without any separation of any kind between them. Now, let's look at the contents of all ten of the files at once that end in .seq. You could write out the path to each of them, but it is much easier to use a wildcard as we did for ls:

```
cat ../examples/*.seq
```

At this point, it is a simple matter to use cat in combination with a redirect to create a new file that contains all the contents of all the other files, joined end to end:

```
cat ../examples/*.seq > chaetognath.fasta
```

Combining files with GUI tools such as the Finder can be a tedious and frequently recurring task. This one-line command just saved you from opening ten different files, copying and pasting the contents of each file into a new blank document, and then saving the file you generated. Best of all, this command would have worked just as well for two files or for a hundred. It is an expandable solution that you only need to learn once for projects of any size. The other thing to note is that the command didn't just join all the files in a directory; it looked for only particular files that ended with .seq. This specificity was important since there are many other files in the examples directory that we didn't want to combine into chaetognath.fasta, and it saved you the added task of sorting through these files.

When using wildcards with cat, do not use the same extension for the output file as the input file. If you try cat *.txt > combined.txt the shell may consider combined.txt to be one of the input files and attempt to continue writing it to itself ...forever. If you accidentally create a never-ending command, like this, remember that you can use ⎡ctrl⎤ C to stop it.

Notice that all the `.seq` file names have the general format `FEC00001_1.seq`, and that the digit after the underscore is either a 1, 2, or 3. You might want to make a combined file from only the original files that have a 1 in this position. This could be done as follows:

```
cat ../examples/*1.seq > chaetognath_subset.fasta
```

The addition of the 1 after the * increased the specificity of the file names that are matched. If you wanted to now add the files with a 2 to `chaetognath_subset.fasta`, you could use a slightly modified redirect which appends to existing files:

```
cat ../examples/*2.seq >> chaetognath_subset.fasta
```

There are two things that are different about this command relative to the one preceding it. The 1 after the * was changed to a 2, and another > was added so that the redirect is now >>. If you had just changed the 2 and rerun the query, `cat` would have found the right files and joined their content, but the > would have written over the existing `chaetognath_subset.fasta` file. It would no longer include the results from the first run. The >> behaves much as >, but it appends the redirected content to the end of a file if it already exists rather than replacing the file altogether. If the file doesn't already exist, it creates it just as > does, so in most cases you can default to using >> and avoid the risk of losing files by accident. You can remember the distinction by noting that the second > is added to the end of the first, just the like operation itself. Later you will learn about other redirects that send data from a file to a program, or from the output of one program directly to the input of another, without ever generating a file.

Regular expressions at the command line with `grep`

Working with a larger dataset

You are now going to dive into a larger dataset, a compilation of measurements taken by Gus Shaver and colleagues on plant harvests in the Arctic. It is available online[2] and is reproduced in the examples folder as a file called `shaver_etal.csv`. The file extension csv usually designates a file of comma-separated values, as it does in this case.

Open `shaver_etal.csv` with TextWrangler to get a sense of its overall layout. You can see that it is comma-delimited text. (We've only dealt with tab-delimited text so far, but any character can in theory be used to separate data fields in a text file.) The rows are different samples, and the columns are various

---

[2]Original source: http://tinyurl.com/pcfb-toolik. Also available at practicalcomputing.org.

measurements. The first row is a header that describes the measurements within each column. You'll also notice that many of the commas don't have data between them. Even when some data are missing (either because they weren't collected or weren't applicable to a particular sample), it is still important to have all the commas in place; this ensures that values coming after the missing data are interpreted as being within the correct column. The last rows don't have any data—they are all commas. (If you are really astute, you'll notice that there are two empty columns at the right side of the file as well.) This isn't uncommon in character-delimited files generated by spreadsheet programs, which will sometime write out empty rows or empty columns that once had data or are formatted differently.

Extracting particular rows from a file

What if you want a file that contains only those records from `shaver_etal.csv` which pertain to Toolik Lake? The command-line program `grep` is an easy-to-use tool that quickly extracts only those lines of a file that match a particular regular expression. This is the first time we will use the regular expression skills that you developed earlier—outside of the context of a GUI-based text editor such as TextWrangler. In this case, no wild cards or quantifiers are needed. The regular expression will be just the literal piece of text you are looking for, "Toolik Lake":

```
host:~/Desktop lucy$ cd ~/pcfb/sandbox/
host:sandbox lucy$ grep "Toolik Lake" ../examples/shaver_etal.csv
```

The first argument (`"Toolik Lake"`) is the regular expression, and the second argument specifies the source file you want it to examine. The `grep` program scans the file and displays only those lines that contain the search phrase. We needed to put quotes around our regular expression because it had a space in it; otherwise, `grep` would have considered `Toolik` to be the search term and `Lake` to be a separate argument.

In the previous example, the results were simply sent to the screen. Now use a redirect to send them to a file: press ⬆ to recall your previous command and add `> toolik.csv` to the end:

```
grep "Toolik Lake" ../examples/shaver_etal.csv > toolik.csv
```

Now you have a file that is a subset of the original, containing only those lines with Toolik Lake. The only issue now is that the new file doesn't have a header. To solve this, you can copy the header and paste it in with `nano` or TextWrangler, or create another file that has just the header and then use `cat` to join it to the new file subset you made.

To tell `grep` to ignore the case of letters in search matches, so that you can find `toolik` as well as `TOOLIK` and `tOOlik`, add `-i` as an argument to the command.

Sometimes when you type a command and hit return, the shell will appear to be frozen, showing only a blank line. This can happen if you made a mistake typing the command, such as forgetting to add a closing quote mark, and it usually just means that the shell is awaiting the rest of the command. Depending on what the status is, you can either try typing the remainder of the command and hitting return again, or using ctrl C to abort the command and start over.

Though you will often need to search a file for the lines that contain a known phrase, the example above didn't leverage the power of regular expressions at all; the search phrase was just a bit of literal text. The command-line version of grep uses slightly different syntax than the regular expressions available in TextWrangler.[3] For instance, grep doesn't understand \d, so you will need to specify the range [0-9] instead. The man file for grep (which you can see with the command man grep) explains some of the command-specific syntax, and you can consult it if something doesn't work as expected.

Now create a file that has only the lines from Toolik Lake that were recorded in August. One way to do this would be to take a two-step approach and just use grep to create a subset of the new file toolik.csv, composed of only those lines that have Aug in them. A more direct strategy would be to simultaneously search for lines that have both Aug and Toolik Lake in them. This requires a wild card and quantifier between the terms to accommodate the intervening text. You will again use the .* formulation from regular expressions, with . as the wildcard and * as the quantifier:

```
grep "Aug.*Toolik Lake" ../examples/shaver_etal.csv > toolik2.csv
```

Some complications arise with grep at the command line when searching for characters that have special meanings for both the shell and regular expressions. For these characters to be taken literally as part of the name and not as special characters, they need to be escaped out with a backslash, so that the shell passes them on to grep. For now, we'll leave it at that and let you explore this a bit more on your own if you like.

When using regular expressions to search for subsequences or other data that might wrap across lines, if there is a line ending in the middle of the pattern it will not match. See the agrep command in Chapter 16 for ways to perform searches which span multiple lines.

You can cause grep to return all the lines that *don't* match your search expression by inserting -v in your command. Thus, if you had the opposite challenge and needed to construct a file with all the records *except* those from Toolik Lake, you could use this command:

```
host:sandbox lucy$ grep -v "Toolik Lake" ../examples/shaver_etal.csv
```

This time the header is included in the output since it doesn't contain Toolik Lake.

---

[3] See Appendix 2 for more information regarding differences in regular expressions syntax.

Redirecting output from one program to another with pipe |

You have already used > and >> to redirect the output of a program to a file. Using the **pipe** redirect, it is also possible to redirect the output of one command directly to the input of another, without ever generating a file. The pipe character is the vertical bar (|) located above the backslash key on your keyboard. In the previous grep example we specified a file to be examined. If you leave off the file name and only specify the regular expression, grep will take the input from another source. In this case, this other source will be the output of the pipe.

To illustrate this, we will use the command history, which displays all your most recent commands line by line. This is a great way to remember what you did, or to find a previous command so that you can reuse it. Try it out:

```
lucy$ history
```

It can sometimes be difficult to find a particular command in the hundreds of results that are displayed. One way to find the right one is if you can remember the argument values or other text you used in issuing that particular command. You can then use history to display your previous commands, and use grep to find those lines that contain the text of interest. Of course, you could redirect the output of history to a file with >, and then run grep on the new file; however, that gets cumbersome and generates a file that you will only need once, cluttering up your system. It is much more convenient to use a pipe to send the output of history directly to grep:

```
lucy$ history | grep Toolik
```

This will display all the commands you have executed that contain the word Toolik. The output of the history command does not depend on what directory you are in when you run it.

You can also use the pipe to construct searches by combining two consecutive grep operations. In the previous example regarding Toolik Lake, you created a regular expression that matched "Aug.*Toolik". If you want to make one of these terms case-sensitive, but the other not, or if you want to invert one search and not the other, you can use a pipe. To do this, first run a grep command for lines containing Aug, and then instead of sending that output to the screen or a file, pipe to another grep command that looks for Toolik. This would give you more flexibility in constructing each particular search:

```
grep "Aug" ../examples/shaver_etal.csv | grep "Toolik" > toola2.csv
```
↳Original grep ↳The searched file   ↳Piped to 2nd grep   ↳Redirected to file

Most but not all programs can accept data from a pipe in the way `grep` did above. It is not enough to just send the output of one program to another; the data must be organized in such a way that the receiving program can make sense of them. Since `grep` can handle any text, this isn't an issue in these particular examples.

Searching across multiple files with `grep`

To search the contents of multiple files in one step for a particular bit of text, you could use `cat` to join the contents of the files together and then feed them to `grep` with a pipe:

```
host:sandbox lucy$ cd ~/pcfb/examples/
host:examples lucy$ cat *.seq | grep ">"
>Fe_MM1_01A01
>Fe_MM1_01A02
>Fe_MM1_01A03
>Fe_MM1_01A04
>Fe_MM1_01A05
>Fe_MM1_01A06
>Fe_MM1_01A07
>Fe_MM1_01A08
>Fe_MM1_01A09
>Fe_MM1_01A10
```

Notice that the > in the command is within quotes. This indicates that > is the character that is being searched for, not a redirect to send the results to a file. This general approach is very useful for summarizing or extracting data spread across multiple files, but disastrous results can occur if you forget the quote marks.

In some cases, you may want to see both the lines with the pattern and the name of the file they were found in. This is the default behavior for `grep` when you designate input files with wildcards:

```
host:examples lucy$ grep ">" *.seq
FEC00001_1.seq:>Fe_MM1_01A01
FEC00002_1.seq:>Fe_MM1_01A02
FEC00003_1.seq:>Fe_MM1_01A03
FEC00004_1.seq:>Fe_MM1_01A04
FEC00005_1.seq:>Fe_MM1_01A05
FEC00005_2.seq:>Fe_MM1_01A06
FEC00006_1.seq:>Fe_MM1_01A07
FEC00007_1.seq:>Fe_MM1_01A08
FEC00007_2.seq:>Fe_MM1_01A09
FEC00007_3.seq:>Fe_MM1_01A10
```

Now you get the filename and contents of every matching line, separated by a colon.

Finally, it is also possible to list just the names of files that contain a particular text pattern. If you specify the −l argument to grep, instead of listing each line that matches, it outputs the name of each file that contains a matching line:

```
host:examples lucy$ grep -l "GAATTC" *.seq
FEC00001_1.seq
FEC00002_1.seq
FEC00004_1.seq
FEC00005_1.seq
FEC00005_2.seq
FEC00007_2.seq
```

Notice that the query pattern has been modified, and that only a subset of the files contain the specified pattern.

In conjunction with a redirect, the above command can generate a new file with these filenames:

```
host:examples lucy$ grep -l "GAATTC" *.seq > ../sandbox/has_EcoRI.txt
```

Refining the behavior of grep

There are a variety of other useful arguments that modify the behavior of grep; you can explore these in the grep manual page with man grep. Many of these are also listed in the table in Appendix 3, and some are listed here (Table 5.1). One of the most helpful is the argument −c, which will cause grep to output the number of lines which contain the specified pattern rather than the lines themselves. You could use it, for instance, to count the number of DNA sequences in a FASTA file: grep −c ">"

TABLE 5.1 Options that modify the behavior of grep

Usage example: grep −ci *text* *filename*

−c	Show only a count of the results in the file
−v	Invert the search and show only lines that do *not* match
−i	Match without regard to case
−E	Use regular expression syntax (as described in Chapters 2 and 3) with the exception of wildcards; use [ ] to indicate character ranges, and enclose search terms in quotes. See Appendix 3.
−l	List only the file names containing matches
−n	Show the line numbers of the match
−h	Hide the filenames in the output

⚡ Be extremely careful if you are constructing a `grep` query that includes > as part of the search term. Be sure that it is within quotes, or else the shell will interpret it as a redirect and replace the contents of the file you wanted to search. When in doubt, use quotes. They are also necessary if the pattern (i.e., the query text) has wildcards or quantifiers.

A NOTE ABOUT `awk` **AND** `sed`

Two powerful commands called `awk` and `sed` are available in most shell environments. These are similar to `grep` in that they let you search and modify the contents of files, but they are like programming languages in that they have even more opportunities for performing complex tasks. Both `awk` and `sed` are tricky, and learning them does not necessarily tie in to other command-line skills. Because of this, and because of our focus on Python programming, we will not cover them here. However, if you are interested, you might read about them online and see if they are especially suited to your needs and inclinations.

will count the header lines, and therefore the number of sequences regardless of their length. It can also quickly ascertain the number of lines of data in a file. One way to do this is with `grep -c $`, which returns the number of line endings and therefore the number of lines. (The command `wc file.txt` will give you a count of the characters, words, and lines in the file.) Remember that `-c` is showing the number of lines that contain the pattern, not the number of times the pattern matches, so if some lines have multiple matches, `-c` will underestimate their total number.

When searching through a long file it can be helpful to see not only the lines that contain your pattern of interest, but also where the lines are in the file. By default, `grep` will output the lines in the order they are encountered, but `-n` adds further location information by prepending the line numbers to the output.

These arguments can be used in combination with the others mentioned above: `-i` for case insensitive and `-v` for inverted searches. The `-v` option is important for some tasks. Instead of trying to figure out the regular expression to search for "lines that don't contain >" (Hmm... `"^[^>]*$"?`) you can just run a `grep` command for `">"` with the `-v` option. (Remember to use quote marks around >.) You also don't have to think about how to make a `grep` search that matches lines with three different items—just chain together three independent searches with the pipe, and the final output will be a consensus.

Retrieving Web content using `curl`

In the course of many projects, it is necessary to fetch data from an online database or other Internet resource. The standard way to do this is to browse to a Web page and then either click to download a linked file, or copy and paste text from the browser window into another document. These methods become impractical when data are spread across multiple pages, or when the data need to be updated from the source on a regular basis. Fortunately, a shell program called `curl` (that is, "see URL"), can directly access Internet files on your behalf, downloading the text content of a Web page without any need of a browser window. Later you will learn how to integrate `curl` with other tools to create automated workflows that can draw on outside data.

Ⓦ Ⓛ▶

See Appendix 1 for alternatives.

The simplest `curl` command consists of `curl` followed by the Web address (also called a URL) of the item you want to retrieve. The URL is the line that shows up in the address bar of a web browser window, usually beginning with `http://` (Figure 5.2). Because this http text is so common as to almost go without saying,

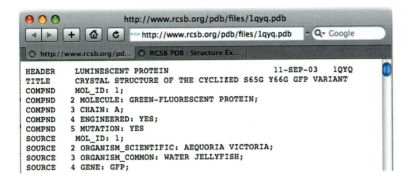

FIGURE 5.2 Example of a URL displayed in the address bar of a web browser

we omit it from many of the URLs in the book, but you should include it in your `curl` commands.

In the shell window, type the following command (or copy and paste it from `~/pcfb/scripts/shellscripts.sh`):

```
curl "http://www.rcsb.org/pdb/files/1ema.pdb"
```

When you press ⟨return⟩, the content of that address (a file describing 3-D protein structure) will be downloaded to your terminal so that it scrolls across the screen. If you want to save the information to a file instead, you can use the redirect > followed by a file name:

```
curl "http://www.rcsb.org/pdb/files/1ema.pdb" > 1ema.pdb
```

To store files without using the redirect operator, use the −o option (output) followed by the destination file name. In this case you can say:

```
curl "http://www.rcsb.org/pdb/files/1ema.pdb" -o 1ema.pdb
```

The `curl` command really begins to become powerful when you use it to download a range of files in one step. For example, to retrieve weather data for each day of a month, specify the range of days in brackets [01-30]. This command will retrieve each day's record from August:

```
curl "http://www.wunderground.com/history/airport/MIA/1992/08/[01-30]/
DailyHistory.html?&format=1" >> miamiweather.txt
```

The command should be entered all on one line. (Again, you can also copy and paste it from the `shellscripts.sh` example file in your `~/pcfb/scripts/`

folder.) It generates thirty separate URLs in sequence, and downloads them one after the other. Because of this, you will need to use the append redirect >> instead of just >, or else each data file will overwrite the previously downloaded file.

An alternative to >> is to save the files using curl's -o option again. By default, curl will only save the last page to that file name, rather than store all the results from the same command end-to-end; but as we shall soon see, this behavior for -o is easily modified to save to multiple files, each bearing a unique name corresponding to the appropriate value in the range you specified. The current value being retrieved within the range is stored in memory as a variable called #1 (reminiscent of the \1 used in regular expression searches). This can be used to generate unique file names for each of the individual results retrieved.

So to download weather data from the first day of each month and save each day as a unique file, use the following command, where the month field is represented by a range instead of the day:

```
curl "http://www.wunderground.com/history/airport/MIA/1979/[01-12]/01/
DailyHistory.html?&format=1" -o Miami_1979_#1.txt
```

Twelve files are retrieved and saved, with the #1 in the file name substituted with 01, 02, on up to 12. Take a look at the names of the saved files:

```
host:sandbox lucy$ ls Mi*
Miami_1979_01.txt    Miami_1979_04.txt    Miami_1979_07.txt    Miami_1979_10.txt
Miami_1979_02.txt    Miami_1979_05.txt    Miami_1979_08.txt    Miami_1979_11.txt
Miami_1979_03.txt    Miami_1979_06.txt    Miami_1979_09.txt    Miami_1979_12.txt
```

Ranges can also use letters [a-z], and they are smart enough to add padding characters, depending on how you write them. So [001-100] will generate URLs in the form file001 rather than file1, with the extra zeros as padding. You can also retrieve data from two ranges simultaneously (e.g., if numbered files exist within numbered directories); in that case, #2 would be the placeholder for inserting the second range of values into a file name.

Rather than provide curl with a sequential range of file names, you may wish to retrieve from a list of URLs where only one portion of the address is changing at a time, but not in a predictable manner. To form this kind of query,

Be sure you **cd** into your **~/pcfb/sandbox** or an appropriate destination folder before running these commands, because they can quickly generate large numbers of files that will clutter your home directory. If the command gets out of hand, perhaps taking longer or generating more files than you expected, you can interrupt with ⌈ctrl⌋ C at any time.

put the list of elements in curly brackets {} separated by commas. For instance, to retrieve a set of four particular protein structures at once, you could use:

```
curl "http://www.rcsb.org/pdb/files/{1ema,1gfl,1g7k,1xmz}.pdb"
```

There are many variations on the queries that you can form with `curl`. Take a look at the `man` page for `curl` for more information on how to tailor your Web downloads.

Other shell commands

The shell commands we have introduced so far will enable you to perform a wide range of tasks, but we have only barely scratched the surface of what is available. In Chapters 16, 20, and Appendix 3 you will encounter a variety of other commands that are very helpful for handling and analyzing data; these include `sort`, `uniq`, and `cut`. With these commands you will be able to do things like count the number of unique entries in a data file—even if it is gigabytes in size, a common challenge that applies to many dataset types. If you find that the command-line skills you have picked up so far are having a big impact on your data analysis abilities, you may want to briefly peek ahead to Chapter 16 and Appendix 3, either now or as you continue with the next chapter; these supplement the material presented here, and provide tips for jumping-off points to learn still more shell commands.

SUMMARY

You have learned how to:

- Edit files at the command line with `nano`
- Send output to a file instead of the screen with > and >>
- Join multiple files together into one file with `cat`
- Use `grep` to extract particular lines from a file
- Pipe output to another command with |
- Download data from the Web at the command line with `curl`

Chapter

SCRIPTING WITH THE SHELL

You are now familiar with some of the powerful commands available within the shell, although to an extent you have been replacing one kind of work (clicking and dragging) with another (typing). You will now learn to gather multiple commands into a file that can be executed as a single command in the shell. This list of commands is a script and is itself a type of custom program.

Combining commands

A shell **script** is just a text file that contains a list of shell commands that you want to run in sequence. Certain properties of the file, which we will describe below, tell the shell how to interpret the commands in the file. When executed, usually by typing the name of the script at the prompt, commands in the file are executed as if you had typed each of them at the command line. Scripts are useful for situations where you want to use the same set of commands on many different occasions, or when you want to execute a set of many commands at once. Here are some examples of possible shell scripts:

- Download twenty files from the Internet and save them to a single large file

- Convert a file from one format to another, process it in a program, and reformat the output

- Rename a set of files or copy them to a different directory

All of these tasks can be accomplished using specialized programs as well as shell scripts, but sometimes the shell approach is the most direct.

Before you write any scripts, we will first walk through a few steps that are necessary for the shell to recognize a text file as a list of commands. In these and subsequent chapters, we will use the formatting convention that text to be entered at the command line (in a terminal window) has `this style` and text in a script (a text editor document) has `this background`. As before, text that you type at the prompt is shown in bold.

The search path

How the command line finds its commands

Each of the commands we've shown you so far (`ls`, `cd`, `pwd`, etc.) is its own little program that the `bash` shell calls when you type in the program name. To find out the directory where a particular program is stored, use the `which` command. For example:

```
host:~ lucy$ which cd
/usr/bin/cd
host:~ lucy$ which ls
/bin/ls
host:~ lucy$ which which
/usr/bin/which
```

Even the `bash` shell that generates the command line is a program with its own location:

```
host:~ lucy$ which bash
/bin/bash
```

These folders (`/usr/bin` and `/bin`) are some of the standard directories where programs are stored on your system.[1] To list the pre-installed programs in these directories, type `ls /usr/bin` and `ls /bin`. If `which` gave you a different answer to the location of a command you tried, take a peek at the contents of that directory as well.

When you create your own script or program, how will the shell know where to find it when you type its name? One way is for you to provide the absolute path. For example, `/bin/date` runs the `date` command. However, it would be inconvenient to have to specify the full path to every program each time you wanted to use it. You also do not want to copy a program to your data directory each time you use it—this is a big mistake that people sometimes make. Copying a program to each directory with data files can lead to the same program being present in dozens of places, and it is a nightmare when a new version of the program comes out and you have to keep track of which version is where. *You want your computer to have a single working copy of each program, and no more.*

---

[1] Even though `/usr` looks and sounds like `/User`, it stands instead for Unix system resource. These commands are available throughout the system, not just to a particular user.

For a variety of reasons, you don't want to put the programs you create in /bin alongside the system programs. First, your computer won't let you unless you grant yourself special administrative privileges. (We'll get to that later.) Second, you don't want to be mucking around deep within the core files of your computer like that. You could end up erasing an important command by mistake (where would you be without ls?) or altering things in such a way that they generate cryptic results.

> **IMPORTANT** The steps described here are critical for setting your computer up for scripting, as well as for programming and other tasks that will be described in later chapters. Follow along with them even if you haven't been trying out other examples.

Better to make yourself a custom directory where you can put your own programs with less risk of damage, and then tell the shell to look for them there. This process is a bit complicated, but it only needs to be done once.

Creating your workspace, the scripts folder

The first step is to make a directory where your program files will go. To begin, cd to your home directory. Then use mkdir to create a folder called scripts. (Remember this command does the same thing as choosing New Folder... from the File menu in the Finder.) Type ls and look at the list of files to make sure that the new scripts directory is there.

```
host:~ lucy$ cd
host:~ lucy$ mkdir scripts
host:~ lucy$ ls
```

You can refer to the new scripts folder anywhere on your system using the full path (/Users/lucy/scripts) or starting with the shortcut to your home directory (~/scripts).

Although you now have a folder for your own custom scripts, your shell still doesn't know to look for programs there. The shell knows to look in /bin and /usr/bin because there is a settings file which lists the places programs may be found. When you log in, this list of places, along with other settings, is loaded into system-wide variables for you as a user. When you type something at the command line that might be a program name, the shell searches through this list to see if it finds anything in those locations that matches. The list itself is stored in a special shell variable called $PATH. To see the current contents of $PATH, type echo $PATH. (Remember, capitalization is important when working in the shell.) The echo command just echoes whatever input you provide it, which in this case is the contents of the variable $PATH.

```
host:~ lucy$ echo $PATH
/usr/bin:/bin:/usr/sbin:/sbin:/usr/local/bin:/usr/X11/bin
host:~ lucy$
```

Your list may be longer or shorter than this; regardless, it shows all the places that the shell will try to find a command corresponding to what you have entered at the command line. You can see that each of the paths is separated by a colon, and that our old friends `/usr/bin` and `/bin` are there, in addition to several other directories.

You can take a look at other shell variables like `PATH` by typing `set`:

```
host:~ lucy$ set
BASH=/bin/bash
COLUMNS=80
HOME=/Users/lucy
HOSTNAME=hosts.local
LOGNAME=lucy
MACHTYPE=i386-apple-darwin9.0
OLDPWD=/Users/lucy/scripts
PATH=/usr/bin:/bin:/usr/sbin:/sbin:/usr/local/bin:/usr/
X11/bin:
PWD=/Users/lucy
SHELL=/bin/bash
USER=lucy
```

This listing only shows a subset of the variables that you will see. The contents of each of these variables is known to the shell and thus available at the command line, and can be retrieved by putting a $ in front of the name. So if someone asks your name, instead of responding `lucy`, you could reply "Hi, I am `$USER`." Instead of typing `cd /Users/lucy` to move to your home directory, you could also type `cd $HOME`. In the upcoming section, we will use the contents of the variable `$HOME`, as well as `$PATH`.

Editing your `.bash_profile` settings file

We now need to show you how to permanently modify `$PATH` so it always includes your new scripts directory.

Your personal settings for the `bash` shell can be modified using a hidden file in your home directory called `.bash_profile`. This file is itself a script, and it contains a list of shell commands that are automatically run each time you log in or open a new Terminal window. The file probably will not exist on your system, but this depends on whether the default preferences have already been modified on your computer. To edit the file, make sure you are in your home directory and type:

```
host:~ lucy$ cd
host:~ lucy$ nano .bash_profile    ← Important step #1
```

For Linux, use
`.bashrc` instead of
`.bash_profile`.

Be sure to include the dot before the name and the underscore in the middle. The `nano` command will open an existing `.bash_profile` if one is present; otherwise it will create a new empty document. If you prefer TextWrangler over nano, you can use the command edit .bash_profile to launch TextWrangler with the file displayed for editing.

Now put a new definition of PATH into this file that includes your scripts directory. Because you want to preserve the other paths too, you will create the new path by adding your scripts folder to the end of the existing paths. One curious property of variables like $PATH in the shell is that they are *read* with a $ before the name, but they are set *without* the $. You will see what we mean in a moment.

Type this line into your file exactly as written:

```
export PATH="$PATH:$HOME/scripts"  ← Important step #2
```

There must be no space on either side of the equals sign. Notice that the first instance of PATH does not have a dollar sign before it, because it is being *set*, not read. This command sets PATH to a combination of its old value (what you saw when you typed echo $PATH) plus a colon, plus the absolute location of your scripts directory. As a shortcut for creating the absolute path of the scripts directory, you read in the value of $HOME, which should be /Users/lucy, and add /scripts to the end of it. You could instead write out /Users/lucy/ scripts, but then this bash profile might not work on another user's system. The whole expression after the equals sign is surrounded by straight quote marks. The export portion of the line just tells the shell to make this PATH variable available anywhere in the shell, not just inside the confines of this particular file.

For Linux or Cygwin, use ${HOME} where it says $HOME.

If your path previously was /usr/bin:/bin, then after executing this command it will be /usr/bin:/bin:/Users/lucy/scripts. In other words, the shell will now include the scripts directory with the other places it looks for commands.

Once the export PATH command looks right, save the file. Because the filename begins with a dot, your .bash_profile will not be visible in a Finder window or with a normal ls command, even though it resides in your home directory. To list

---

⚠ **CHANGING SYSTEM SETTINGS** While you have your .bash_profile open, you have the option of turning on a safety setting. Remember how the cp and mv commands and the > redirect function will overwrite existing files without asking permission? You can turn on a warning to yourself by adding the following line to .bash_profile:

```
set -o noclobber  ← Add this line when you edit your .bash_profile
```

This will make your system reluctant to "clobber" or wipe out existing files. You may sometimes work on systems that do not have this safety net, so you should not become dependent on it. However, especially when starting out, it is easy to have a typo cause you misery. At this point, then, we advise you to stay vigilant, but do turn on **noclobber** in your **bash** settings.

the file, type `ls -a ~/` in a terminal window, or type `cat ~/.bash_profile` to see its contents.[2]

Checking your new $PATH

The moment of truth: seeing if your new settings worked! The configuration file is only read when `bash` first launches, so you will need to open a new terminal window. In the new window, again type `echo $PATH`. This time you should see the absolute path of your scripts folder attached to the end of the default $PATH. If you see your `scripts` directory listed as part of the new path, congratulations! You are all set up now to start writing and using your own software. If you do not see a `scripts` directory in your path after creating a new terminal window, check for typos and then seek help in the forums at `practicalcomputing.org`. If you are working on an operating system other than OS X, also consult Appendix 1.

In the future, you can add other script and project directories to your path by appending them to the end of the `export` line, inside the quotes. It is a good idea, though, to only have a small number of such folders. It is not fun to spend time trying to find problems in a script, only to realize that the copy you are editing is not the same one that the shell finds and executes when you type the command. The order of directories that the shell searches is based on the order they are listed in your path, so you could have a script right in your current working folder, yet when you type its name at the command line, the version that is actually being executed is elsewhere on your computer.

At this point you have created a `~/scripts` folder, and edited your `~/.bash_profile` settings file so that the shell looks in this folder to find commands you type. This part of the setup only needs to be done once. Now you are ready to add as many scripts and other custom programs as you like to this special directory.

Turning a text file into software

You have a place to put your scripts, but how do you actually write a script to put in it? There are a few operations involved:

1. Type the commands into a file.

2. Tell the operating system which program it should use to interpret the commands.

3. Give the script the permissions it needs in order to be executed by the shell.

All that is required to make a working script is a text editor and a shell—both of which you are now familiar with.

Open up your text editor and enter the following text exactly as written (except for the note of warning, of course):

---

[2] If you use **MATLAB**, your `startup.m` file is similar to `.bash_profile`.

```
#! /bin/bash
ls -la  ← Note: this is lowercase L, not the number one
echo "Above are the directory listings for this folder:"
pwd
echo "Right now it is :"
date
```

This file will print a complete directory, then echo a message, and then print the name of the working directory. It then prints the date. Save the file as `dir.sh` in your `~/scripts` directory, with the `.sh` indicating a shell file (Figure 6.1).

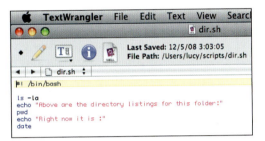

    If you are editing in TextWrangler or in an advanced command-line text editor such as `emacs`, you will notice that the text colors change when you save the script. This is because when you save a file with a `.sh` extension, the editor realizes that it is a shell file and gives you color-coded hints about the content. You will see this behavior when working through the examples in other chapters as well, since TextWrangler is set up to recognize many types of program and data files.

FIGURE 6.1 The bash script `dir.sh`

Control how the text is interpreted with #!

The first line of your script is very special. When the shell executes a text file and encounters `#!` at the beginning of the first line of a file, the shell will send the entire contents of the file to the program named immediately after `#!`. This means that you can write a file that automatically feeds a set of commands to any shell program. In this case, the program we are feeding the commands to is the shell itself, `/bin/bash`.

    When you execute this text file a bit later, the shell will look at the first line, see `#! /bin/bash`, and send the rest of the contents to `bash`. You may have seen the hash mark (#) used to indicate comments in programs (that is, text to be ignored during execution), but it does something different when it occurs like this at the very top of a file.

    The combination of `#!` is called a **shebang**, a name that comes from the "sh" in "hash mark" and the word "bang," which is a Unix-y way of describing the excla- mation mark. It is sometimes hard to remember which order the characters go in (`!#` or `#!`), but if you remember shebang, you'll always get it right. The space after `#!` is not usually included, but we have used it here because it makes it easier to see the path of the file, and helps make sure you don't have an error (such as for- getting the leading slash and typing `#!bin/bash`).

> You have now accomplished one of the key require- ments to make a text file into a script: you've told the shell which program should interpret the contents of a particular file. This step—adding a `#!` to the beginning of the text file that contains a script—is necessary for each script you write, even in other languages.

Making the text file executable by adjusting the permissions

Try your new script by typing `dir.sh` and hitting ⎡return⎤. You can do this from any working directory:

```
host:~ lucy$ dir.sh
-bash: /Users/lucy/scripts/dir.sh: Permission denied
```

It didn't work! This is because you still have to explicitly make the text file executable. The `#!` tells the shell what to do with the text when it is executed, but you still need to give the shell permission to execute the file as a command. By default, the creator of a file can read and write to the file, but cannot execute it. The reason for this is very simple: security. If any text file could be executed, then seemingly innocuous text files given to you by someone else or downloaded from the Web could cause malicious damage to your system.

There is a shell command to make a file executable. First, check where things stand right now:

```
host:~ lucy$ cd ~/scripts
host:scripts lucy$ ls -l ← Again, lowercase L
-rw-r--r--  1 lucy   staff       45 Dec 18 13:52 dir.sh
```

Remember that calling `ls` with `-l` causes it to list files and directories, one per line, and also to provide a bit more information about each listing. This information includes a compact representation of the permissions for the file (the dashes and letters at the beginning of each line), the owner of the file, the group for the file, the size of the file, the date it was last modified, and its name. Within the permissions section, the `r`'s indicate who can read to it, the `w`'s indicate who can write to it, and the `x`'s indicate who can execute it. These are grouped in order of who they apply to: user, group members, and all other users. A dash indicates that permission is not granted to that category of user. See Figure 6.2 for a closer look at how these are arranged.

In the permissions section of the listing for `dir.sh`, you may see that there are no `x`'s for now, which explains why the text file couldn't be executed as a program. You would need to modify the permissions of the file with the command `chmod` (**ch**ange **mod**e) to remedy this. You will do this with each script you write:

```
host:scripts lucy$ chmod u+x dir.sh
```

Check the results with another `ls -l` command:

```
host:scripts lucy$ ls -l
-rwxr--r--  1 lucy   staff   45 Dec 18 13:54  dir.sh
```

The permissions for `dir.sh` now include an `x`, indicating that the file is now executable by the user (in this case, you). The `u+x` argument for `chmod`, which comes between the command name and the filename, tells the shell, "For the main user/owner of this file, add executable permission." You can also subtract permissions using a minus in the first argument (e.g., `chmod o-x dir.sh`); this includes modifying the ability for others to read (`r`) and write (`w`) to a directory or file. Be careful not to take away your own permission to modify a file. For now, you will almost exclusively be using the `u+x` modifier.[3]

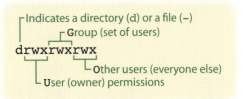

FIGURE 6.2 **Permissions as shown by `ls`**

Try your program again. Type `dir.sh` at the command line and press return . You should see a listing of all the files in your scripts directory, followed by the folder path and the current date:

```
host:scripts lucy$ dir.sh
total 8
drwxr-xr-x    3 lucy   staff   102 Dec 18 15:03 .
drwxr-xr-x+  19 lucy   staff   646 Dec 18 15:03 ..
-rwxr--r--    1 lucy   staff   118 Dec 18 15:03 dir.sh
Above are the directory listings for this folder:
/Users/lucy/scripts
Right now it is :
Sat Dec  18 15:03:34 PST 2010
```

You have created and run your first script! Now go somewhere else in your filesystem and try it again:

```
host:scripts lucy$ cd ~/Desktop
host:Desktop lucy$ dir.sh
```

By making the file executable, you have completed the final step of converting a text file to a script that can be run from anywhere on your system. This sets the properties of the file so that the shell knows it is safe to interpret the file contents as a set of instructions. You will need do this to make each script you write executable.

Generating scripts automatically

Now that you know how to make an ordinary text file into a script, you can begin writing programs that will help ease your workload. In the course of processing

---

[3] There are other ways to specify permissions using a binary-type format. See Appendix 6 for a more detailed explanation, or type `man chmod`.

your data, you will probably use regular expressions—the search and replace operations covered in Chapters 2 and 3—more than any other single tool. In addition to manipulating and reformatting datasets directly, regexp searches can convert a plain text list into a working script, with minimal drudgery. You will use this trick in the next three examples to create useful scripts.

Copying files in bulk

Imagine you have a folder with hundreds of files in it, and you want to copy a subset of those files to another directory based on their contents, rather than on their names. It is not possible simply to use cp with wildcards in this case; you will need to look within the files to see if they contain the text of interest, make a list of those files which qualify, and then use cp on that list.

In this example, you will write a script to copy only those files which contain the words `fluorescent` or `fluorescence` to another directory. The files in question will come from a series of 3-D structure files in PDB format, which are in your examples folder. Start by generating a list of the files you want—specifically, those containing the text fragment `fluor`. Use the grep command, with the option `-l` to return matching filenames rather than matching lines:

```
host:Desktop lucy$ cd ~/pcfb/examples
host:examples lucy$ grep -li fluor *.pdb
structure_1ema.pdb
structure_1g7k.pdb
structure_1gfl.pdb
structure_1xmz.pdb
```

Notice that by using the text fragment for our search word, we can match both `fluorescent` and `fluorescence` with the same query. Note too that in addition to `-l`, we added the case-insensitive flag, `-i`. The result is a list of all the files in ~/pcfb/examples/ with the extension `.pdb` and containing the text `fluor`.

The goal now is to reformat this list of filenames into a series of copy commands. If the list is short, you can copy it from the terminal window and paste it

---

PATHS AND GREP SEARCHES You can modify the scope of the grep search by using * in the path description. For example, to search inside .pdb files in all subfolders of a specific directory, you could write `grep fluor ~/pcfb/*/*.pdb`. The extra */ after **pcfb** means search *all* folders in the **pcfb** folder, not just the **examples** folder. This is a powerful technique to remember for the ls command as well. (For example, `ls ~/*/siphs*.fta`). When you add extra path elements in this way, though, the files must be located in the specified subdirectory. An asterisk here represents one or more subdirectories, not zero or more.

You may see an absolute or relative path in your results, depending on the way you call grep.

into a blank text document. If the list is long, rerun the `grep` command (use ⬆ to step back through your command history) and redirect it into a file by appending `> ~/scripts/copier.sh`:

```
grep -li fluor *.pdb > ~/scripts/copier.sh
```

Instead of displaying the list to the screen, this will put the results into a file called `copier.sh`. Open this file in TextWrangler. (If you are not in the examples subdirectory, you can specify the full path as part of the `grep` command as `~/pcfb/examples/*.pdb`, but the results that are returned will also contain the full path, and not just the filenames by themselves.)

Here is where regular expressions come in. Instead of typing out the command for each filename, you will use regular expressions to semi-automatically transform each of these filenames into a copy (`cp`) command. These files are all located in the same directory, so the beginning of each copy command will be:

```
cp ~/pcfb/examples/
```

You want to insert this text without spaces before each filename. To do this, open the Find dialog, make sure **Grep** is checked, and search for the special boundary character `^` which, though not visible, indicates the beginning of each line. Remember that this marker represents the position just before the first character in a line. For the replacement text, type the `cp` command above. Figure 6.3 illustrates what your search box should look like.

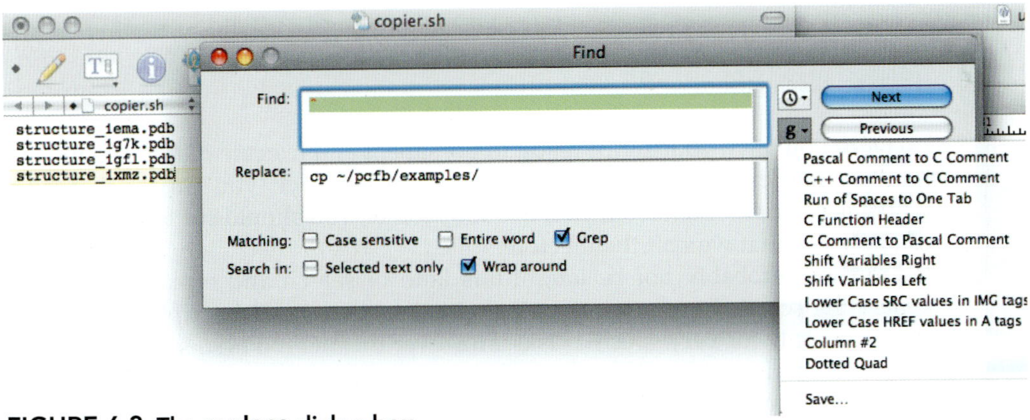

FIGURE 6.3 The replace dialog box displaying its **Save Search** menu g

After you perform this operation, you will have a file with the beginnings of several copy commands:

```
cp ~/pcfb/examples/structure_1ema.pdb
cp ~/pcfb/examples/structure_1g7k.pdb
cp ~/pcfb/examples/structure_1gfl.pdb
cp ~/pcfb/examples/structure_1xmz.pdb
```

Because the ^ searched for *any* line beginning, if you had a blank line at the beginning or end of your document, you will have an extra `cp` command in your file.

Now you need to append the rest of the copy command—the destination directory—to the end of each line. Search for the boundary character signifying the end of each line, `$`, and replace it with the name of the destination directory. In this case, you will copy the desired files to `~/pcfb/sandbox/`. Use this name, with a space before the initial tilde, as the replacement text. Each copy command is now complete:

```
cp ~/pcfb/examples/structure_1ema.pdb ~/pcfb/sandbox/
... lines omitted ...
cp ~/pcfb/examples/structure_1xmz.pdb ~/pcfb/sandbox/
```

This is a somewhat trivial example, because you could have pasted the commands together rather quickly. However, it is not hard to imagine cases where you are dealing with one hundred files instead of four. With two simple search and replace operations, you have turned a plain file list into a script.

At this point you should be protesting that it's not a working program quite yet. We have to indicate what kind of script this is by adding a shebang line (`#! /bin/bash`) at the beginning, and then change the permissions to make it executable:

```
chmod u+x copier.sh
```

Once you have your script ready to go, save it and execute it at the command line by typing `copier.sh`. It won't give you any feedback (you can add that capability later), but if you look in your `sandbox` folder, you should see four duplicates of `.pdb` files that weren't there before.

You will probably not be taught this kind of search and replace method of generating scripts by many computer scientists, because the scripts it generates are useful only for very specific tasks. However the method is so straightforward, adaptable, and practical, that we find ourselves using it very frequently. As you will see later in Chapter 9, you can also use this method to rapidly convert tables of information into program elements.

Flexible file renaming

Renaming files is generally more difficult than just moving or copying them to another directory. This is because `mv` and `cp` cannot automatically generate new filenames from portions of old filenames. For example, to rename a series of protein structure files ending in `.pdb` to files ending in `.txt`, you cannot just type `mv *.pdb *.txt`, since the `*` in the destination filename (`*.txt`) is not replaced with the characters matched by the `*` in the source filename (`*.pdb`). People have written a wide array of file-renaming utilities, but in this section you will learn to generate a shell script to rename an arbitrary list of files however you wish.

In TextWrangler, open the basic list of `pdb` files from the previous example. (You will need to regenerate this file if you already modified it.)

```
structure_1ema.pdb
structure_1g7k.pdb
structure_1gfl.pdb
structure_1xmz.pdb
```

This time, you will modify the filenames as you make new copies, removing `structure_` at the start and also changing the extension from `pdb` to `txt` at the end. When writing scripts which rename files, it is best to copy them with `cp`, rather than move and rename them with `mv`. Renaming is a risky process: even a single error in your script during the testing process could end up deleting all the files.

The basic command to copy a single file to one with the new name format would be:

```
cp structure_1ema.pdb 1ema.txt
```

Again, you can use a series of search and replace commands to convert the list into the necessary commands to perform this operation. First we'll show it as a series of sequential replacements, and then we'll combine them into a single operation.

To generate the new filenames from the old ones you will need to think back to the chapters on regular expressions. The regular expression you construct will need to reuse part of the text found in the search to generate the replacement. In this case there are three parts to the name:

`(structure_)(1ema)(.pdb)`   ← Parentheses denote the text to be captured

Now replace the text in the second element with a wildcard and quantifier, and escape the period in the third element. This is the search query:

`(structure_)(\w+)(\.pdb)`

Check the search term by doing Find without replacement. If it does not highlight the line, then check for trailing spaces or other inconsistencies in the format.[4]

All three parts are needed for the source filename, while only the second part, which falls between the underscore and the period, is reused for the destination filename. The replacement text that generates the destination filename will consist of the second element of captured text, (denoted by \2) plus the new file extension.[5]

`\2.txt` ← Backslash 2 is the contents of the second parentheses from the search

To create the entire replacement string in command format, add `cp` at the beginning, stitch back together the original filename with `\1\2\3`, and generate the replacement filename at the end with `\2.txt`:

Search term	Replacement term
`(structure_)(\w+)(\.pdb)`	`cp \1\2\3 \2.txt`

Run this replacement on the list of filenames, and you should have a series of command lines to rename files (move them) within a directory. You could either copy and paste these commands into a terminal window or save them into a script, as described below.

```
cp structure_1ema.pdb 1ema.txt
cp structure_1g7k.pdb 1g7k.txt
cp structure_1gfl.pdb 1gfl.txt
cp structure_1xmz.pdb 1xmz.txt
```

To turn this list of commands into a script, add the line `#! /bin/bash` to the top of the file, save it as something like `~/scripts/renamer.sh`, and then make it executable with `chmod u+x ~/scripts/renamer.sh`. As the script is currently written, nothing is specified about the paths except for the filenames; therefore, you must run it in the folder containing the files to be renamed. Note that the paths to the files are not relative to where the script is located, which is `~/scripts` in this and most other cases, but rather, relative to where you run the script (the working directory).

This script will only rename filenames that start with `structure_` and end with `.pdb`. If you wanted to write a single script that could handle filenames with other starts and endings, you could add further wildcards and quantifiers:

`(\w+\_)(\w+)(\.\w+)`

> 🔵 **SAVING SEARCHES** If you find a particular search to be especially useful or complex, you can save it in the **TextWrangler Replace** dialog by clicking on the pop-up menu labeled with a **g** next to the **Find** box (see Figure 6.3). The terms will then be accessible from any document you edit.

---

[4] If the concept of captured text is not familiar, review Chapter 2 on regular expressions.
[5] For advanced regular expression operators, it is possible to capture text in nested parentheses. The outer pair becomes \1 and the inner pair becomes \2. So a simpler and equivalent search-replacement pair for this would be (\w+_(\w+)\.\w+) replaced by cp \1 \2.txt.

In (somewhat) plain English, this matches `word_word.word` and captures each word in sequence as `\1`, `\2`, and `\3`. If you have names that contain punctuation (and especially periods), then your search would have to be adjusted to fit that situation.

Automating `curl` to retrieve literature references

Recall from Chapter 5 that the shell program `curl` retrieves documents over the Internet, much like a web browser does. This command already has the ability to retrieve a range of files at once, but it is still often useful to generate a script which contains a series of `curl` commands. This approach is more flexible than using `curl`'s wildcard ranges. In this case, we will start with a list of literature citations and from it generate a script to retrieve a more complete bibliographic record, including the DOI (digital object identifier, a universal address for electronic records). This script starts off with modest goals, but can potentially be made into a very useful reference retrieval device.

The CrossRef registration agency (`www.crossref.org`) provides a means of searching for published literature. The basic format for a CrossRef query is:

```
http://www.crossref.org/openurl/?title=Nature&date=2008&
volume=452&spage=745
```

The variable portions of the address have been highlighted here. This won't work quite yet, though. The system requires verification, so you can either register your e-mail address for free,[6] or temporarily use the a demonstration ID we have created. In addition, the default is for the URL to find the reference and redirect you to the journal's Web site, but we want to retrieve the full set of information associated with that reference. These can be gathered by adding more options to the URL, including a redirect field, a format field, the PCfB identifier for CrossRef:

```
http://www.crossref.org/openurl/?title=Nature&date=2008&volume=452&
spage=745&redirect=false&format=unixref&pid=demo@practicalcomputing.org
```

Open the file `~/pcfb/examples/reflist.txt`, and you will see that the file contains the Web address above, a series of reference entries, and the (very long) search and replacement strings you will use. Copy the first URL line (beginning with `http:`) into a browser's address bar (not the search box) and hit ⌜return⌝. If it works, you can try plugging in some of your own search terms to see how customized queries can be formed. You will use this basic command to retrieve full reference information from several citations.

The references in the example file are in the format:

```
JournalName△    Year△    Volume△    StartPage    ← Remember, △ indicates a tab character
```

---

[6] `http://www.crossref.org/requestaccount/` and see also `http://labs.crossref.org/site/quick_and_dirty_api_guide.html`

A nice feature of CrossRef searches is that author names, titles, and end pages are not required, and even the year is expendable. The actual entries are listed below:

```
American Naturalist△ 1880△   14△    617
Biol. Bull.△ 1928△   55△      69
PNAS△   1965△    53△   187
Science△ △   160△   1242
J Mar Biol Assoc UK△   2005△   85△   695
Biochem. Biophys. Res. Comm.△ 1985△ 126△ 1259
Gene△   1992△   111△ 229
Nature Biotechnology△     △   17△   969
Phil Trans Roy Soc B△   1992△   335△   281
```

Cut and paste these reference lines into a new document, and save it with the name `getrefs.sh` in your `~/scripts` directory.

Now you will convert these entries into a `curl` command to retrieve the entries rather than view them one at a time. Because of the peculiarities of Web addresses (URLs), you will first need to replace any spaces that fall between parts of the journal names, using the replacement `%20`—that is, a percent marker followed by the ASCII code for the space symbol.[7] Search for spaces (not `\s`, and not tabs, but an actual space character) and replace all with `%20`.

Now generate a search query that will capture each of the four fields of the source file. Each field is separated by tabs; thus you can search for "any character" in the first field, and digits in the remaining fields. Some fields may be missing (for example, the year is missing in some references), but the corresponding tabs will still be there; because of this, you should use `\d*` instead of `\d+` for the first digit field, to allow for empty `\t\t` combinations:[8]

```
(.+)\t(\d*)\t(\d+)\t(\d+)
```

This search will store the journal name, year, volume, and starting page of the reference as `\1` through `\4` for use in the replacement.

To construct the replacement term, you can copy it from the example file:

```
curl "http://www.crossref.org/openurl/?title=\1\&date=\2\&volume=\3\&
spage=\4\&redirect=false\&format=unixref\&pid=demo@practicalcomputing.org"
```

This URL is so long that it is split onto two lines here, but in using it, make sure you keep it as a single unbroken line. Note that `\1` is located where the journal title should occur, `\2` is a placeholder for the year, and so on. The ampersands are escaped with backslashes to avoid being misinterpreted as captured text in the replacement string.

---

[7] See Appendix 6 for information on ASCII.

[8] Again, these regular expression search conventions should be familiar to you. If not, review Chapters 2 and 3 or Appendix 2.

When you perform this replacement, your list of bibliographic information will be transformed into a series of `curl` commands. Add a line containing `#! /bin/bash` to the start of the file, save it, and make the file executable with the command `chmod u+x getrefs.sh`. Try the script out by typing the name at the command line. It should print out a list of details about all the references. Once you have confirmed this, you can capture its output by redirecting to a file to retain your new bibliography:

```
host:sandbox lucy$ getrefs.sh > references.xml
```

The new `references.xml` file has much more information on each reference than the original file did. To find out how to automatically import this data file as a record in a bibliography program, look in the `reflist.txt` example file, or download the CrossRef importer plugin from `practicalcomputing.org`.

General approaches to `curl` scripting

For your own purposes in the future, you can take a general approach to making batch retrieval files:

1. Explore the Web site in a browser to find the exact search you are interested in. Note that in some cases it may be necessary to View Source for a page in your browser to find the actual link to data of interest.

2. From the address bar or page source, find the URL for a Web query that gives the result you want. Look for link names in the source, usually starting with `href=` . At times, these might have been visible from the preceding page, but not once you follow them to the data page itself.

3. Determine how the URL is constructed and what parts change. These changing elements will often come after a ?, =, or & symbol.

4. Separate the URL into parts that can be used in your scripts. Sometimes the part that changes is just at the end, but sometimes there are several points where you will want to insert information.

5. Generate the regular expressions needed to convert the data you have into a `curl` command of the appropriate format.

6. Fold these commands into a `bash` script file and make it executable.

7. Save the output of your script with redirection `>>` if needed.

Aliases

You've seen how to join shell commands together into a script. There are also other ways to make multiple operations occur with a single command. One is to create a shortcut called an **alias**. With an alias, you can type a short simple command, and have the system substitute a much longer command in its place. Aliases are defined using the following general format, with no spaces on either side of the equals sign:

```
alias shortcut="longer commands with options"
```

We mention aliases here because they can potentially save you time as you work through the upcoming chapters, but they are explained at length in Chapter 16. For example, you will probably spend a lot of time within the terminal window changing to your scripts or pcfb/sandbox directories. Even with the [tab] shortcut for completing directory names, this takes quite a few keystrokes. With an alias, you can make a shortcut which lets you easily change to that directory:

```
alias cdp='cd ~/pcfb/sandbox'
```

If you type this example at the command line, you will be able to type cdp at any time during your current terminal session and move to your pcfb folder from anywhere in the system.

Aliases defined at the command line only last until that terminal window is closed. To create a more permanent alias, edit .bash_profile, and add the alias line somewhere in that file. Each alias you create in this manner will be loaded for your use at the start of every session.

You can also use aliases to form the beginning of a command, as well as use them in combination with any terms that follow. For example, if you define this shortcut:

For other shells, see Appendix 1 for the format.

```
alias cx='chmod u+x'   ← Give a file executable permission
```

then you can just type:

```
cx myscript.sh
```

and the shell will respond as if you had typed out the entire chmod command. You can also use aliases with wildcards; in this case, for example, you can make all the .sh files in a folder executable with cx *.sh.

Other useful aliases include:

```
alias la="ls -la"   ← Directory listing with hidden files and permissions
alias eb="nano ~/.bash_profile"   ← Edit your .bash_profile
```

Later, as you find yourself working on a project or performing operations repeatedly, you can add your own personal shortcuts to your .bash_profile.

SUMMARY

You have learned how to:

- Set up your command line for scripting
- Manually write scripts specifying a list of commands
- Automatically generate scripts from a list of filenames or data
- Use `curl` in a script
- Use aliases to streamline commonly used command-line operations

Moving forward

Appendix 3 provides a reference table for the shell commands we have covered. Although we will be moving our focus away from `bash` commands for a while, more `bash` commands and ways to use them in your workflows are explained in Chapters 16 and 20. Other powerful shell commands you will encounter include `sort`, `uniq`, `cut`, and `wc`, which can be used in conjunction with each other and with other commands such as `grep`.

## PART III

PROGRAMMING

Chapter **7**

COMPONENTS OF
PROGRAMMING

The idea of "writing a computer program" can sound like a difficult and complicated job, full of strange terminology and complex rules. There are many elements, though, that are universal to nearly all programming languages. Some of these common building blocks are the subject of this chapter. By understanding these components in general terms, you will be better able to think of the structure of your program independent from the syntax and grammar of particular languages.

What is a program?

If you followed along with the previous chapter, you have already written a program in the form of a shell script. There are many programming languages, each optimized for different trade-offs between ease of use, computational efficiency, portability, and specific tasks. The suite of popular languages has evolved over time, so that most of the widely used languages today were not even in existence twenty years ago. However, many key programming concepts and basic building blocks have remained largely the same.

Goals of the next few chapters

At first, writing a program may seem like taking a spelling test, a foreign language exam, a math test, and a car repair class—all at the same time. Learning the syntax of a programming language is the most conspicuous impediment to becoming a programmer, but it is actually more important to learn to think about your data analysis problems as a series of programmable tasks.

As an analogy, imagine you want to instruct someone in how to bake a pie. You might write instructions in English with U.S. units of measure—but you would be describing a set of underlying *tasks* that are themselves independent of language and measurement conventions. That is, the tasks required to bake the pie could be articulated in any language, regardless of syntax, vocabulary, or grammar. You can think of the physical acts that go into cooking as similar to the building blocks of programming, and think of the words used to describe these acts as the terms of a programming language.

This book is not a cookbook—we do not provide you with lists of instructions for different tasks. This book is an introduction to how to make your own recipes. It might seem like a big step to go from not being able to cook at all to cooking from scratch without a recipe, but you can ease into it by starting simply, with only a few ingredients. You will grow familiar with programming by reusing the new skills you acquire and recycling and adapting bits of programs you write.

COMPILED VERSUS INTERPRETED PROGRAMS Programs are rarely written in a language that microprocessors can understand directly. Programs written in some languages, such as C and C++, are translated from human-writable and human-readable code (called **source code**) to computer-understandable instructions. Such translated programs are said to be **compiled**. Once the program has been compiled, it can be run again and again without being retranslated. The majority of programs that you use regularly are compiled in this way. The compiled program file is specific to a particular family of microprocessors and an operating system, and is not **portable** to another operating system. Also, other programmers can't modify the program or recompile it for a different type of computer unless they have the source code.

However, another category of programming languages exists in which programs are not typically compiled, but instead processed by an **interpreter** each time they are run. The interpreter is itself a compiled program that runs directly on the microprocessor. These languages are called interpreted languages, or **scripting languages**, and include Python, R, MATLAB, and Perl, among many others. There can be additional computational overhead to reinterpreting a program each time it is run, but even so interpreted programs have several advantages, especially for scientists. One such advantage is portability: the program file itself is the source code, so it is easier to modify, and can usually be run on any computer that has the correct interpreter. The shell scripts you wrote in previous chapters were interpreted programs, with the interpreter being `bash`. In the following chapters you will write programs for the `python` interpreter.

The distinctions between compiled and interpreted programming languages are discussed in greater detail in Chapter 21, along with instructions for compiling simple programs.

Practical programming

The scope of the next few chapters is modest compared to books dedicated to learning programming; we hope to convey a minimal set of skills that will get you started writing your own programs. These chapters are not intended as a comprehensive introduction to programming, or as a grounding in programming theory, but as a taste of practical programming. This will give you a footing for solving some of the simple problems you already face, as well as provide a departure point for getting started on more complex projects. The next step to taking on these projects would likely include more in-depth instruction, via either online resources or any of the many excellent books available for beginners. At the end of the next few chapters you should have a much better sense of which tasks are best approached by writing new software, what's entailed in writing a new program or repurposing an old one, and what specifically you would need to learn to take on more complex challenges.

The present chapter focuses on the basic building blocks of programs and introduces some commonly used terms (Table 7.1). Most of this discussion is not specific to any particular computer language, but we do lean somewhat towards Python, the language we will focus on in the following chapters. If you plan to learn programming with a language other than Python, it is still important to become familiar with the concepts described here. In Appendix 5, you can see examples of how these elements are applied in a variety of programming languages. Although this treatment is neither comprehensive nor general to all computer languages, it will form a foundation for translating data-analysis tasks ("I need to reformat each line of this file into tab-delimited decimal values") into programming terms ("Open file, read in each line with a `while` loop, parse variables, save formatted output"). After this overview, the subsequent chapters will explain how to use these building blocks specifically in the context of Python.

TABLE 7.1 Glossary of program-related terms

Term	Definfition
Arguments	Values that are sent to a program at the time it is run
Code	*Noun*: A program or line of a program, sometimes called source code *Verb*: The act of writing a program
Execute	To begin and carry out the operation of a program; synonymous with *run*
Function	A subprogram that can be called repeatedly to perform the same task within a program
Parameters	Values that are sent to a function when it is called
Parse	To extract particular data elements from a larger block of text
Return	In a function, the act of sending back a value; the value can be assigned to a variable by referring to the function name (e.g., in $y = \cos(x)$, the function cos calculates and returns the value of the cosine of x, which is assigned to y)
Run	To execute the sequence of commands in a program; can also refer to the processing of a file by a program that finds instructions within it
Statement	A line of a program or script, which can assign a value, do a comparison, or perform other operations

Variables

The anatomy of a variable

Anyone who has taken algebra knows about variables. In essence, each variable is a name that holds a value. The value can be a number, a series of characters, or something even more complex. As the name suggests, variables can vary, by changing what pieces of information they hold or point to.

Each variable has several attributes. First, there is its **name**, usually a plain word that gives some indication of the variable content, such as `Sequence`, `Plot_Num`, or `Mass`. The programmer designates the name of the variable at the time the program is written. Some languages require a special symbol or symbols to be associated with the name of the variable; in Perl, for instance, variable names begin with `$`. Other languages, such as Python, forbid punctuation within variable names. Almost all languages make it illegal for the first character of a variable name to be a digit. Sometimes single letters alone are used as names, especially when the variable is keeping track of some internal value just for controlling program flow. In general, though, it is best to pick variable names that are distinctive and memorable, even if it means a bit more typing. This is one of the simplest steps you can take to improve the readability of your programs so that they can be understood later, whether by you or by someone else. To improve clarity, most variable names in this book will begin with a capital letter, so that they can be easily distinguished from other program components. (The exception is variable names only a single letter long; we have left these lowercase, since they can't be confused with anything else.)

The second variable attribute addressed here may initially seem a bit more abstract. This is its **type**, a designation of the kind of information the variable contains. Variables in modern computer languages each have a type, accommodating numbers, strings made up of text, images, lists of data, and so on. It is also possible to design and use entirely new types.

A third attribute for a variable is its **value**, that is, the actual piece of information that it holds. Not only can a variable be assigned a value upon creation, but it can be reassigned different values throughout the course of a program. A close relationship exists between a variable's type and its value, such that a variable of a certain type can only contain certain kinds of values.

Variables have other attributes that we won't discuss in detail here, but they can become important as programs get more complex. One of these is **scope**, which designates where in a program a variable can be accessed.

Basic variable types

There are a few variable types that you will use again and again; examples of these are shown in Table 7.2. You will notice that there are several variable types for numbers. This is because there are trade-offs between the size and precision of a number and the computational resources needed to store and manipulate the number. Having several number types may seem unnecessarily confusing, but it allows programmers to write code that is more computationally efficient and uses less memory; it also helps ensure that the program is behaving as expected.

Integer   One of the most basic variable types is **integer**, which holds whole numbers without any fractional component (e.g., 0, –1, and 255). The limits on integer size are language-dependent, but in most languages integers can range from –2,147,483,648 to 2,147,483,647.[1] This range of possible values for an integer may seem so large that there is no chance that you could ever exceed it—and yet a timer that counts elapsed time in milliseconds would do so within a month. If you find yourself in a situation where the regular integer type is insufficient, you can go with one of the more specialized integer types: a variable of type **long** can hold much larger integers, but will take more memory, while a variable of type **unsigned** can go twice as high as a normal integer, but cannot hold negative integers.

TABLE 7.2 Commonly used variable types

Type	Example
Integer	98
Float	98.6
Boolean	False
String	'Bargmannia elongata'

Floating point   Many scientific applications require numbers that either are extremely large, or else have decimal fractions. A useful type for these situations is **float**, shorthand for "floating point." This name signifies that the decimal point can float to any position, and not just lurk at the very end of the number. Like scientific notation, a float can represent a huge number or a tiny number using just a few digits (e.g., 4×10^{22}, the number of bacteria in a bioluminescent milky sea, or 1×10^{-15}, the diameter of a proton in meters). Floats are sometimes written as 4e22 or 4E22. The limitations of floats mainly have to do with the number of significant digits they can accurately retain, rather than the size of the number. If a standard float isn't precise enough for your program, you can use a **double**, which occupies twice as much memory (hence the name) and has higher precision.

Boolean   A variable of type **Boolean**[2] can have one of two values, True or False. Booleans are used for logical operations, such as the response to the question, "Is the value of variable x greater than variable y?" Booleans can also be interpreted as numbers. In this context, True is equivalent to 1 and False is equivalent to 0.

Strings   Most languages have a data type called **string** to handle a sequence of text characters. A string can include letters, digits, punctuation, white space (spaces, line endings, etc.), and other characters. Within a program, string values are almost always bounded by a pair of straight quotation marks, either single (') or double ("). This is to differentiate text that is placed in a string from variable names and commands. Here are some examples of assigning strings to variables:

---

[1] Such limits are determined by the amount of memory used to store a variable; in the case of an integer with these limits, this amount is 32 bits. See Appendix 6 for more specifics on how numbers are stored and how this affects the way computers can work with them.

[2] The name "Boolean" is derived from the 19th century mathematician George Boole.

```
SequenceName = "Bolinopsis infundibulum"
Primer1 = 'ATGTCTCATTCAAAGCAGG'
DateString = "18-Dec-1865\t13:05"
Location = "Pt. Panic, Oahu, Hawai'i"
```

In each case, the text within the quotation marks on the right is placed in the variable named on the left. Some programming languages allow strings to hold nearly any character, while others cannot include certain characters. Unfortunately for international users, support for extended character sets (for example, ø, ü, °, ß, and even カナ) can require special kinds of string variables that support Unicode text encoding. Strings can be of arbitrary length—that is, they can contain no characters at all, a single character, or many characters.

Variables as containers for other variables

Arrays and lists

An **array** or matrix is a collection of data that lives under the roof of a single variable. Arrays can consist of one dimension, such as the set of integers from 0 to 9; two dimensions, such as a grid describing the pixels of a black-and-white photograph, or a record of depth, temperature, and salinity over time; or three or more dimensions. One-dimensional arrays are often referred to as **lists** or **vectors**. Depending on the computer language, an array or list can contain a mixture of types, including other lists. For example, a list called morphology could include floats, integers, string descriptions, and lists:

```
Morphology=[1, 0, -2, 5.27, 'blue', [4,2,4]]
```

Lists of lists can store multidimensional data, such as two-dimensional matrices with columns and rows. Lists come with useful tools for finding particular values, extracting elements, sorting data, performing some function on each datum, and other similar tasks.

Each item in an array can be accessed by referring to its location within the matrix, usually with an index number contained in brackets. In many languages, the first element is index 0, rather than 1. For a one-dimensional list such as [A,G,T,C], the index is just the number referring the position in the list (in this case, 0 for A, 1 for G, 2 for T, and 3 for C). With a two-dimensional matrix that looks like a checkerboard (Figure 7.1), you can specify an entire row or column of data, or just a single element by giving its row and column indices. To change elements in a list, you can usually use the same syntax that is used to access the values, with the list item on the left and the value to be assigned on the right. For example:

$$A = \begin{bmatrix} 2 & 7 & 6 \\ 9 & 5 & 1 \\ 4 & 3 & 8 \end{bmatrix}$$

FIGURE 7.1 A multi-dimensional array

```
MyArray[2] = 5
```

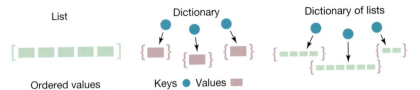

List   Dictionary   Dictionary of lists

Ordered values   Keys ● Values ▨

FIGURE 7.2 The structures of a list, a dictionary, and a dictionary of lists

Arrays are very important in scientific programming, making the code more efficient and easier to write and understand. If you find yourself repeating the same operation or calculation on a long series of values—especially if you are copying and pasting names like `Species1`, `Species2`, `Species3`—then this is the perfect time to learn how to store and process such variables as lists.

Dictionaries and associated arrays   A **dictionary**, also known as an associative array, hash, or map, is another type of container for more than one variable. Unlike a list, which is simply a sequence of ordered values, a dictionary is a collection of names, or **keys**, with each key pointing to an associated **value** (Figure 7.2). The keys can be numbers, strings, or other types of variables. The keys for a dictionary must be unique, since each key can point to only one value, but different keys can have the same value. Here is some example Python code showing the creation of a dictionary, followed by the assigning of key-value pairs:

```
TreeDiam={}  ← Create an empty dictionary with {}
TreeDiam['Kodiak'] = [68]
TreeDiam['Juneau'] = [85]
```

True to the dictionary analogy, values in dictionaries are looked up according to their keys, rather than by their position, as would happen in a list. For example, if you used a dictionary type to create and populate an actual dictionary, you could find the value for "cnidarian" using the word itself as the key, rather having to know that it was the 1024th item in the alphabetical list of definitions. In fact, the elements of data in a dictionary are in no particular order, and can be accessed *only* by looking up their key word.

Dictionaries are especially useful for gathering together data that describe some properties of a series of entities. For instance, you could create a dictionary containing all the molecular weights of amino acids, allowing you to easily retrieve the values for each amino acid by its name. Another handy trick is to combine dictionaries and lists, making a dictionary of lists. For instance, you could associate lists of DNA sequences from several different genes with particular specimen numbers.

Converting between types

Some programming languages, including C and C++, require the programmer to explicitly state variable types and then strictly enforce these types thereafter. In other words, somewhere in the program before you use x, you have to specify that x is going to hold an integer. This makes for robust programs, since the computer never has to guess what type was intended, but it tends to reduce flexibility and require more code. Other languages, such as Python, set the type of the variable according to the values that are assigned to them, and only then strictly enforce these types. This is also quite robust, and means that the programmer still needs to think about variable types even though they aren't explicitly specified. Still other languages, for example Perl, try to take care of all type-related issues behind the scenes. The problem with this less restrictive approach is that code can break in very confusing ways when things go wrong. Although it can be convenient to have types handled entirely in the background, in many cases your intentions may be ambiguous and the computer's best guess is not what you expected.

This is particularly evident when you have a string containing a series of characters that could represent a number. For instance, take the string "123". (Remember that the contents of a string are always contained in quotes.) To the computer, this is a sequence of characters, just like any other sequence, and not a number at all. A variable that contains this set of three text characters is very different from the integer 123; the latter is stored in the computer's memory as a binary representation of the number 123, not as a sequence of characters. If you were to add the integer 5 to the string "123", for instance, would you expect to have the "123" treated as a number, resulting in 128, or the 5 treated as a string, resulting in "1235"?

In most programming languages, you would need to explicitly convert the variable types in this example to clarify the intended operation. That is, if you wanted the result to be the number 128, you would write the equivalent of, "Add 5 to a numerical interpretation of the string '123'"; whereas if you wanted the result to be the string '1235', you would write the equivalent of, "Append a string interpretation of the integer 5 to the end of the string '123'." Because this kind of thing occurs often, there are commands built into nearly all languages to perform such conversions.

Variables in action

Programs need to do a lot more than just store information in variables. They also need to be able to do something useful with them. Often this consists of performing some kind of calculation to generate a new value. The tools for modifying or calculating new values include **operators** and **functions**.

Mathematical operators

Operators in computer programs include the mathematical operators that you are already familiar with: addition, subtraction, multiplication, division, power (x^y), and modulo (finding the remainder left after division).

The actions that operators perform depend on the language and the type of data they are operating on. Usually this is quite intuitive. Take the + operator:

If it is applied to two integers, it will return their sum. In many languages, + can also be applied to strings, as alluded to in the previous section, to produce a new string that contains the joined contents of the other strings. Although the operator is written as + in both cases, it is performing different tasks in each context.

The data that an operator acts on don't necessarily have to be of the same type. The + operator can add a float to an integer, for example, but not all combinations of data types are supported by all operators: the + operator cannot combine a string and an integer.

In some languages, including Python, operations on integers alone will often generate integers, perhaps even when you were expecting a float. The most trouble comes with the division operator, /. The precise result for 5/2 would of course be 2.5, which needs to be stored as a floating point number. When the result is shoe-horned into an integer variable, however, it truncates the decimal portion, leaving the somewhat confusing result of 2, with the remainder having been discarded.

The operators discussed so far all do a calculation and return some new value as the result. One operator that we have taken for granted so far is the = operator, which assigns the value on the right side of = to the variable on the left side. This allows you to copy the value of one variable into another, or to store a new value.

Comparative and logical operators

Other operators make comparisons of variables and then return a Boolean value in the form of True or False. A common example is to test if one variable is greater than another. The results of operators can be assigned to a variable, or just used directly to help the program decide what to do.

Comparison operators can report whether two entities are the same or not. The equality operator is often written as ==, and returns True if the entities on the left and right side of the operator have the same value. This is different from a single =, which assigns values rather than comparing them. Though it is common to think of = alone doing both of these tasks, most languages will not allow its use in both contexts.

In some languages, equality operators can even apply to strings. The statement GenusName == 'Falco', for example, would return True if the variable named GenusName contains a string with the value 'Falco'. These kinds of equality comparisons have to be exact, including whether the characters in the string are uppercase or lowercase. If you want to test whether a bit of text partially matches another string, there are other string-related functions which work more like the regular expressions of Chapter 2 and the grep command in Chapter 5.

Another useful operator which returns a Boolean value is the in operator (some operators are words, not punctuation marks). For example, the expression x in A returns True if the value of x is contained among the items in list A. If A is a list of several lists, then x must be a list as well, not one of the elements of those lists.

Other logic-related operators work with Boolean variables themselves. The names of these operators (and, or, not) describe their mode of operation pretty well. These are important in testing several comparisons simultaneously, for example (`Depth > 700 and Oxygen < 0.1`).

Operators are evaluated according to a hierarchical order, per normal algebraic rules. Usually the order is exponents; multiplication and division; addition and subtraction; comparisons of equality; and finally, logical operators. Within these

TABLE 7.3 Common operators and their symbols	
Operator*	**Common symbols**
Mathematical	
Addition	+
Subtraction	–
Multiplication	*
Division	/
Power	**
Modulo (remainder after division)	%
Truncated division (result without remainder)	//
Comparative	
Equal to	==
Not equal to	!=, <>, ~=
Greater than	>
Less than	<
Greater or equal	>=
Less or equal	<=
Logical	
And	and, &, &&
Or	or, |, ||
Not	not, !, ~

*Not all languages support all operator symbols, and in some cases the name of the operator is used instead of a symbol.

categories, operators are evaluated left to right. This order can be modified with parentheses, grouping operations that should be performed first. The most commonly used operators are listed in Table 7.3.

Functions

Functions can be thought of as little stand-alone programs that are called from within your own program. Many functions for common tasks come built into programming languages. You can also write your own custom functions; these can be reused as often as you want, both within a program and potentially later on by other programs that you write. Functions can be defined locally in your program, or they can be stored in external files and loaded into your program as needed. Functions will be an important part of all but the simplest of the programs you write.

Functions can accept variables, which are then referred to as **parameters** of the function. These are usually sent to the function within parentheses, `()`. Functions can also return values. For example, the `round()` function, as used in the context of `y = round(2.718)`, takes a number with a fractional portion (in this case the floating point value `2.718`) and returns the value after the decimal is rounded off (here, rounded number is assigned to the variable `y`). Even when functions aren't designed to take any parameters, a set of trailing empty parentheses, e.g., `PrintHelpInfo()`, is still usually needed to differentiate the name as pointing to a function, rather than to a variable or other code element.

Flow control

Decisions with the `if` statement

At this point you know a bit about variables and how to obtain or create values based on their state, through comparisons and calculations. With just these tools alone, though, you couldn't write much more than a glorified calculator. The real power of programming becomes apparent with conditional decision-making—the ability for a program to follow different courses of action depending on the state of variables.

The `if` statement is the most widely used building block for such decision-making. It is like a sign at a fork in the road. If the statement on the sign is true, then you take one fork, and if it is false you take another route, eventually rejoining the main sequence of commands within the program. In programming terms, the `if` statement evaluates an expression that returns a Boolean. If that expression is true, it then executes a specified set of commands. You can use an `else` statement in conjunction with an `if` statement to execute another set of commands when the expression is false. You can also chain together a series of `if-else` statements to combine several of the logical conditions into a complex sequence of events. Figure 7.3 depicts both an `if` and an `if-else` statement.

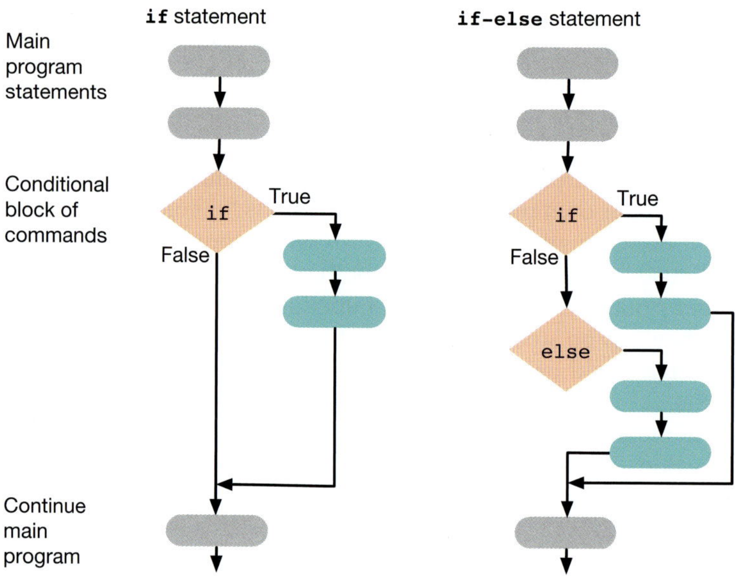

FIGURE 7.3 The flow of a program through conditional statements The ovals represent program statements and the diamonds represent decision points. Arrows represent the operation of the program under different circumstances. In an `if-else` statement, for example, if a statement is evaluated as `True`, you take the `if` fork; but if it evaluated as `False`, you take the `else` fork.

Below is an example of an `if-else` statement as it would be written in Python; other languages will have slightly different ways of stating the same thing:

```
A = 5
if A < 0:
    print "Negative number"
else:
    print "Zero or positive number"
```

Looping with `for` and `while`

The building blocks presented so far allow you to write a program consisting of a linear sequence of events to be executed from top to bottom, with the power to control which events in this sequence are executed. A linear program like this can automate a complicated sequence of analyses, but it isn't well-suited to repetitious tasks that require the same computations to be made over and over. This, however, is the exact kind of task you usually want to write a program for. The power of a program to plow through hundreds of calculations in only a few lines of code is realized by **looping** repeatedly through the same basic set of commands. A `for` loop and a `while` loop are illustrated schematically in Figure 7.4.

The `for` loop   Loops are portions of programs that are repeatedly executed until some condition is met. Of these, by far the most widely used is the `for` loop. A `for` loop cycles through each item of a predefined collection, performing a series of commands each time it does so. The items it cycles through can be drawn

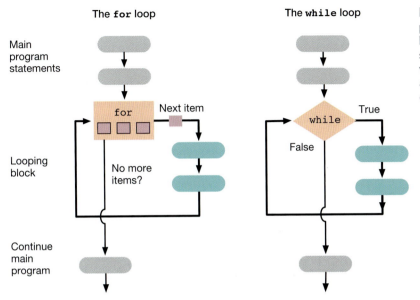

FIGURE 7.4 **The flow of a program with a for or while loop** The orange box represents a collection of objects which are taken one-by-one and operated on by the blue ovals.

from a variety of collections, such as a range of numbers from 1 to 100, a list of species names, an archive of protein sequences, the lines of a data file, and so on.

The power of a `for` loop comes from being able to repeat the same operation on each item in a long list. In fact, some programming languages have a specific `foreach` statement emphasizing this item-by-item capability. Loops can also be nested, or placed one inside the other. For instance, you might use three nested `for` loops to step through each amino acid of each protein sequence in each file of a folder.

As much as any programming element, the syntax used in `for` loops varies from one language to another. Many examples of this variation can be seen in Appendix 5. Two examples in particular are shown in Figure 7.5, one from Python and the other from C; note how verbose the C code is compared to the Python code. This may help you understand why we prefer Python for our work, and why we chose it as the programming language for this book.

(A) `for` loop in Python

```python
for Num in range(10):
    print Num * 10
```

(B) `for` loop in C

```c
for (Num=0; Num < 10; Num++) {
    printf("%d",Num * 10);
}
```

FIGURE 7.5 **Code snippets showing a for loop in Python and in C**

Innumerable situations can be tamed with `for` loops: processing every file that meets a certain criterion, splitting each line of a data file into its component parts, translating each gene sequence in a FASTA file, or performing a unit conversion on each entry of a column of numbers in a data-logger file.

The `while` loop   In some situations, you enter a loop without having a predetermined list to cycle through. When a loop is open-ended like this, you can use a `while` loop to continue cycling through until some logical condition is evaluated as `False`. Like an `if` statement, the `while` loop begins with a logical test, but unlike the `if`, when the conditional block of code is complete, the program does not return to the main path, but loops back up and checks the condition again.

At least one variable involved in the test expression must be modified by the statements in the loop, or else the loop will keep cycling forever. This is the infamous "infinite loop." To avoid this situation, programmers sometimes link in a second fail-safe condition. Typically this is a counter—that is, a variable which keeps count of how many times the loop has been traversed. The loop can be made to exit when the count exceeds the expected maximum iterations.

Situations where `while` loops are useful include improving an approximation until it reaches a sufficient level of precision, waiting until a temperature or environmental condition is reached, continuously monitoring a sensor until a stop command is given, and reading lines from a file until the end is reached.

Using lists and dictionaries

Lists

Think of a list as a series of numbered boxes that start with box 0. You can create a list with a given number of boxes, and later append more boxes to the end of this series if you need more containers. A list that includes a box numbered 16 will necessarily have a box numbered 10. If you try to access box number 94 when there are only 70 boxes, though, an error will be generated. The data in a list can be accessed one at a time, or ranges of data can be accessed at once. You can peek inside any box you like, replace the contents of a box, remove or insert boxes into the beginning or end of the list (shifting the remaining boxes), or even reorder the boxes. After such modifications, the number used to access each box depends only on its position.

When do you use lists as opposed to singular variables? Pretty much any time you are handling more than one piece of closely related information. Imagine, for instance, a field site at which you measure the diameter of 5 trees. You might first think to load these values into a series of 5 variables named `Diameter1`, `Diameter2`, `...`, `Diameter5`, make calculations on each (`crosssection1=`, `crosssection2=`), and then print each result with five separate `print` statements. This is a cumbersome solution, since you would need to repeat a nearly identical set of commands to act on each variable independently for each tree. It is cleaner and easier to create a list variable called `Diameter` which holds the five values in a list. By using a `for` loop, you can then easily write one set of commands which performs the same operations on each measurement in turn.

Lists also have the advantage of making your code more general. In our example, you may later want to use the same program to analyze a dataset with 6 measurements per site. If you used a different variable name for each measurement, you would have to add a variable to your program to accommodate the new dataset. If you use a list, however, your program would require no additional modifications. A list can handle thousands of values just as easily as it can handle ten.

Dictionaries

Dictionaries may seem a bit less intuitive than lists at first, but it is likely that you will find them to be equally useful, if not more so. Recall that the data entries of a dictionary are looked up with unique labels known as keys. If you wanted to perform an operation using each entry in a dictionary, you would first get a list of the keys, and then use those to retrieve each dictionary item one by one. Since in a dictionary, data elements don't have a fixed position, sorting the entries within the dictionary is not meaningful the way it would be with a list. However, sorting the keys and then retrieving the associated values in a particular order is often useful.

The unordered nature of dictionaries makes them more flexible than lists in several important ways. In a dictionary, you can create an entry for key 16 without having any data corresponding to keys 0 to 15. This lets you fill in data as they become known. Although list indices must always be integers, dictionaries allow you to associate values with keys of many different variable types. For instance, you could use a dictionary where the keys are strings, each string being a specimen name:

```
TreeStat = {}  ← Create an empty dictionary to build upon
TreeStat['Kodiak'] = [ 68, 57.8, -152.5]
TreeStat['Juneau'] = [ 85, 58.3, -134.5]
TreeStat['Barrow'] = [133, 71.3, -156.6]
```

Dictionaries and lists can be combined in useful ways. Imagine that each field site where you are measuring tree diameters has a unique name. You could create a dictionary `TreeStat` that uses the name of each site as the key, with a whole list of measurements and coordinates as the associated value for each entry. Such dictionaries of lists provide an easy way to store and return a large set of related measurements.

Other data types

One of the other basic data types, the **string**, is actually quite like a list in many respects. It is, after all, an ordered sequence of characters. You can step through each character of a string with a `for` loop, and you can even extract substrings or certain elements from a string in the same way you would extract subportions of a list. A difference is that you can't always modify individual elements of a string the way you can those belonging to a list. Fortunately, there are other ways to carry out these kinds of operations on strings, as you will learn in later chapters.

Finally, there are other data types that hold collections of objects, such as sets and tuples in Python; however, these will not be addressed here.

Input and output

User interaction

Most programs provide for some means of user interaction. The vast majority of computer users have interacted with programs only through graphical user interfaces. Programs that are run at the command line (including the programs that you will soon write) also provide for several means of user interaction. Program options specified at the command line when a program is launched are called **arguments**. The command `ls`, for instance, can be run with several arguments that control what files are displayed, and in what format. These arguments are specified by listing them after the command when it is executed, for example `ls -a` or `ls *.txt`.

Arguments are a mode of user input for programs that first gather all the information they need, and then go about their business in one big push. This is good for the user because it isn't necessary to babysit the program while it runs, or to respond to queries during the execution. This is especially important for programs that take a long time to finish. If your program is configured with arguments, then it is also easier for other programs to control it. This will be important if you want to build up workflows that automatically shuttle data through a series of programs. Python and many other languages have built-in tools for passing arguments from the command line to variables within the programs you write.

Some command-line programs are designed to respond to user input while they are running. Command-line text editors, for example `nano`, are an elaborate example of this. Simple interactive interfaces can be desirable in some cases, and like arguments, they are simple to implement in your own software. For instance, you will create an interactive interface for the DNA analysis program described in the next chapters; this will allow you to calculate basic molecular properties of short sequences that you enter at the interactive prompt.

User interaction isn't just about getting input from the user; it is also about generating output, as well as giving the user feedback about the progress and results of the program. Typically, the command used to send output to the screen is either called `print` or some variation of that word. Output can also go directly into a file, a process we discuss next.

Files

Like variables, files associate some chunk of data with a name. Unlike variables, which have a fleeting existence while the program is running and thus live only in the computer's volatile memory (RAM), files reside on the disk drive. They are still there after the program stops, and even after the computer is turned off. While variables are only accessible from within the program that created them, files are available to any program with access to the disk and permission to read them. Files therefore serve both as a long-term repository of data and a way to shuttle data between programs. Reading and writing files is an essential part of the programming you will do for your scientific research.

There are two separate levels of operations that must be completed to access data in a file. The first level consists of all of the mechanical details of gaining access to the contents of the file from within a program, such as finding it on the disk and loading its contents into memory. Fortunately, programming languages provide tools and commands that take care of all this work behind the scenes. In order to gain access to the contents of a file, you usually need to indicate only the path to the file and whether you want to read from it or write to it.

The second level of operations required to access data from a file is to establish relationships between the data in the file and variables in the program. This works in both directions. That is, when reading a file, you are **parsing** its data from a file into program variables, whereas when writing to a file, you are packaging data from program variables.

For example, a data file might contain lines of text like the following:

```
ConceptName     Depth   Latitude  Longitude Oxygen TempC   RecordedDate
Thalassocalyce  348.7   36.7180   -122.0574  1.48  7.165   1992-03-02 17:40:09
Thalassocalyce  520.3   36.7491   -122.0368  0.52  5.826   1992-05-05 20:01:10
Thalassocalyce  118.4   36.8385   -121.9676  1.52  7.465   1999-05-14 17:48:27
Thalassocalyce  100.9   36.7270   -122.0488  2.30  9.497   1999-08-09 20:12:11
Thalassocalyce  1509.6  36.5846   -122.5211  0.95  2.774   2000-04-17 20:04:23
```

This information, stored on the disk, will be most usable in the programming environment if each column can be loaded into list variables of appropriate types (strings, integers, floats, etc.). Conversely, once a calculation or conversion is performed, the variables will likely be written out to a file and saved.

Parsing and packaging file data usually comes down to string manipulation. The text is read from a file as strings, usually line-by-line, and the data are then extracted from these strings and placed in the appropriate variables. Regular expressions, the first tools you learned in this book, are often used for this process. Writing data to text files is the opposite process: data from variables are converted to strings, joined together with the appropriate formatting characters, and then written to the text file, once again line-by-line. The specifics of how you parse and package data will depend on the format of the data within the file, meaning that data from different sources, databases, or instruments will often require that you write or modify a small program to handle conversions and calculations. This, then, is where the heavy lifting of file handling occurs when you are programming.

In addition to storing data for analysis, files can also be a means of controlling program behavior. You have already seen how a text file can store a list of program commands as a script. Many programs can take their raw input and marching orders from a file, after which they send their output directly to another file. Text files can also serve as a record of what a program did; in such cases, they are called log files, or just plain logs.

Libraries and modules

Computer languages have many built-in tools available out of the box. These include some of the basic building blocks we have mentioned—predefined variable types, basic functions, and simple mathematical and logical operators. For more specialized tasks, additional functions can be written and then bundled together in **modules**. By **importing** from one of these modules, you can gain access to the suite of functions stored within. Some of the most commonly used modules provide tools for more advanced mathematical functions (e.g., sin and log), interacting with the operating system (e.g., listing directory contents and executing shell commands), retrieving content from the Internet, performing regular expression searches, working with molecular sequences, and generating graphics. Python modules are explored in Chapter 12.

Comment statements

Programming languages provide for comments—that is, portions of text that are helpful to human beings but are to be ignored by the computer when the program is run. Comments are usually marked by starting the line with a special character, such as #, //, or %.

Comments serve several important functions for programmers. Commented blocks of text are used to provide documentation and a description of both what a program does and how to run it. Commented text at the top of a script usually includes a list of any arguments the program may take, and a revision history of changes, so that updated versions of the script can be distinguished.

Within the program, comments may be associated with individual lines or with subsections of code, to explain how they operate. These internal notes are important for future readers of your program (including yourself), who typically are trying to determine (or maybe just remember) how the program operates.

Comments are also useful for troubleshooting your program. You can isolate problems by turning portions of the code on and off with comments, to see if an error still occurs in the same way.

Objects

A full treatment of **objects** is beyond the scope of this book. However, objects are essential components of contemporary programming, and aspects of these components are woven into the fabric of many programming languages; thus, even a simple treatment needs to mention them.

An object is a sort of "super variable" that can contain several other variables within it as part of its definition. In fact, not only can objects contain variables in the traditional sense, but functions as well; in this context, these are called **methods**.

Consider a bicycle as an object. Your own personal bicycle may be just one example or instance of the platonic ideal of a bicycle. Likewise, your bike has many properties associated with it. In computer terms, you access the nested properties of an object using a **dot notation**, where the name of the property is joined to the name of the object with a period:

```
MyBike.color  ← Color of your bike
MyBike.tires  ← Properties of your tires
```

Used in this way, these statements might report the status of those properties, returning `'blue'` for color, as well as various values for the brand, air pressure, diameter and tread of your tires. These are examples of **nested** properties:

```
MyBike.tires.pressure  ← A nested property of your tires
```

Methods (that is, functions) can also be associated with the object; if so, these are accessed with dot notation as well. These methods can be little programs, which take in values and apply them to the bicycle object:

```
MyBike.steer(-4)  ← Steer 4° to the left
MyBike.color('red')  ← You can often set and read with the same notation
MyBike.pedals.pedal(100)  ← With the pedals, pedal at a speed of 100
```

The reason that these object-related ideas are important is that in many languages, even simple variables are objects and have their own methods and properties. Sometimes instead of using a separate function and sending your variable to it as a parameter, you can access a method from within the variable itself.

For example, imagine you have a string variable and you want to convert it to uppercase letters and print it. You might find a function `uppercaser()` and send your string into it:

```
MyString='abc'
print uppercaser(MyString)
```

Alternatively, the string variable itself might contain an uppercase function which you can access:

```
print MyString.upper()
```

There are many more implications and unique properties of object-oriented programming that are beyond the scope of this book. The main significance of objects for the topics we will address is that some functions you will use are methods of variables rather than stand-alone functions.

SUMMARY

You have learned:

- The differences between compiled and interpreted programs
- That variables have several attributes, including name, type, and value
- Some of the variable types widely used among programming languages
- That groups of variables can be stored in ordered lists and unordered dictionaries
- The basics of using variables with operators
- Key elements of the structure of programs, such as loops and functions

Chapter 8

BEGINNING PYTHON PROGRAMMING

Now that you have a passing familiarity with some of the basic components of programming, it is time to put them into practice. This will require applying some general programming concepts in the context of a particular language: Python. Within the next few pages you will learn how to make and use basic Python programs.

Why Python

In this book, you are going to write your programs in the popular scripting language Python. There are many programming languages available to scientists, and a brief explanation of our choice of Python will help you understand a bit about the language and the goals of these next chapters. Python is a relatively new language, but it has rapidly grown in use. It is now widely taught in introductory programming courses, and many students and researchers will likely encounter it in this and other contexts, reinforcing the information we present here. Many of the programs written and distributed by scientists are in Python, which makes Python directly relevant to the real-world computing environment in biology. Python is also widely used in industry, with considerable intellectual and financial support for further development coming from Google and others.

Beyond a large and growing user base, the reasons most often cited for learning and using Python are the relative clarity of the code, the ease of manipulating text data, and its native support for advanced features such as object-oriented programming. Newcomers tend to get up and running quickly in Python, yet it is also a fully functional and mature language suitable for complex tasks.

In key respects, Python is most directly comparable to Perl, another language that is also well-suited to text manipulation and which has been widely used by biologists. Emotions can run high when discussing the relative merits of the two languages. Python does have some shortcomings, and we ourselves have used Perl in our past research. Despite this, we have selected Python for this book for two primary reasons: First, Perl code can be difficult to read and understand, especially for newcomers; this is due in part to its heavy reliance on punctuation characters with special meanings. Second, the growing momentum behind Python provides more opportunities to take advantage of other computational and scientific resources, such as existing code and documentation, and integration with ongoing projects and course work.

Among other essential tasks, Python programs can reformat and organize data files, perform mathematical and statistical analyses, and assist with visualizing data and results. This book, however, will emphasize data handling over analysis and visualization. As discussed in *Before you begin*, most of the day-to-day computational challenges encountered by biologists consist of reformatting and reorganizing data files. This is a natural consequence of the explosive growth in dataset size across the field, which precludes manual manipulations that might have worked in the past. Not only are these manual operations tedious and time consuming, but they can't legitimately address the sophistication of interdisciplinary analyses, which require modifying data files to enable cooperation between programs that weren't designed to work with each other.

Even if you are already familiar with Matlab, R, or some other language, Python's simple power will complement those tools, as you wrangle your data into an optimal format for further analysis. So while you could build on the data manipulation skills we present here to write sophisticated new Python analysis and visualization tools, why reinvent those technologies? A better use of your new skills will be to prepare your datasets for further analysis, performed by packages dedicated to those tasks.[1]

Writing a program

If you have skipped through the last chapters, this is a good place to rejoin the narrative, but you first need to make sure that your programming environment is set up correctly. If you haven't already done so, create a scripts directory and edit your PATH variable as described in Chapter 6.

In the coming pages, it is helpful to follow along by testing the examples as you read. You can type these in as you go; the final programs and data files are also available in the examples package, which you can download. The added value to typing them in is that because you will have built up a set of programs that you understand intimately, you will be more easily able to modify them later for your own purposes.

Appendix 1 describes how to install and use Python on Windows.

Getting a program to run

If you are running Mac OS X, Linux, or most any other flavor of Unix, your system is already set up to run Python programs. You will be able to create a blank text file

---

[1] If you *really* end up liking Python and want to use it for analyses also, you might look at matplotlib, briefly introduced in Chapter 12, which provides **MATLAB**-type graphing commands to the Python environment.

and quickly set it up to operate as a Python script. If for some reason you have to install a new copy of Python on your system, use Python version 2.6, even for Mac OS X. Remember also that as you work through examples, a `blue background` will indicate shell commands entered in a terminal window, and a `tan background` will indicate scripts edited in a text editor.

From Chapter 6 you are already familiar with creating scripts—that is, text files containing a series of commands to be executed by your computer. Instead of being interpreted as a series of `bash` shell commands, as your shell scripts were, these next scripts will be read by the `python` program. Like a `bash` script, the first line of the file needs to tell the system where to send the rest of the file contents. In this case, you will begin the file with the shebang line (`#!`) followed by the location of the `python` program.[2] You can find the location of the Python program with the command `which python`:

```
host:~ lucy$ which python
/usr/bin/python
```

This gives the absolute path to `python`, and it would work fine on your computer. However, hardcoding absolute values for paths and other variables into your scripts, as this practice is called, can be a problem if you share those scripts with colleagues. For example, they may have `python` installed somewhere else on their computer. You will therefore want to use a more flexible method to indicate where the `python` program is located.

Nearly all systems have another program called `env` (similar to `which`) that can find `python` or any other program for you. Instead of sending your scripts to `python` directly you will send them to `env`, and ask it to pass them along to `python`. To do this, the first line of the file will read:

```
#! /usr/bin/env python
```

Notice that there is no slash after `env` (because it is a program, not a folder) and that there is a space between `env` and `python` (because you are telling `env` to find and run `python`, wherever it might be).

To begin a new Python script, open your text editor, create a new empty document, and type the shebang line above into this document.

Constructing the `dnacalc.py` program

As the first exercise you will build up a program which takes a string of text representing a DNA sequence, made up of the bases A, G, C, and T, and prints out information about the sequence, including the percent composition within the se-

---

[2] If this doesn't make sense to you, go back and read Chapter 4 on the shell and Chapter 6 on scripting. You will need to have your system set-up properly and be familiar with the basic skills which we described in the earlier shell chapters to function effectively here.

quence of each base. To do this you will need to determine the length of the sequence, count the number of each of the bases, and do some simple arithmetic. To present the results, you will print the numbers to the screen.

To get you up and running as fast as possible, this example draws on portions of many different aspects of Python. If you get stuck at any point, you can consult with other readers of the book at `practicalcomputing.org`. You can also refer to Chapter 13, which covers general debugging practices and approaches, as well as specific `python` error messages and their likely causes.

Simple `print` statements

In your script, type the short program that follows. Be sure to include the shebang line, and to make sure the capitalization of each command is right. You can add blank lines between sections of the script, but be sure not to add any extra spaces at the beginning of the lines:

```
#!/usr/bin/env python

DNASeq = 'ATGAAC'
print 'Sequence:', DNASeq
```

Now save this file in your ~/scripts folder with the name dnacalc.py. The colors of the text should change when you save it, indicating that the editor (for example, TextWrangler) has recognized it as a script. Because of the modifications to the $PATH that you completed in Chapter 6 on shell scripting, the scripts folder has special properties, so be sure to save the file here.

Open a terminal window, `cd` to your ~/scripts directory and then use `chmod` to make the file executable:[3]

```
host:~ lucy$ cd ~/scripts
host:scripts lucy$ chmod u+x dnacalc.py
```

Remember that you can use the [tab] key to auto-complete a file name after you have typed the first few characters. Also, if you are not sure how a particular shell command works, you can look up the manual page using the `man` command, for example `man chmod`.

You now have a file named dnacalc.py which is ready to execute. Type its name, press [return], and see what happens:

---

[3] If for some reason you find yourself working and storing your script within a folder that is not part of your path, you will have to run the program using the command `./dnacalc.py` or else it won't be found. The dot indicates the current directory (similar to how `..` means the enclosing directory). Even though you may be working in the folder which contains your program, the shell will not know where to find the command you are typing if it is not in your path.

```
host:scripts lucy$ dnacalc.py
Sequence: ATGAAC
host:scripts lucy$
```

Did it print out the text `Sequence: ATGAAC`? If so, congratulations. If not, then an error message may have been generated that will help you pinpoint the problem. If there is an error, does it indicate that the program was not found, or does it perhaps say "permission denied"? If so, there is probably something wrong with the way your path is configured, where the program is saved, or the permissions that were set with `chmod`. If the error says something like `Traceback:`, then that is actually a good sign. It means that it tried to run your program, but there was an error in the code itself. Check for typos, including extra spaces. If you get `IndentationError: unexpected indent`, then check that each line has no extra spaces at the start.

This program first creates a variable named `DNASeq` and puts the string `'ATGAAC'` in it. The print statement then displays three different things all on one line—the literal string `'Sequence:'` (this text is used directly and never placed in a variable), then the contents of the variable `DNASeq`, and then an end-of-line

NAMING VARIABLES Even for experienced programmers, it can be difficult to read through a new program and figure out what words are variable names created by the programmer and which ones are built-in parts of the programming language. Text editors can help with this by automatically displaying variable names as a particular color. The best way to avoid this problem, though, is to choose your names well, or use a convention which distinguishes them from the style of the reserved words. Because nearly all Python functions start with a lowercase letter, you can capitalize the first letter of your variable names, as we have done here.

character. You didn't have to tell Python to add the end-of-line character, for it does that on its own (though this behavior can be turned off). A comma separates the two pieces of text that are given to the `print` command for display. Later, we will explore other ways of formatting and joining pieces of text that provide a bit more control. Also, just as the line ending is added automatically, a space is added between the pieces of information that are separated by a comma.

Python interprets any text within straight quote marks, either single (') or double ("), as a string. This is to differentiate the contents of the string from the computer code itself. The advantage of having both types of quotes at your disposal is that it is easy to include a quote mark within your string without inadvertently terminating it. For example, this works:

```
MyLocation = "Hawai'ian archipelago"
```

but this would not work:

```
MyLocation = 'Hawai'ian archipelago'
```

The second single quote, between `i` and `i`, is interpreted as the close of the first one. The characters following the second quote are thus considered to be outside of the string. Most text editors designed for programming will help out by showing open strings in a conspicuous color.

FURTHER TROUBLESHOOTING If the `print` operation seems to be the source of your problem, check what version of Python you have installed. You can do this in a terminal window by typing `python -V` (make sure to use a capital V). The scripts presented here are designed to work with versions 2.3 through 2.7; there are no versions 2.8 or 2.9. Although Python 3.0 has been released, it contains some fundamental changes in the language and has not yet been widely adopted at the time of this writing. If your version is 3.0 or greater, try putting parentheses around the material that follows the print statements. If you reach an impasse, visit the forum at `practicalcomputing.org`.

You did not have to explicitly state that the `DNASeq` variable is a string when you created it. Python figured that out when you assigned a string to the variable. Likewise, `print` recognized `'Sequence:'` as a literal string rather than a variable or function name. In both cases, this is because the text is within quotes.

The `len()` function

Now that you have a variable holding a DNA sequence, you will do some calculations to summarize some attributes of the sequence. These calculations will take advantage of built-in functions. Functions are miniature programs or commands that are available within your program. A function can take input from variables sent to it as parameters (usually contained within the parentheses that follow the function name), perform some task (often a calculation involving the parameters), and return one or more results. The names of functions are case-sensitive, so it is important to use the correct capitalization. A space between the function name and the parentheses is optional in Python, but is not typically used.

The built-in `len()` function takes as a parameter an object such as a string, list, or dictionary; it then returns the number of elements in the object, which by convention is considered to be the object's length. In this case, it will return the number of characters in a string you give it. Back in the editor window, add the following two lines to the end of the program, save it, and then run it again (using the ⬆ in the terminal window to recall your previous command):

```
SeqLength = len(DNASeq)
print 'Sequence Length:', SeqLength
```

When you re-execute, you should get two lines of output: the first line again showing the sequence, and the second line indicating its length. If not, make sure that you saved the changes you made to the program before running it.

In your program, the variable `DNASeq` already stores a string, so it is printed as-is. In the second case, the `SeqLength` variable holds an integer, and the Python `print` command translates it into a string for display. This automatic translation by `print` works with numbers, letters, and even calculations, converting them to strings on the fly. However, it only works when a single data type is involved, or when different types are separated by commas. Other formatting strategies require the explicit conversion of non-string data to strings, as you will learn next.

Converting between variable types with str(), int(), and float()

Python has the ability to do **mathematical operations** on numbers using typical operators for addition (+), subtraction (–), multiplication (*), and division (/). You can also do exponents (**) and many other operations.

The command `print 7+6` returns `13`, and `print 7+3*2` also returns `13`. Just as with algebra, the sequence of operations in your formula gives precedence to exponents, then multiplication and division, then addition and subtraction. Also as in algebra, you can control the order of these operations using parentheses, so `print (7+3)*2` returns `20`. Parentheses in expressions such as `7+(3*2)` can often help clarify what a long formula or logical expression is actually doing, even though they are not necessary.

Python also allows addition and multiplication with strings. In this context, `print '7'+'6'` returns `'76'`, and `'a'+'b'` returns `'ab'`. Note that there are quotes around the 7 and 6; the numbers in this case are not naked, as they were in the previous examples. The quotes indicate that Python should interpret the numbers not as integers (actual numbers) but as strings (bits of text), in this case single characters. Where adding integers gives their sum, adding strings returns the joined strings.

The + operator is a useful and easy way to join together multiple strings into one. However, it is not legal (that is, not permitted by Python) to add together a mixture of strings and numbers, because the correct interpretation is unclear. The statement `print '7'+3` will fail because it is impossible for the python interpreter to know whether you want the output to be the integer `10` or the string `'73'`. If you want to use + to build up a string from variables of different types, you must first convert each variable to a string, and then join the strings together. The `str()` function helps with this by converting other data types, including numbers, to strings. In a program, the statement:

```python
print '7'+ str(3*2)
```

would display the string `'76'`.

Here, 3 and 2 are both integers, so they are multiplied by *. Their product, 6, is then converted to a string by `str()` and joined to `'7'` to produce a new string, `'76'`.

Just as it is often necessary to convert numbers to strings when formatting output, it is often necessary to convert strings to numbers when you gather input from a user or a text file. The `float()` function converts a string or integer to a floating point number (defined in Chapter 7). For example, this statement in a script:

```python
print float('7.5')+3*2
```

would return `13.5`.

The `float()` function is flexible, in that it can interpret scientific notation as well. So this statement:

```
print float('2.454e-2')
```

would return `0.02454`.

The `int()` function converts a string or float to an integer. If the value being converted contains any decimal fraction it is truncated, not rounded. Most mathematical operations that combine floats and integers, including addition, subtraction, multiplication, and division, return a float.

The built-in string function `.count()`

As described in Chapter 7, some variable types have functions built into the variables themselves. These built-in functions, known as methods, are accessed using dot notation. Strings have a method called `.count()`, which counts how many times a particular substring occurs in a string. The statement `print DNASeq.count('A')`, if inserted into your program would output the value 3. Modify the program to add the following lines:

```
NumberA = DNASeq.count('A')
NumberC = DNASeq.count('C')
NumberG = DNASeq.count('G')
NumberT = DNASeq.count('T')
```

 Using four different variables that each need to be independently analyzed is not a very efficient way to do this operation, but for the moment it is sufficient. In later chapters, you will learn how to store similar data like these in a list, and then write your analysis process once and use it on each element of the list.

With these commands, you create four new integer variables. Each contains the count of how many times the corresponding nucleotide is found in your sequence string. You could provide the user with these raw counts, but they are typically more informative when displayed as fractions. To determine fractions, you will next divide each count by the total number of nucleotides in the sequence, which you already determined and stored in the `SeqLength` variable above.

Math operations on integers and floating point numbers

Remember from Chapter 7 that numbers with decimal components are called floats. This brings us to one of the foibles of the Python language: in Python 2.7 or earlier, an integer divided by another integer is truncated to give an integer result. So in Python, `5/2` gives the answer `2`. A mixture of integers and floating point will return a floating point number, so `5/2.0` gives the correct answer of `2.5`. We admit that this is potentially troublesome behavior, and it has been modified in Python 3.0. For the moment, the solution is to just make sure that there is at least one floating-point element in your statement so that the answer is provided as a float as well.

In addition to converting strings to floats, `float()` can also convert integers to floats. Modify the `SeqLength` line of your code to include the `float` function as follows:

```
SeqLength = float(len(DNASeq))
```

Functions can be nested—that is, placed inside one another. In this case, the result from the function `len()` is evaluated first and used as the input to the function `float()`. The output of *that* function is what is assigned to the variable `SeqLength`. This is a bit like the pipe at the command line: data can be processed through several steps without writing them to variables or files at each stage. With this modification, the variable `SeqLength` is now a float rather than an integer. If you run the program at this point, notice that the length is now reported as `6.0` instead of `6`, indicating that it can have a fractional part (even though that fractional part is currently 0).

Since `SeqLength` is used in each calculation, changing it to a float is sufficient to ensure that all of the subsequent division operations will also result in floats. To avoid having to use `str()` to convert the resulting fraction to a string, we'll combine the print operation with the calculation into one line. At the end of the existing script, add in a print statement like this for each of the four nucleotides:

```
print 'A:', NumberA/SeqLength
```

Here is the whole program up to this point:

```python
#!/usr/bin/env python

DNASeq = 'ATGAAC'
print 'Sequence:', DNASeq

SeqLength = float(len(DNASeq))

print 'Sequence Length:', SeqLength

NumberA = DNASeq.count('A')
NumberC = DNASeq.count('C')
NumberG = DNASeq.count('G')
NumberT = DNASeq.count('T')

print 'A:', NumberA/SeqLength
print 'C:', NumberC/SeqLength
print 'G:', NumberG/SeqLength
print 'T:', NumberT/SeqLength
```

Run the program at the command line and verify the output:

```
host:scripts lucy$ dnacalc.py
Sequence: ATGAAC
Sequence Length: 6.0
A: 0.5
C: 0.166666666667
G: 0.166666666667
T: 0.166666666667
host:scripts lucy$
```

You might also verify for yourself what happens if you try the division using an integer value of SeqLength. Copy the line where it is defined and paste it below. On this copy, remove the float function from the equation, and run it again. Because the more recent assignment overrides the first one, you should see a different output, with all zeroes for the percentages. In each case, the division result was truncated. Be sure to delete this test line after you are done playing with it.

Adding comments with

It is often useful to add notes within your programs to remind yourself, as well as inform others, how they are meant to work. These comments are indicated with a hash mark (#), also called a pound sign. Any text following this symbol to the end of the line is simply ignored by Python. The hash mark can come at the beginning of a line, or it can occur within a line after a code statement. If it occurs within quote marks, it will be interpreted as part of a string, but otherwise it signals the start of a comment.

A large block of comments is often placed at the beginning of a program, after the shebang line, to describe what the program does and how it is used. Other comments describe what major subsections of the program do, and particularly complex lines may be individually annotated for clarity. Most editors have a Comment/Uncomment Lines command which can insert # at the beginning of each line of a selection of text. You can also comment a block of lines with triple quotes (" " ") placed before and after the block. This can be more convenient than placing a # before each line.

Commenting also serves an important function in debugging. Debugging is the process of finding and eliminating errors in your script. This process is described in detail in Chapter 13, but it is good to begin thinking about it as soon as you begin writing programs. (You won't have a choice but to think of it if things don't work as expected.) Comments can be used in debugging to effectively turn off certain portions of code, thereby helping to isolate problems. Another trick is to insert extra print statements which report on the status of your program as it executes; you can comment these extra statements out when they are not being used.

The example code in this book will include comments from this point on. As you follow along, you can insert your own comments in the code you are writing, with observations of how and why the various portions of the programs function. In general, comments are appreciated by anyone who reads your code—even your future self when you try to revisit an old program. It is hard to have too many comments.

Controlling string formatting with the % operator

Although you have a functioning program, it is less than ideal for the task at hand. It would be clearer to present the fractions as percentages, and display the percentages with a fixed number of significant digits. The first problem just requires multiplying the fractions by 100. Controlling the number of significant digits displayed requires an additional way to format text when printing. You have already seen how to use the default behavior of the print command to display strings and numbers separated by commas, and how to create custom strings from mixed data types using the plus sign and the str() function.

A more flexible way to build strings from different data types is with the string formatting operator, %. It inserts the values present on the right side of the % symbol into the positions marked by placeholders in the strings to the left. Placeholders indicate whether the substitution should be interpreted as an integer digit (%d), a floating point number (%f), or a string (%s). The placeholder symbols also control the type conversion, so you don't need to use the str() function to convert numbers before placing them into the string. This will be much clearer with an example:

```
print "There are %d A bases." % (NumberA)
```

```
There are 3 A bases.
```

The value of the integer NumberA to the right of the lone % is inserted in the position held by %d in the string. Multiple values separated by commas can be simultaneously substituted into a series of placeholders, and they are substituted in the order that they occur:

```
print "A occurs in %d bases out of %d." % (NumberA, SeqLength)
```

```
A occurs in 3 bases out of 6.
```

One of the most common uses of this operator is to control the number of decimal places that are printed, using the floating point placeholder modified by the insertion of a decimal precision specifier. This consists of a dot followed by a number indicating the number of decimal digits to display, nested within the %f:

```
print "A occurs in %.2f of %d bases." % (NumberA/SeqLength, SeqLength)
```

```
A occurs in 0.50 of 6 bases.
```

The `%.2f` placeholder inserts the corresponding float into the string, but only with two digits of precision to the right of the decimal point. Values inserted in this way are rounded, not truncated.

Now put this into practical use in your `dnacalc.py` program. Modify your `print` statements with `%` operators as follows:

```
print "A: %.1f" % (100 * NumberA / SeqLength)
print "C: %.1f" % (100 * NumberC / SeqLength)
print "G: %.1f" % (100 * NumberG / SeqLength)
print "T: %.1f" % (100 * NumberT / SeqLength)
```

When you run the program now, the value in parentheses is calculated and inserted at the position of `%.1f`, giving you the following output:

```
host:scripts lucy$ dnacalc.py
  Sequence: ATGAAC
  Sequence Length: 6.0
A: 50.0
C: 16.7
G: 16.7
T: 16.7
```

Other ways to control the output of the `%` operator are listed in Table 8.1. These are used across many programming languages, including MATLAB, C, and Perl. There is also a `%s` placeholder, which places a string (specified either with quotes or with a variable name) within the string being formatted.

The `%` operator doesn't need to be used in conjunction with the `print` function. The formatted string could be assigned to a variable, for example:

```
pctA = "%.1f" % (100*NumberA/SeqLength)
```

With the padding feature (triggered by inserting a number after the `%`, as in `%2d` or `%4f`) you can create and print strings which are aligned along their right edges, no matter what the number of digits involved in the value. So using `"A:%3d"` would return output as follows, where each number occupies three spaces:

```
A:  2
A: 10
A:100
```

When using the `%` operator, if you actually want a percent character to show up in your string (as we might here), you have to escape your code with two consecutive percents: `'%%'`. The backslash character does not work as an escape character in this context, so `'\%'` doesn't display a percent sign.

TABLE 8.1 String formatting options		
Given the string s = '%x' % (4.13) where %x is a placeholder listed below:		
Placeholder	**Type**	**Result**
%d	Integer digits	'4'
%f	Floating point	'4.130000'
%.2f	Float with precision of 2 decimal points	'4.13'
%5d	Integer padded to at least 5 spaces	' 4'
%5.1f	Float with one decimal padded to at least 5 total spaces (includes decimal point)	' 4.1'

Play with the program you've written to change the way the text is displayed. The example file `dnacalc1.py` includes the version of the code discussed above.

Getting input from the user

Gathering user input with `raw_input()`

Using this example script, you can analyze different sequences by pasting them into the program file and re-running the file each time. Later you will learn to read data from a file and perform calculations on them. Sometimes, however, it is useful to have a utility program which performs calculations directly from user input—that is, from what you type at the command line. To add this feature to our example script, you can use Python's `raw_input()` function. Add this line immediately below the existing `DNASeq` definition:

```
DNASeq = raw_input("Enter a DNA sequence: ")
```

This function presents a prompt to the user, consisting of the string within the parentheses. When the user types a response and presses return, the response can then be assigned to a variable. In this case, it is assigned to the `DNASeq` variable, writing over any previous value assigned to it. The original value is forgotten.

Now when the program is run, it will wait to receive a typed or pasted value. Try different sequences to see how it works. You no longer need to edit the program to change the sequence that it considers. In Chapter 11 you will see other ways to gather input when a program is run, using the `sys.argv` function.

Sanitizing variables with `.replace()` and `.upper()`

Before doing anything with user input, it is a good idea to check and make sure the values entered are appropriate. As it is written, your program will count the number of uppercase A's but not lowercase a's. This could lead the program to behave in ways that aren't expected by the user, who might not know that it is case-sensitive. Instead of placing the onus on the user to keep track of this, it is better to design the program so that it counts both uppercase and lowercase letters. Rather than count uppercase and lowercase letters separately, it is easier just to convert the entire string to uppercase before analyzing it. If it is all uppercase already, this will do nothing. If all or some of it is lowercase, it will change the offending characters to uppercase.

This can be accomplished using the built-in string function `.upper()`. When this method is called for a string, it returns an uppercase version of that string. This function doesn't require any parameters in the parentheses. The data it uses are in the string it is part of (that is, the string just before the period). However, since it is a function, you do need to include parentheses, even though in this case they are empty. Below your raw input, add this line:

```
DNASeq = DNASeq.upper()
```

Note that in this line of code, you call the method `.upper()` on the string `DNASeq`, and then immediately rewrite the contents of `DNASeq` with the method's result. It is usually okay to access and rewrite the same variable in one statement like this. It is necessary to do so anyway in this case, because `.upper()` alone doesn't make the string uppercase, but merely returns a new uppercase string. You can assign this new string to a new variable, or if you won't be needing the original variable again, you can overwrite it with the modification, as you did here. Rewriting variables can be a good strategy to avoid creating unneeded new variables in your programs.

Case isn't the only thing that might be a problem with our user input, however. It might also be pasted from another document and could therefore contain spaces. These spaces will not affect the count of nucleotides, but they *will* alter the total calculated length of the DNA sequence. This would result in incorrect calculations. Within a string like that generated by `raw_input()`, spaces are just another character.

To remove the spaces, you can use another built-in string function: `.replace()`. This function takes two arguments separated by commas: the item to be removed, and the item to replace it with. In this case, we will remove a space, `" "`, and replace it with an empty set of quotes, `""`. Below the `.upper()` command, insert this `.replace()` function:

```
DNASeq = DNASeq.replace(" ","")
```

This is just a simple search and replace, not a regular expression. You will learn how to use regular expressions within a Python program a bit later. Also note that we again had to reassign the result of this function to the variable `DNASeq`, since it doesn't directly search and replace within the original string, but instead generates a new string.[4]

Now you have a moderately sanitized version of the DNASeq string, ready to work with the rest of the program you already wrote. Before sending your program out into the real world for others to use, you would want to incorporate further checks—for instance, making sure the resultant string length is not zero, and that there are only legal characters in it (C, G, A, T). For the moment, though, you have a nice utility for performing calculations based on the character composition of a string. This same kind of functionality can be used to make a great many useful calculations, such as the molecular weight of a protein or molecule, or the melting temperatures of oligonucleotides. Here is the final script:

```python
#!/usr/bin/env python
# This program takes a DNA sequence (without checking)
# and shows its length and the nucleotide composition

DNASeq = "ATGAAC"
DNASeq = raw_input("Enter a DNA sequence: ")
DNASeq = DNASeq.upper()  # convert to uppercase for .count() function
DNASeq = DNASeq.replace(" ","")  # remove spaces

print 'Sequence:', DNASeq

SeqLength = float(len(DNASeq))

print "Sequence Length:", SeqLength

NumberA = DNASeq.count('A')
NumberC = DNASeq.count('C')
NumberG = DNASeq.count('G')
NumberT = DNASeq.count('T')

# Old way to output the Numbers, now removed
# print "A:", NumberA/SeqLength
# print "C:", NumberC/SeqLength
# print "G:", NumberG/SeqLength
# print "T:", NumberT/SeqLength

# Calculate percentage and output to 1 decimal
print "A: %.1f" % (100 * NumberA / SeqLength)
print "C: %.1f" % (100 * NumberC / SeqLength)
print "G: %.1f" % (100 * NumberG / SeqLength)
print "T: %.1f" % (100 * NumberT / SeqLength)
```

---

[4] You can also nest methods by putting consecutive dot notation on the same line:
DNASeq.upper().replace(" ","")

Reflecting on your program

The program you've constructed could easily be modified to be a general calculator applied to other purposes. As we mentioned, though, it is not written in a very optimal way, because instead of using a loop to do the repetitive commands, we have just duplicated the .count() and print lines and manually edited them. This is officially Bad Programming Practice. In the next chapter you will see a simpler way to work through the elements of a string or a list of variables using a for loop. At the end of Chapter 9, we will revisit this program and show how to accomplish most of its capability in a script of only four lines.

SUMMARY

You have learned how to:

- Create a Python program and execute it
- Use the function len()
- Change variable types with int(), float(), and str()
- Add comments to your program with #
- Print strings and numbers together
- Format strings to your own specification with %d, %.2f, and %s
- Get user input with raw_input()
- Use the built-in string functions .count(), .replace(), and .upper()

Chapter 9

DECISIONS AND LOOPS

You now are able to construct and run a basic program in Python starting from only a blank text file. This chapter takes a brief break from writing programs to show the interactive prompt—an area that functions like a scratchpad, for testing small bits of code—and then describes finding help for different elements of the Python language. You will then augment the program you wrote in the previous chapter so that it can make decisions and work with lists of data. You will also write a protein calculator, which requires converting a web-formatted table so it can be used by your program. Finally, the chapter will explain lists in detail, so that you become comfortable working with this important type of variable.

The Python interactive prompt

Sometimes you just want to quickly try out a few different Python commands to see what they do. You could write a whole program that just contains those few lines, but most of your time would go into getting the file set up to run. As an alternative, you can use the Python programming language through an **interactive prompt**. This interface allows you to try Python code right at the command line without ever generating a program file. We often work with one terminal window open to the shell for executing programs, and a second terminal window open to a Python prompt to test pieces of code and check results before we place them into the actual program file.

> **INTERACTIVE PROMPTS VERSUS EXECUTABLE TEXT FILES**   Using a program interactively versus sending it a series of commands in a text file is not as different as might seem. When you execute a text file containing a shell script or a Python program, the shebang line at the very start of the file instructs the system to send the rest of the contents of the file to the appropriate program (in this case, **bash** or **python**). For the most part, the programs don't even know whether the commands they are running are coming from the user or from a file.

To launch the Python interactive prompt, open a new terminal window and type **python** at the shell prompt:

```
host:~ lucy$ python
Python 2.6.2 (r262:71600, Apr 16 2010, 09:17:39)
[GCC 4.0.1 (Apple Computer, Inc. build 5250)] on Darwin
Type "help", "copyright","credits" or "license" for more information.
>>>
```

You will see some diagnostic information, including the Python version number. If the version reported is between 2.3 and 2.7, you should be okay, although some specific commands may differ slightly in versions older than 2.5. After the diagnostic information has been printed, the prompt then changes to >>>, indicating you are now in Python's interactive mode.

Try the following Python commands at the interactive prompt to get a feel for how it works. If you initialize a variable with a value, then enter the variable name either by itself or as part of an operation, the resulting value is printed to the screen:

```
>>> x='53'   ← No output is generated; the variable x is initialized with a string
>>> x
'53'   ← The value of the variable x is printed to the screen
>>> x+2
Traceback (most recent call last):   ← Fails because of type mismatch
  File "<stdin>", line 1, in <module>
TypeError: cannot concatenate 'str' and 'int' object
>>> float(x)+2
55.0   ← Note the decimal indicating the result is a float; the value of x remains unchanged
>>> int(x)/2
26   ← Note lack of decimal since the result is an integer
```

 One useful feature of the Python prompt is the ability to get information about the variables in your workspace. The `dir()` function lists all the variables and methods that are nested within a specified variable. If you run `dir()` on a string, you will see some familiar functions, such as .replace() and .upper(). You will also see some names that begin and end with ___, but these are internal names that we have trimmed from this screen capture:

```
>>> DNASeq='ATGCAC'
>>> dir(DNASeq)
['capitalize', 'center', 'count', 'decode', 'encode',
'endswith', 'expandtabs', 'find', 'index', 'isalnum',
'isalpha', 'isdigit', 'islower', 'isspace', 'istitle',
'isupper', 'join', 'ljust', 'lower', 'lstrip',
'partition', 'replace', 'rfind', 'rindex', 'rjust',
'rpartition', 'rsplit', 'rstrip', 'split', 'splitlines',
'startswith', 'strip', 'swapcase', 'title', 'translate',
'upper', 'zfill']
```

In this case, we have defined the string called DNASeq, and then run dir() to see all of its components. These are the methods and variables nested within a string. Any of these can be accessed using dot notation:

```
>>> DNASeq.isdigit()
False
>>> DNASeq.lower()
'atgcac'
>>> b=DNASeq.startswith('ATG')
>>> print b
True
>>> DNASeq.endswith('cac')    ← Case–sensitive
False
```

A variation of the dir() command is to use the help() function on a variable. If you type help(DNASeq), it will print out an extended version of the commands that are available for this variable type. This doesn't work for all variables and for older versions of Python, so you might have to enter the name of the variable *type* instead, as in help(list) or help(str).

You can also use the Python interpreter to check your program: just copy and paste portions of a script into the window, press ⟨return⟩, and it will attempt to run. (This works best when the script does not involve reading files or asking for user input.) The variables will be loaded into memory so that you can interact with them, to check that things are operating as you expect. As you will see shortly, indentation makes a difference in Python, so you have to pay attention to both extra space and missing space at the beginning of the lines that you paste into the interactive prompt.

When you are done with the Python interpreter, you can return to the bash shell prompt by typing quit() or ⟨ctrl⟩ D.

Use ⟨ctrl⟩ Z.

Ⓦ

Getting Python help

The built-in Python documentation has many shortcomings. To begin with, it isn't installed by default on many computer systems. Even if it is present, there are still problems. The documentation is difficult to search, and once you find a promising document, the majority of the document typically explains the history of the command and why it was constructed as it is, rather than explaining what it does and showing examples of how to use it. For this reason we are reluctant to recommend the `help()` function at the Python prompt, or even the `pydoc` program available at the `bash` command line.

As with most things now, the quickest way to answer your Python-related question is through a Web search. This doesn't help if you are on an airplane or sitting at the beach, but in general it will give you the fastest results. You can often do a general Web search including the word `python` (maybe also `-monty -snake` to remove spurious results), plus the concept you are interested in, and get your question answered. Some helpful sites dedicated to Python include:

```
http://python.about.com/
http://docs.python.org/tutorial/
http://rgruet.free.fr/PQR26/PQR2.6_modern_a4.pdf
http://rgruet.free.fr/PQR25/PQR2.5.html
http://www.diveintopython.org/
```

The Python commands discussed in this book are summarized in Appendix 4 for easy reference, and Chapter 13 offers help on debugging.

Adding more calculations to `dnacalc.py`

You will now continue building on the `dnacalc.py` script from the previous chapter. The next challenge is to estimate the melting temperature for the sequence (the temperature at which the strands of a DNA molecule with this sequence would separate from their complement, in half of the sample). This calculation will be based on the counts of various nucleotides in the sequence, and you will employ a different formula depending on the overall length of the sequence. To get started again, reopen your `dnacalc.py` script in TextWrangler, or open `dnacalc1.py` from the `pcfb/examples` folder.

Although the ability to accept user input at the command line makes your program more convenient to use, during the development process it is usually faster to hardcode a fixed value for testing. Find the `raw_input` statement around line 10 and comment it out by adding # at the beginning of the line. This turns off the input statement and retains the hardcoded value for the `DNASeq` variable in the preceding line:

```
DNASeq = "ATGTCTCATTCAAAGCA"   ← Now using a longer sequence
# DNASeq = raw_input("Enter a sequence: ")   ← Not executed anymore
```

Now you can add the ability to calculate the melting temperature, also known as Tm. One simple approximation for Tm is based on the number of strongly binding nucleotide pairs (G+C) and the number of weak ones (A+T). At the end of your existing program, add these lines to calculate the Tm:

```
TotalStrong = NumberG + NumberC
TotalWeak = NumberA + NumberT
MeltTemp = (4 * TotalStrong) + (2 * TotalWeak)
print "Melting Temp : %.1f C" % (MeltTemp)
```

Run the program and see that the result of the new calculation is displayed.

Conditional statements with `if`

In reality, this formula for the melting temperature is only recommended for short nucleotide sequences. If sequences are 14 or more nucleotides long, you should use another formula rather than the one shown above.[1]

To implement this "conditional behavior" where different formulas are employed under different circumstances, you will use the `if` statement. The `if` statement in Python has this general syntax (shown in pseudocode[2]):

```
if logical condition == True:   ← Notice the colon
    do this command   ← Indented block
    do this command
continue with normal commands   ← Main thread of the program
```

The commands that are indented under the `if` statement constitute what is sometimes called a **block** of code—that is, the commands are not only contiguous, but constitute a functional unit, with the entire set of lines executed one after another. In this case, this particular block is only executed if the condition is logically true. Otherwise, if the condition is false, the block is skipped and the program continues executing below.

Designating code blocks using indentation

The method of indicating which commands are controlled by the `if` statement brings us to one of the most unique aspects of Python: *indentation is used to indicate blocks of code*. The programs you have written so far have consisted of only a single block of code, run from top to bottom. From here on out, though, this strictly linear sequence of events will often be modified with conditional statements that only execute some blocks of code if certain criteria are met, as well as with blocks of code whose commands are reiterated multiple times within a loop. In each case, blocks are nested within other blocks. Therefore you need to be able to signal

---

[1] According to `http://www.promega.com/biomath/calc11.htm`.

[2] Pseudocode is a combination of computer code and plain English, used for illustrative purposes.

which groups of lines go together, forming a block, and thus indicate how the flow of the program will proceed.

In many other programming languages, nested blocks of code must be designated with brackets, with indentation being technically superfluous. Even in these other languages, however, programmers typically indent blocks anyway, to improve the readability of their code. Just remember that in Python, indentation of nested blocks is not an option or an aesthetic decision, but a requirement.

Indentation can be produced by either tabs or spaces. If you choose to use spaces, you must always use the same number of spaces to designate the same block. (Each block is typically indented four spaces relative to the block it is nested within.) We find it easier to just stick with tabs, and so they are used throughout this book and in the sample files. When copying and pasting code from programs you find on the Web, be prepared for inconsistencies in indentation. One program might use tabs, another three spaces, another four spaces. Web examples can also contain invisible characters that are not normal spaces. This can lead to strange behavior if you combine code of disparate origins into one program. It is important to go back through all the copied code and make sure the indentation characters are consistent, since in most environments, you can't mix tabs and spaces, nor different numbers of spaces. In TextWrangler, you can turn on Show Invisibles to see such characters in your document.

This formatting via visual indentation corresponds to the flow of the program. For example, the flow of a program takes a detour when it reaches an `if` statement with conditions that are true, executes the indented block of code, and then continues on. If the condition is false, the detour is avoided and the program goes on as if the indented code didn't even exist.

The line leading to an indented block of code ends with a colon, whether the line is an `if` statement, as described here, or controls a loop, which we will discuss a bit later. Unlike other languages, there is no specific marker at the end of a block of code in Python. The code present where normal operation resumes after the end of a nested block is aligned with the indentation of whatever statement started the block, be it an `if` statement or some other statement.

Logical operators

Conditional statements used in `if` statements and throughout programming are usually either simple comparisons such as `SeqLength >= 14`, tests of equality such as `SeqType == 'DNA'`, or else some logical combination of these, such as `(Latitude > 30 and Latitude < 40)`. This last example can also be written as `30 < Latitude < 40`.

TABLE 9.1 Python comparison operators

Comparison operators These operators return `True` (1) or `False` (0) based on the result of the comparison.

Comparison	Is `True` if...
`x == y`	x is equal to y
`x != y`	x is not equal y
`x > y`	x is greater than y
`x < y`	x is less than y
`x >= y`	x is greater than or equal to y
`x <= y`	x is less than or equal to y

Be sure that when comparing the equality of two entities, you use two consecutive equal signs (==) as the operator. A single equal sign would assign the value of the second entity to the first. This is a common slipup when first programming. See Tables 9.1 and 9.2 for comparison and logical operators in Python.

The `if` statement

To put all this into practice, you will add an `if` statement that will execute one block of code if the sequence is 14 or more characters long, and another block of code if the sequence is less than 14 characters long. In the final version of this program, one block or the other, but never both blocks, will be run.

Insert these lines in the program just above where the `MeltTemp` is calculated:

```
if SeqLength >= 14:
    #formula for sequences 14 or more nucleotides long
    MeltTempLong = 64.9 + 41 * (TotalStrong - 16.4) / SeqLength
    print "Tm Long (>14): %.1f C" % (MeltTempLong)
```

TABLE 9.2 Python logical operators

Logical operators In this table, A and B represent a true/false comparison like those listed Table 9.1.

Logical operator	Is True if...
A and B	Both A and B are true
A or B	Either A or B is true
not B	B is false (inverts the value of B)
(not A) or B	A is false or B is true
not (A or B)	A and B are both false

You can either type this or copy it from the `dnacalc1.py` example file.

Go through these new commands line-by-line. First, the `if` statement itself is just a test of whether the `SeqLength` variable is 14 or greater. If this condition is true, it executes the next two indented lines and then goes on to also complete the Tm calculation designed for short DNA sequences. If it is false, the program skips the calculation for long sequences and goes straight to the Tm calculation designed for short DNA sequences.

The `else:` statement

As the program stands, the Tm derived from the short-sequence formula will be printed no matter what the sequence length is. This is a bit annoying, since the number resulting from the short calculation is irrelevant if the sequence is 14 or more nucleotides in length. In many situations where you are using the if statement, including this one, you want one block of commands to be executed *only* if the condition is `True`, and another *only* if it is `False`. To accomplish this, use the `else:` statement, which provides an alternative route, or detour, to be executed only when the main condition is `False`.

Add an `else:` statement immediately before the formula for sequences less than 14 nucleotides long. Then indent the `MeltTemp` calculation line and the `print` statements so that they form a nested block of commands under the control of the `else:` statement. Remember, even though it is just a single word by itself, an else: statement must always have a colon at the end of the line, just like an if statement.

In TextWrangler, you can easily indent existing lines by highlighting them and pressing ⌘]. Similarly, ⌘[moves a block of highlighted commands to the left. The entire modified program should now look like this:

```python
#! /usr/bin/env python
# This program takes a DNA sequence (without checking)
# and shows its length and the nucleotide composition

DNASeq = "ATGTCTCATTCAAAGCA"
# DNASeq = raw_input("Enter a sequence: ")
DNASeq = DNASeq.upper()  # convert to uppercase for .count() function
DNASeq = DNASeq.replace(" ","") # remove spaces

print 'Sequence:', DNASeq

# below are nested functions: first find the length, then make it float
SeqLength = float(len(DNASeq))

# SeqLength = (len(DNASeq))
print "Sequence Length:", SeqLength

NumberA = DNASeq.count('A')
NumberC = DNASeq.count('C')
NumberG = DNASeq.count('G')
NumberT = DNASeq.count('T')
# Calculate percentage and output to 1 decimal
print "A: %.1f" % (100 * NumberA / SeqLength)
print "C: %.1f" % (100 * NumberC / SeqLength)
print "G: %.1f" % (100 * NumberG / SeqLength)
print "T: %.1f" % (100 * NumberT / SeqLength)

# Calculating primer melting points with different formulas by length

TotalStrong = NumberG + NumberC
TotalWeak = NumberA + NumberT

if SeqLength >= 14:
    #formula for sequences > 14 nucleotides long
    MeltTempLong = 64.9 + 41 * (TotalStrong - 16.4) / SeqLength
    print "Tm Long (>14): %.1f C" % (MeltTempLong)
else:
    #formula for sequences less than 14 nucleotides long
    MeltTemp = (4 * TotalStrong) + (2 * TotalWeak)
    print "Tm Short: %.1f C" % (MeltTemp)
```

Now you can assign DNA sequences of varying lengths to the `DNASeq` variable and test the effectiveness of your conditional expressions. You can also reinstate the `raw_input` command, to once again make the program work interactively.

Choosing from many options with `elif`   When you are using combinations of `if` and `else:` to choose from a long list of items, you can get into a deeply indented series of statements:

```
if X == 1:
    Value = "one"
else:
    if X == 2:
        Value = "two"
    else:
        if X == 3:
            Value = "three"
```

A cleaner way to write this is by using the special `elif` command, short for else if:

```
if X == 1:
    Value = "one"
elif X == 2:
    Value = "two"
elif X == 3:
    Value = "three"
```

In this trivial example, an even better way to solve the problem would be to think like a programmer and use a list, with `X` as an index to the list, rather than as part of a logical test. You will learn about lists later in this chapter:

```
ValueList=["zero","one","two","three"]
Value = ValueList[X]
```

Introducing `for` loops

In the next section, you will create a simple script called `proteincalc.py`. This program will take an amino acid sequence as input. Amino acids and their corresponding molecular weights will be stored in a dictionary. You will use a `for` loop to calculate the total molecular weight of the protein by adding up the molecular weight of each of its amino acids.

Remember from Chapter 7 the basic premise of a `for` loop: it cycles through each object in a list or other collection of variables, performing a block of commands each time through. In Python pseudocode the syntax is similar to an `if` statement:

```
for MyItem in MyCollection:
    do a command with MyItem
    do another command
    # return to the for statement and move to the next item
    resume operation of main commands
```

This `for` statement says: "For each item in `MyCollection`, assign that value to `MyItem`, and run the block of commands below. Then go through the next item in `MyCollection` and run the commands again."

The first time through the loop, the variable `MyItem` temporarily takes on the value of the first item in `MyCollection`. Within the loop, any calculations or print operations that involve `MyItem` will be done using this first value. When the last statement of the indented block is complete, the execution returns to the `for` line, and `MyItem` takes on the second value of `MyCollection`. Finally, after `MyItem` has been assigned the last value and the last of the indented statements has been executed, the program continues with the following unindented line.

A brief mention of lists

Items in `MyCollection` are typically of the type list. In Python, lists are created with square brackets enclosing a group of items, and with each item separated by commas:

```
MyCollection=[0,1,2,3]  ← List of integers
MyCollection=['Deiopea','Kiyohimea','Eurhamphaea']  ← List of strings
MyCollection=[ [34.5, -120.6] , [33.8, -122.7] ]   ← List of lists
```

In the last example, where `MyCollection` is a list of lists, in the course of cycling through the `for` loop, `MyItem` itself would be a two-element list, so that a pair of values could be used within the loop.

There is also a built-in Python function called `list()` which can convert the contents of a string into list elements. For example:

```
MyCollection= list('ATGC')
```

is the same as:

```
MyCollection = ['A', 'T', 'G', 'C']
```

 This command can save you a lot of typing of commas and quotation marks, but in the context of a `for` loop, unless you want to sort the list, such conversions are not necessary: in Python, `for` loops can operate directly on each character of a string without conversion to a list.

Writing the `for` *loop in* `proteincalc.py`

Begin writing a new script called `proteincalc.py` in the editor. (The completed script is included in the **examples** folder as well.) Prepare your script for execution by performing the now-routine steps of adding a `#!` line at the top, saving the file to your **scripts** folder, and doing a `chmod u+x` to make it executable. Then add a string definition and the lines of code below:

```
#!/usr/bin/env python
ProteinSeq = "FDILSATFTYGNR"
for AminoAcid in ProteinSeq:
    print AminoAcid
```

Now try running your program. If you get an error, make sure that you remembered the colon after `ProteinSeq` at the end of the `for` statement. Notice that this loop treats the variable `ProteinSeq` as a list: it steps through each letter of the string, assigns it to `AminoAcid` in turn, and prints each on its own line. So the first time through the loop, it is as if there is a statement saying `AminoAcid = 'F'` before the `print` statement.

> For quick debugging, many text editors have a tool to execute code without switching to the terminal window, or they can automatically switch to the terminal for you. In **TextWrangler**, the menu with this option is easy to miss since it is labeled with a shebang (`#!`).

One potentially confusing aspect of `for` loops is that variables like `AminoAcid` are never mentioned or defined before appearing in the `for` statement. This is because they are defined each time the `for` statement is executed.

Now that you have a way to cycle through each amino acid in the protein sequence (each character in the string), you will use a Python dictionary along with that bit of information to look up a corresponding value in a table of molecular weights.

Generating dictionaries

Recall from Chapter 7 that dictionaries are collections of objects, with the objects being looked up by associated keys, rather than by order of occurrence in the collection. In our example, we want to associate the amino acid's single letter code (such as `'A'`) to its corresponding molecular weight (a floating point value such as `89.09`).

There are several ways to create dictionaries. One is by defining the pairs of keys and values inside curly brackets. Keys and their values are separated by colons, and each pair is separated by a comma, with the basic format of:

```
MyDictionary = {key:value, nextkey:nextvalue}
```

For example:

```
TaxonGroup={'Lilyopsis':3, 'Physalia':1, 'Nanomia':2, 'Gymnopraia':3}
```

In this example, the keys are strings, and the values are integers.

Although dictionaries are *created* using {}, values are *extracted* from the dictionary using square brackets [ ], which is the same way that values are extracted from lists. So given the `taxongroup` dictionary above, the command:

```
print TaxonGroup['Lilyopsis']
```

would print out the value 3.

OTHER WAYS TO CREATE DICTIONARIES There are several other ways to create dictionaries. Which method you use depends on the format of the information you have to begin with. In addition to the {} method described in the text, another useful method is when you have two lists of equal length, one containing keys and the other containing values; in such a case, you can use the `dict()` and `zip()` functions to associate them. For example, if you have these lists:

```
TaxonKeys=['Lilyopsis', 'Physalia', 'Nanomia', 'Gymnopraia']
TaxonValues=[3, 1, 2, 3]
```

you can create an association between them using the command:

```
TaxonGroup=dict(zip(TaxonKeys,TaxonValues))
```

First the `zip` command pairs the two lists together into a list of pairs, and then the `dict` command takes these pairs and connects them as dictionary entries.

Despite its strict indentation rules, Python does allow a statement to be split across lines if the splits occurs within ( ), [ ], or {}. This is often useful for large entries, such as long lists or dictionaries. As an example, the definition above could also be written as:

```
TaxonGroup={
'Lilyopsis' :3,
'Physalia'  :1,
'Nanomia'   :2,
'Gymnopraia':3 }
```

For your `proteincalc` program, you will be making a comparable dictionary for amino acids. We won't ask you to type out a dictionary definition of twenty amino acids and their molecular weights. This kind of busywork wastes time and also has a high probability of introducing errors. Instead, you will use your regular-expressions skills to convert data derived from a web page into Python code.

It will often be convenient for you to gather tables of information from the Web, other spreadsheets, or documents you already have. In all likelihood, these data will not have been intended for use in a computer program, but converting them into programming syntax is usually a matter of doing a few searches and replacements.

In this case, you will create a dictionary where each line has an amino acid as the key and its molecular weight as the value:

```
AminoDict={
 'A':89.09,
 'R':174.20,
 ...several more lines...
 'X':0,
 '-':0,
 '*':0 }
```

Drag the file `aminoacid.html` from your examples folder into a web browser, or obtain the file from `http://practicalcomputing.org/aminoacid.html`. You should see a table of amino acid names, abbreviations, and molecular weights (Figure 9.1).

The first step is to get these data into a blank text document so that you can edit them. One way to do this would be to copy and paste them from the web page. This would work fine in many situations. However, tables copied from some browsers end up as lists rather than tables when they are pasted into a text document. There might also be hidden characters on the Web site that you don't see, but which can cause problems in the text file.

You will usually get more consistent results when you pull web data directly from the page source code. You can view the source code of any Web page by selecting **Page Source** or **View Source** from your browser's **View** menu. (The exact wording and location of this command will depend on your browser, but with a bit of digging you should be able to find it. You can also right-click on the page and see if there is an option to view the source.) Take a look to see where the data of interest start and end in the file. Even if you don't understand a word of HTML (the language used to encode most web pages) you will quickly be able to identify the letter code and molecular weight of each amino acid within this text. Most web pages will have more extraneous text than this one, but the essential information will be embedded somewhere within.

In this case, the useful information is all in lines that start with `<tr><td>`, signifying the start of a table row boundary. Every line that starts with `<tr><td>`, except the header line that starts `<tr><td>Name`, contains the properties and name of an amino acid, as it would be indicated in a protein sequence. Copy these lines, from the one that begins with `<tr><td>Alanine` through the one that begins with `<tr><td>Stop`, and paste them into a blank text file in your text-editing program.

Name	Abbreviation	Single-Letter	Mol Wt
Alanine	Ala	A	89.09
Arginine	Arg	R	174.20
Asparagine	Asn	N	132.12
Aspartic acid	Asp	D	133.10
Cysteine	Cys	C	121.15
Glutamine	Gln	Q	146.15
Glutamic acid	Glu	E	147.13
Glycine	Gly	G	75.07
Histidine	His	H	155.16
Isoleucine	Ile	I	131.17
Leucine	Leu	L	131.17
Lysine	Lys	K	146.19
Methionine	Met	M	149.21
Phenylalanine	Phe	F	165.19
Proline	Pro	P	115.13
Serine	Ser	S	105.09
Threonine	Thr	T	119.12
Tryptophan	Trp	W	204.23
Tyrosine	Tyr	Y	181.19
Valine	Val	V	117.15
Unknown	Xaa	X	0
Gap	Gap	-	0
Stop	End	*	0

FIGURE 9.1 The table of amino acid names, abbreviations and molecular weights contained in `aminoacid.html`

The data will look like this:[3]

```
<tr><td>Alanine</td><td>Ala</td><td>A</td><td>89.09</td></tr>
<tr><td>Arginine</td><td>Arg</td><td>R</td><td>174.20</td></tr>
<tr><td>Asparagine</td><td>Asn</td><td>N</td><td>132.12</td></tr>
...omitted lines...
<tr><td>Unknown</td><td>Xaa</td><td>X</td><td>0.0</td></tr>
<tr><td>Gap</td><td>Gap</td><td>-</td><td>0.0</td></tr>
<tr><td>Stop</td><td>End</td><td>*</td><td>0.0</td></tr>
```

Remember your goal is to reformat those four text fields so they look like this:

```
'A':89.09,
'R':174.20,
```

You will now use regular expressions to reformat each of these lines into Python code that you can copy and paste into your program file. (If you don't recall how to use regular expressions, turn back to Chapters 2 and 3). Look through the raw data to find the minimal chunk of text that has all the information you need. In the first line this would be:

```
A</td><td>89.09
```

This is the single-character code that would be used in a protein sequence (A, R, to – and *), followed by a few formatting characters that are the same in every line, followed by the molecular weight value.

The first step in constructing the regular expression is to figure out the search term. To capture both fields of interest, you can use:

```
.+(.)</td><td>([\d\.]+).+
```

This search term has the following components:

> `.+`   stands for any series of one or more characters

> `.`   stands for any single character by itself, which you want to capture

> `()`   mark the portions of the search you want to capture

> `</td><td>`   is the separating text, which is the same on each line

> `[\d\.]+`   is any series of one or more digits and decimal points

The reason we use a dot in this search instead of `\w` for the first character we are capturing is that there are some symbols at the bottom of the list which are not letters or numbers. The leading and trailing `.+` are necessary because we want the search term to match the entire line, not just the part of interest, so that this flanking text is removed in the search and replace. This matches only the correct portion of the line because there is only one place where the numbers occur.

---

[3] To get colored formatting, tell **TextWrangler** that this is an XML or HTML file using the pop-up menu at the bottom of the page. (It will say **None** by default.)

For a replacement string, use:

```
'\1':\2,
```

This will put the first bit of captured text, designated by \1, between a pair of single quotation marks. It will then add a colon and the second bit of captured text. Execute the search and replace, and you should get the following:

```
'A':89.09,
'R':174.20,
'N':132.12,
...omitted lines...
'X':0.0,
'-':0.0,
'*':0.0,
```

To finish off the dictionary statement, add a line above this list which begins the definition, then delete the comma after the final item in the list, and finish it off with a closing curly bracket as follows:

```
AminoDict={
'A':89.09,
'R':174.20,
'N':132.12,
...omitted lines...
'V':117.15,
'X':0.0,
'-':0.0,
'*':0.0
}
```

Your text file now contains legal Python code which is ready to be copied and pasted into your `proteincalc.py` script. Paste these lines into your existing script just above the definition of the variable `ProteinSeq`.

This dictionary will let your program look up values using the single-letter amino acid name in square brackets. For example, to use the molecular weight of methionine in your script, you can type `AminoDict['M']` and it will give you the corresponding numerical value. As an intermediate test, you could modify the current `print` statement so that it prints both the name and the associated value:

```
print AminoAcid, AminoDict[AminoAcid]
```

Using this dictionary and the `for` loop you have already created, you can now step through each amino acid in `ProteinSeq`, look up its value in the dictionary, and add it to a variable `MolWeight` to sum up the full molecular weight. Modify your existing code to replace the `print AminoAcid` statement in the loop with the running total of the `MolWeight` as shown below. Before adding to the `MolWeight` variable, you need to initialize it to zero:

```
MolWeight = 0
for AminoAcid in ProteinSeq:
    MolWeight = MolWeight + AminoDict[AminoAcid]
```

This loop takes each character of the string `ProteinSeq` and assigns its value to the `AminoAcid` variable. Using this letter as a key, it retrieves the corresponding value from the dictionary called `AminoDict`. It adds this value to the value of `MolWeight`, which grows in value each time through the loop.

Once the loop is completed, print `ProteinSeq` and the molecular weight:

```
print "Protein: ", ProteinSeq
print "Molecular weight: %.1f" % (MolWeight)
```

You could also modify this program by adding user input, so that the user is asked to provide a sequence when the program is run. If you did this, you would want to use the `.upper()` function in the same way it was used in `dnacalc.py`, to make sure the user input is in uppercase. This is important because dictionaries give an error when you try to look up a key (such as a lowercase `'a'`) which is not among the defined entries. (There is a way around this using the `.get()` function to retrieve variables, as discussed in the next section.)

The entire program is summarized below. To save space, the dictionary has been reformatted here to occupy fewer lines. It will still run in this format:

```
#! /usr/bin/env python

# This program takes a protein sequence
# and determines its molecular weight
# The look-up table is generated from a web page
# through a series of regular expression replacements

AminoDict = {
'A':89.09,  'R':174.20, 'N':132.12, 'D':133.10,
'C':121.15, 'Q':146.15, 'E':147.13, 'G':75.07,
'H':155.16, 'I':131.17, 'L':131.17, 'K':146.19,
'M':149.21, 'F':165.19, 'P':115.13, 'S':105.09,
'T':119.12, 'W':204.23, 'Y':181.19, 'V':117.15,
'X':0.0,    '-':0.0,    '*':0.0     }
```

```
# starting sequence string, on which to perform calculations
# you could use raw_input().upper() here instead
ProteinSeq = "FDILSATFTYGNR"

MolWeight = 0

# step through each character in the ProteinSeq string,
# setting the AminoAcid variable to its value
for AminoAcid in ProteinSeq:

    # look up the value corresponding to the current amino acid
    # add its value of the present amino acid to the running total
    MolWeight = MolWeight + AminoDict[AminoAcid]

# once the loop is completed, print protseq and the molecular weight
print "Protein: ", ProteinSeq
print "Molecular weight: %.1f" % (MolWeight)
```

Other dictionary functions

Given that you will probably be working extensively with dictionaries, a few other commands will be useful to you.

The .get() function   In addition to using square brackets to extract values from a dictionary, you can use the `.get()` function. For retrieving the value associated with the key `'A'`, the equivalent statement to `AminoDict['A']` would be:

```
AminoDict.get('A')
```

This function operates just like square brackets, except that you can specify a default value to be returned if the entry doesn't exist. For example, instead of defining `0.0` as the molecular weight for stop codons (`*`) and for dashes, you could access your dictionary with the statement:

```
AminoDict.get(AminoAcid,0.0)
```

where the parameter after comma is the default value to use. Be careful if you use this formulation, because you won't get an error if your input sequence is somehow scrambled and includes improper characters or other punctuation.

Listing keys and values   A list of the keys in a dictionary can be extracted using the `.keys()` function. Remember that there is no intrinsic order to the keys or values in a dictionary. You can't count on them being alphabetical, or even in the same order that they were entered. If you want to loop through each key of a dictionary in some kind of sorted order, use the `.keys()` command, along with the `sorted()` function (described at length later in this chapter) to produce a separate list:

```
SortedKeys = sorted(AminoDict.keys())
```

You can loop through this list using:

```
for MyKey in SortedKeys:
```

A list of all of the values of a dictionary can be retrieved using the `.values()` method:

```
AminoDict.values()
```

Although the keys and values will not be returned in a predictable order, the output from the `.keys()` and `.values()` methods will occur in the same order relative to each other.

Applying your looping skills

Although this program is doing something more intricate than your `dnacalc` program from earlier in the chapter, it is nonetheless accomplishing its task in fewer steps. This is because it is programmed in a more efficient and flexible manner.

You could rewrite the first part of the `dnacalc.py` program, which calculates the percentage of each nucleotide in a DNA sequence, in a similar fashion:

```
#! /usr/bin/env python
DNASeq = "ATGTCTCATTCAAAGCA"
SeqLength = float(len(DNASeq))

BaseList = "ACGT"
for Base in BaseList:
    Percent = 100 * DNASeq.count(Base) / SeqLength
    print "%s: %4.1f" % (Base,Percent)
```

This example is available as `compositioncalc1.py`. In this case, instead of looping through the sequence, we are looping through the list of bases that we want to count within the sequence. The `print` line has also been formatted with `%4.1f` to pad the output to four total spaces, producing output where the values are aligned along their right edges:

```
A: 15.4
C:  7.7
G: 30.8
T: 30.8
```

The power of this approach is something you should use frequently in your scripts. To add support for counting ambiguity codes in the sequence, like `'S'` (strong nucleotide pairs G or C) and `'W'` (weak nucleotide pairs A or T), you could just add them to the `BaseList` string, and their percentages would also be calculated and printed without adding any additional lines of code.

Lists revisited

Lists are an integral part of many programs, particularly those designed to analyze and convert large datasets. While we have introduced aspects of lists in several places, here we will take a break from developing programs, and look in more detail at ways to use lists in Python in particular, and at the related commands for doing so (see also Appendix 4).

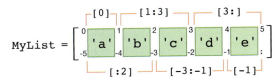

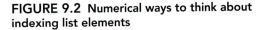

MyList =

FIGURE 9.2 Numerical ways to think about indexing list elements

Indexing lists

Lists are collections of values. These collections are defined using [ ], for example `MyList = ['a','b','c']`. Unlike in some languages, in Python the items in a list can be a mixture of data types, including strings, numbers, and even other lists. In addition to being defined with square brackets, list elements are also retrieved from a list using square brackets, as in `MyList[1]`. Between the brackets, there can be a single number, or a range of numbers separated by a colon, for example `MyList[1:3]`. List indexing can be confusing for several reasons: first, the index numbers start at zero, not one; second, indices sometimes don't seem to line up with their values; and third, Python indexing is handled differently than in some other programming languages. A diagram can help, in this case showing a list of 5 characters `'a'` to `'e'` (Figure 9.2).

The natural inclination is to think of the indices in brackets as corresponding to particular boxes in the list. If you want to extract `b,c`, you might think to try `[1:2]` or `[2:3]`, but neither is correct! In Python, it is better to imagine your list as a series of boxes with numbers placed *between* them. Indices can be positive, counting up from zero at the beginning of the list, or they can count back from the end of the list using negative numbers. If a single index without a colon is included within the brackets, it specifies the element to the right of the indicated boundary. A colon turns the brackets into tongs that reach into the list at particular numbered locations to grab the elements between them, such as `[1:3]`, which grabs elements b and c. Another way to think of it is that the first element in the brackets is inclusive, meaning that this element is included in the range, and the element after the colon is exclusive and is not part of the recovered range.

THE DUAL ROLE OF BRACKETS You have probably noticed that square brackets, [ ], are used in a couple of unrelated ways with lists. First, they are used to define lists with particular elements, in much the same way that quotes are used to define strings. In the statement `MyList = [33,12,89]`, for example, the brackets are used to define a list that is then assigned to the name `MyList`. If the brackets immediately follow the list name, however, they are interpreted as specifying particular elements in the list by their indices; thus `MyList[1]` returns the second element of `MyList`, because `[1]` is not defining a list but specifying an index—in effect, a range within the list to extract.

Both ends of the range may be specified, providing for left and right boundaries within the list. If one of the numbers in a range is omitted (for example `[:3]` or

[2:]), then this grabs from the beginning of the list (if the number to the left of the colon is omitted) or up to the end of the list (if the number to the right of the colon is omitted). A colon by itself within the brackets specifies all elements from the beginning to the end of the list; thus MyList[:] returns a full copy of all the elements in MyList.

Open the Python interpreter at the command line (just type python at the shell prompt) and try out some list commands:

```
lucy$ python
>>> MyList = ['a','b','c','d','e']
>>> MyList[:]
['a', 'b', 'c', 'd', 'e']
```

MyList[:3] returns the first three elements of the list, the same as MyList[0:3]:

```
>>> MyList[:3]
['a', 'b', 'c']
```

This can get confusing, since the first three elements have indices of 0, 1, and 2. The element with the index after the colon is not included in the result. This differs from the number before the colon. MyList[2:] returns all the elements from index 2 to the end of the list, the same as MyList[2:5]:

```
>>> MyList[2:]
['c', 'd', 'e']
```

The element with index 2, the 'c', is included in the result. Just imagine those tongs, and keep in mind that the index before the colon is inclusive, and the index after the colon is exclusive.

Using negative indices to count from the end of the list works the same as counting from the beginning of the list; it is merely that the reference point of the index has changed:

```
>>> MyList[-2:]
['d', 'e']
>>> MyList[:-2]
['a', 'b', 'c']
```

An additional colon followed by an integer can be added to the end of the range of indices. This last value, if specified, indicates the **step size** of the slice. By default, if it isn't specified, the step size is 1, and all of the values in the range are returned. If a value larger than 1 is specified, then every *n*th element in the range will be returned:

```
>>> MyList[0:5:1]    ← same as [0:5]
['a', 'b', 'c', 'd', 'e']
>>> MyList[0:5:2]
['a', 'c', 'e']
>>> MyList[::2]
['a', 'c', 'e']
```

A negative step size can also be specified, which reverses the order in which the elements are returned:

```
>>> MyList[::-1]
['e', 'd', 'c', 'b', 'a']
>>> MyList[::-2]
['e', 'c', 'a']
```

This allows you to quickly reverse your lists.

Unpacking more than one value from a list

Sometimes you want to create a new list when you are retrieving values from an existing list, but at other times the reason you are extracting particular elements from the list in the first place is because you want to put each in its own variable. It is simple to place the value of a single element at a time into a new variable, just by specifying its index, as with i = x[0]. You can also extract more than one variable at a time:

```
i,j = x[:2]
```

This unpacks the first two elements of the list x into the variables i and j. If x doesn't have at least two elements to begin with, this code generates an error. You can unpack any list in this way, as long as the number of elements you retrieve from the list matches the number of variables you are trying to put them into.

The range() function to define a list

The function range() takes parameters that are similar to those used to index lists, and generates a list of integers. The range starts at the first parameter and ends just before the last parameter:

```
>>> RangeList = range(0,6)
>>> RangeList
[0, 1, 2, 3, 4, 5]
```

The range() function works with negative numbers, but these behave differently than when referring to a subset of a list. For instance, range(-5,6) will generate a list of numbers from -5 to 5.

Use a third parameter to indicate a step size if you want the numbers in the list to be incremented by a value other than 1:

```
>>> range(1,10,2)
[1, 3, 5, 7, 9]
>>> range(10,0)    ← A lower limit of 10 doesn't work with 0 as the upper
[]
>>> range(0,10,-1)  ← Likewise, stepping backward from 0 to 10 doesn't work
[]
>>> range(10,0,-1)  ← Now you're talking
[10, 9, 8, 7, 6, 5, 4, 3, 2, 1]
>>> range(-5,-11,-1)
[-5, -6, -7, -8, -9, -10]
```

Notice that if the step size goes in the wrong direction, such as being negative when the end value is greater than the starting value, an empty list is returned. This is what happened in the second and third examples here.

The range() function simplifies a variety of tasks. Imagine that you want to create a vertical list of labels corresponding to the wells of a 96-well plate. By convention, the columns of these plates are designated with the numbers 1–12 and the rows with the letters A–H:

A1	A2	. . .	A12
B1	B2		B12
. . .			
H1	H2	. . .	H12

A quick way to do this is with a tiny program consisting of two **nested loops** that cycle through the eight letters and twelve numbers, printing all combinations. This program will take advantage of the chr() function, which returns the ASCII character based on its special number (see Chapter 1 and Appendix 6). In ASCII, capital A is chr(65), B is chr(66), and so on:[4]

```
#! /usr/bin/env python
for Let in range(65,73):   ← Step through character number 65 to 72
    for Num in range(1,13):  ← For each letter, step through numbers 1 to 12
        print chr(Let) + str(Num)
```

---

[4] The inverse function for chr(), to find out the ASCII number corresponding to a letter, is ord(), so ord('A') is 65 and ord('a') is 97. These functions end up being useful in creating text that would be arduous to type.

The result of this miniature program is to print out a column of ninety-six labels which you could use to label the rows in a spreadsheet:

```
A1
A2
A3
...
H11
H12
```

In the first `for` loop, the variable `Let` steps through 65 to 72, the integers corresponding to ASCII code for characters A through H.[5] The nested loop (that is, the loop within the loop) goes through twelve times for each of the values of `Let`, printing the combined letter and number each time through the loop. To print the labels in table format, add a comma to the end of the `print` line without any other characters after it. This tells the `print` command to suppress the end-of-line character that it usually adds. Also add another `print` statement by itself at the same indentation level as the `for  Num` loop. This will print a line break each time the letter (equivalent to the row) increments:

```python
#! /usr/bin/env python
for Let in range(65,73):
    for Num in range(1,13):
        print chr(Let) + str(Num),    ← The comma is important
    print    ← Prints a line end once for each letter, after 12 numbers have passed
```

The output of this modified script will be eight lines of twelve elements:

```
A1 A2 A3 A4 A5 A6 A7 A8 A9 A10 A11 A12
B1 B2 B3 B4 B5 B6 B7 B8 B9 B10 B11 B12
...
H1 H2 H3 H4 H5 H6 H7 H8 H9 H10 H11 H12
```

A comparison of lists and strings

Indexing can also be used with strings, which in some respects behave as if they were lists of characters. Both list and strings can be combined using the plus (+) operator, sorted, and iterated within a `for` loop. The most important difference between strings and lists is that individual elements of a string cannot be directly modified, as the elements of a list can be. See the box on the next page on copying and modifying Python variables for more details. Elements can be retrieved from a list and a string in the same way, but trying to change the value of a character in a string will result in an error, while modifying an element in a list is acceptable:

---

[5] For international characters, the corresponding function is `unichr()`, and fortunately, for the UTF-8 version of Unicode, the values corresponding to A-Z are the same as they are in ASCII.

```
lucy$ python
>>> SeqString = 'ACGTA'
>>> SeqList = ['A', 'C', 'G', 'T', 'A']
>>> SeqString[3]
'T'
>>> SeqList[3]
'T'
>>> SeqString[3]='U'
Traceback (most recent call last):
  File "<stdin>", line 1, in <module>
TypeError: 'str' object does not support item assignment
>>> SeqList[3]='U'
>>> SeqList
['A', 'C', 'G', 'U', 'A']
```

THE SUBTLETIES OF COPYING AND MODIFYING PYTHON VARIABLES In Python, copying a variable doesn't create a new variable with a new name and a new value; instead, it creates a new name that refers to the old value. After you copy a variable, both names point to the same place in computer memory. However, the value of most variable types in Python, including integers, floats, and strings, can't be changed. When you assign a new value to an existing variable, what really happens is that a whole new variable is created, with a different value but the same name. For example, when you create an integer **x=5** and set **y=x**, both names refer to the same 5. If you then assign **x** a new value, say **x=8**, a whole new variable with the value 8 and the name **x** is created, and the old **x** is deleted. The name **y** still points to the original 5. In the end, this means that you can change the value of **x** without worrying about what this will do to the value of **y**, and forget everything mentioned in this box.

On the other hand, the values of some Python variables, including lists and dictionaries, can be changed. When a new value is assigned to such a variable, no new variable is created and the old variable really has a new value. When you define **A** as the list **[1,2,3]**, and then define **B=A**, you have created a potentially confusing situation: if you change the contents of the list by altering **A**, the values of **B** change as well, because **B** is still pointing to this same now-modified collection.

To break such linkages when they are not wanted, you can copy the actual values of a list, rather than just its name. This is done by selecting the full range of the list (using a colon within brackets, without any specified beginning and end point), and then putting these into a new list. This copies the elements to a new list. In place of **B=A**, which just copies the list name, the statement for creating an actual copy of the list is **B=A[:]**. If the copy is created in this way, any modifications to **B** won't affect **A** and any modifications to **A** won't affect **B**.

Converting between lists and strings

There are several ways to convert between lists and strings. For example, using the list function it is simple to create a list of the characters present in a string. This function will try to convert any variable into a list format:

```
>>> MyString = 'abcdefg'
>>> MyList = list(MyString)
>>> MyList
['a', 'b', 'c', 'd', 'e', 'f', 'g']
```

A list of strings or characters can be joined together into a single string with the `.join()` method. This method can be a bit confusing, because it seems to operate backwards. Rather then being a list method that takes a string argument, it is a string method that takes a list argument. The string that it acts on is inserted between each of the elements of the list when they are stitched together:

```
>>> MyList = ['ab', 'cde', 'fghi']
>>> ''.join(MyList)
'abcdefghi'
>>> '\t'.join(MyList)
'ab\tcde\tfghi'
>>> ' '.join(MyList)
'ab cde fghi'
```

In the first example above, the string that `.join()` acts on is empty—that is, defined with an empty set of quotes—and so the strings in the list are joined together without any characters in between them. The `.join()` method is particularly useful for building up lines of tab-delimited text from lists of data, or for creating a single string from a list of characters.

Adding elements to lists

Earlier, you modified a list by setting one of the elements to a new value. You may be tempted to do something similar: *add* elements to a list by directly accessing their index within the list. However, given a 2-element list, defined by:

```
x = ['A','B']
```

you can't add a third element `x[2]='C'`. The assignment operator is used to change the value of existing list elements, but can't be used to create elements that don't already exist. You have to use the `.append()` function to build lists:

```
x.append('C')
```

You can't add element number 20 directly to a list without defining the intervening values. Nor can you begin a list using `.append()` on a new variable name, if that variable hasn't already been defined. You must instead create an empty list, with `x=[]`, and then build up a list by appending to that starting point.

To insert items between existing list elements, use two matching indices to specify the location of insertion:

```
>>> MyList=['a','e']      ← Define a list with two elements
>>> MyList[1:1] = ['b','c','d']      ← Insert into position 1
>>> MyList
['a','b','c','d','e']
```

Lists are convenient for organizing data that correspond to sequential integer values, for example, consecutive field site numbers from 0–19. This kind of sequence, though, isn't always useful for your application, since it might be missing some numbers: there might be twenty field sites, but they could be numbered 0–9 and 12–21 (maybe a storm took out sites 10–11 and you added a couple more). A list would not be ideal for organizing these data, because you can't add a 12th element without a 10th and 11th element. One solution would be to use a list and pad unused elements with a placeholder value, but then you need to be sure to account for these placeholders in later steps. Another solution would be to use a dictionary that has integer keys. Then the keys could be any value at all, as long as they are unique.

Removing elements from lists

To delete elements from a list, use the del() function, or else just reassign an empty list to those elements:

```
>>> MyList = range(10,20)
>>> MyList
[10, 11, 12, 13, 14, 15, 16, 17, 18, 19]
>>> MyList[2:5]=[]
>>> MyList
[10, 11, 15, 16, 17, 18, 19]
>>> MyList = range(10,20)
>>> MyList
[10, 11, 12, 13, 14, 15, 16, 17, 18, 19]
>>> del(MyList[2:5])
>>> MyList
[10, 11, 15, 16, 17, 18, 19]
```

Checking the contents of lists

Often it's important to know if a particular element is contained in a list, regardless of its exact position. The in operator makes this test and returns True or False:

```
>>> MyList = range(10,20)
>>> MyList
[10, 11, 12, 13, 14, 15, 16, 17, 18, 19]
>>> 11 in MyList
True
>>> 21 in MyList
False
```

Sorting lists

There are a couple of different ways to sort lists in Python. One is the list method `.sort()`, available since Python version 2.4:

```
>>> MyList = [4,3,6,5,2,9,0,8,1,7]
>>> MyList.sort()    ← You don't have to assign the output to a new variable
>>> MyList
[0, 1, 2, 3, 4, 5, 6, 7, 8, 9]
```

Notice that `.sort()` doesn't return anything, not even a sorted list. This is because the list is sorted in place—in other words, the list that `.sort()` acts on is itself changed. Most Python variables can't be directly modified, but lists can be (see the previous box on copying and modifying Python variables).

At times you may want to get a sorted copy of your list, and leave the original list unchanged. In these cases, you will want to use the `sorted()` function with the original list as a parameter in `()`:

```
>>> MyList = [4,3,6,5,2,9,0,8,1,7]
>>> NewList=sorted(MyList)
>>> MyList    ← The original list is unchanged
[4, 3, 6, 5, 2, 9, 0, 8, 1, 7]
>>> NewList   ← The sorted list has been placed here
[0, 1, 2, 3, 4, 5, 6, 7, 8, 9]
```

Identifying unique elements in lists and strings

Often you aren't interested in all the elements in a list—you just want those which are unique. You may, for instance, want to know which character states are possible for a set of specimens. The `set()` function[6] can be used to summarize the unique elements in lists and the unique characters in strings. We won't work with the output of the `set()` function directly, but instead immediately convert it to a list:

---

[6] The `set()` function actually returns a special set-type variable, but we are skipping that distinction. The function became built-in to Python installations starting with version 2.4. To use it in version 2.3, add this command to the beginning of your program: `from sets import Set as set`.

```
>>> Colors = ['red','red','blue','green','blue']
>>> list(set(Colors))
['blue', 'green', 'red']
>>> DNASeq = 'ATG-TCTCATTCAAAG-CA'
>>> list(set(DNASeq))
['A', 'C', '-', 'T', 'G']
```

This approach to identifying unique elements can be applied to make much of the code you write more general. As an example, take the `dnacalc.py` program you wrote in Chapter 8. In its original incarnation, this program had dedicated code for each of the four nucleotides (`A`, `T`, `G`, and `C`). Earlier in this chapter you modified the program (the new version was called `compositioncalc1.py`) to loop over a list of the nucleotides, which was a much cleaner way to approach the problem, since you could use the same code to analyze each of the nucleotides in turn. Still, though, the nucleotides to be analyzed were hardcoded in the definition of `BaseList`. Now, with the `set()` function, you can extract the list of characters from the string itself. This program would give the same result as the previous version when applied to DNA, but could also be used if there are nonstandard nucleotides, or if you wanted to count the frequency of amino acids in proteins. In fact it is so general now that it can be used to calculate the frequency of characters in any string. This program is available in the scripts folder as `compositioncalc2.py`:

```
#!/usr/bin/env python
DNASeq = "ATGTCTCATTCAAAGCA"
SeqLength = float(len(DNASeq))

BaseList = list(set(DNASeq))
for Base in BaseList:
    Percent = 100 * DNASeq.count(Base) / SeqLength
    print "%s: %4.1f" % (Base,Percent)
```

List comprehension

Many analyses require batch modifications or calculations for each item in a list. For example, you might want to square each element in a list, or make each element in a list of strings uppercase, or make a new list containing the length of each word in an existing list. You can't just say `WordList.upper()` or `NumberList*2` to transform each element individually. Transformations applied to a list aren't automatically applied to each item. Sometimes such an attempt will result in an error, while at other times it will modify the list as a whole rather than the elements. The `*` operator, for instance, creates a new list that consists of multiple copies of the original list placed end-to-end:

```
>>> MyList = range(0,5)
>>> MyList
[0, 1, 2, 3, 4]
>>> MyList * 2
[0, 1, 2, 3, 4, 0, 1, 2, 3, 4]
```

There are modules, such as the numpy module described in Chapter 12, that do provide more sophistication for applying operations to elements in a list. A general solution would be to write a for loop that goes through each element in the list and applies the desired transformation:

```
>>> Values = range(1,11)
>>> Values
[1, 2, 3, 4, 5, 6, 7, 8, 9, 10]
>>> Squares = []
>>> for Value in Values:
...        Squares.append(Value**2)
...
>>> Squares
[1, 4, 9, 16, 25, 36, 49, 64, 81, 100]
```

The code above[7] produces a list called Squares containing the squares of 1–10. (Note that indented code works fine at the Python interpreter: the prompt changes to ... to let you know you are in a nested code block, and you can hit an extra return to get out of the code block.)

There is a shorthand construct called **list comprehension** that lets you perform this same kind of operation on each element in a list, but with a single command. This is a little bit complex to understand, but it can be a big time-saver in your programs. For example:

```
>>> Values = range(1,11)
>>> Values
[1, 2, 3, 4, 5, 6, 7, 8, 9, 10]
>>> Squares = [Element**2 for Element in Values]
>>> Squares
[1, 4, 9, 16, 25, 36, 49, 64, 81, 100]
```

The calculations are made in a single line, rather than with a multiline loop. The list comprehension statement loops through the list Values and performs some operation (**2 in this case) on each item (Element), and returns the list of results. Notice that the entire construct is within square brackets.

---

[7] The ** operator is for doing exponents, so x**y is x to the y power.

List comprehension is a useful way to extract columns of data from a two-dimensional array or a list of strings. You can specify a single index to get a column, or a range to get a subset of each list element. For example:

```
>>> GeneList = ['ATTCAGAAT','TGTGAAAGT','TGTATCGCG','ATGTCTCTA']
>>> FirstCodons = [ Seq[0:3] for Seq in GeneList ]
>>> FirstCodons
['ATT', 'TGT', 'TGT', 'ATG']
```

In this case, the variable `Seq` takes on the value of each entire string in the list, and then the first three characters are extracted into a new list. Several operations can even be combined together in this stage. Here the first three characters of each string are extracted and concatenated to a string with the + operator:

```
>>> Linker='GAATTC'
>>> Start = [(Linker + Seq[0:3]) for Seq in GeneList ]
>>> Start
['GAATTCATT', 'GAATTCTGT', 'GAATTCTGT', 'GAATTCATG']
```

Here is another example of a function used inside a list comprehension, reminiscent of your first Python program. (The resulting list isn't stored in a variable, so the interactive prompt displays it directly):[8]

```
>>> [ Seq.count('A') for Seq in GeneList ]
[4, 3, 1, 2]
```

Although you can convert a string to a list of characters using the `list()` function, it is sometimes hard to go from a list of numbers `[1,2,3]` to the equivalent strings `['1','2','3']`. You can't just use `str(ListOfIntegers)`. List comprehension again comes to the rescue:

```
>>> [ str(N) for N in range(0,10) ]
['0', '1', '2', '3', '4', '5', '6', '7', '8', '9']
```

The format for list comprehension can be a bit confusing (try things out in the interactive prompt before putting them in your program), but if you work with matrices of data, you will probably find them to be very useful. If you need even better manipulation and access to the components of an array, see the section on the `numpy` and `matplotlib` modules in Chapter 12.

---

[8] You can even nest list comprehension loops. For example:
```
[ [ Seq.count(Base) for Seq in GeneList] for Base in "ACGT"]
```

SUMMARY

You have learned how to:

- Start the Python command-line interpreter
- Use `dir()` to see functions within a variable
- Go to the Web for Python help
- Create logical expressions
- Write `if` statements
- Use the `else:` command
- Store data in a dictionary as `{key: value}` pairs
- Retrieve dictionary entries using `[]` or `.get()` after the dictionary name
- Convert data from the Web into programs by using regular expressions
- Write `for` loops to work with strings
- Use loops to look up values in dictionaries
- Work with lists as follows:

   Define lists with `MyList =[1,2,3]`

   Extract elements with `[]`

   Add elements with `.append()`

   Define numerical lists with `range()`

   Convert strings to character lists with `list()`

   Convert lists to strings with `' '.join()`

   See if an item is in a list with the `in` function

   Identify unique list elements with `list(set())`

   Sort lists with `.sort()` and `sorted()`

   Remove elements with `del()` or `[]`

- Use list comprehension, for example
  `Squares = [Val**2 for Val in MyList]`

Moving forward

- Indent the `print` statements in `proteincalc.py` so that they occur inside the `for` loop, and see how it affects the output of your program.

- Working at the interactive prompt, try constructing and sub-sampling multi-dimensional lists.

- Build a list-comprehension statement that returns a list of the characters `'a'` to `'z'`.

## Chapter 10

READING AND WRITING FILES

The programs you have created so far rely on information being written into the program itself, or else provided by the user at a prompt. In most cases, though, you will want to process data that are stored in files. In this chapter you will learn how to open text files, parse the data, and then use that information to generate new text files. You will develop these skills by building a file converter program that reads in a tab-delimited file containing latitudes and longitudes, then writes out a file that can be visualized with **Google Earth**. In the course of building up this example, you will get experience in Python with file handling, regular expressions, several new functions, and a variety of other tools.

Surveying the goal

This chapter will focus on building up a program that can read location data from one text file format, convert it to another format, then rewrite it in another file. Specifically, this program will reformat an input file with a series of locations to create an output file that can be read and displayed by the geographical viewer Google Earth. This is a very typical challenge for biologists—data are in hand and you know what you want to do with them, but the format of the data isn't understood by the program you want to use.

The input file, `Marrus_claudanielis.txt`, contains the latitude, longitude, depth, and associated data for several specimens that were considered when a new species of siphonophore, *Marrus claudanielis*, was described. The first challenge is to simply read the text from the file. Next, the individual components of the data must be parsed—that is, extracted from each line. Finally, these data must be repackaged into a new format and written to an output file.

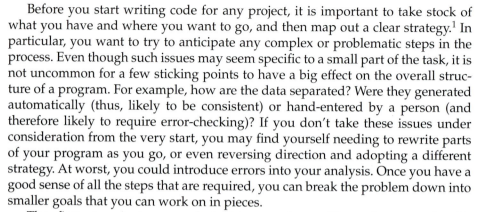

Before you start writing code for any project, it is important to take stock of what you have and where you want to go, and then map out a clear strategy.[1] In particular, you want to try to anticipate any complex or problematic steps in the process. Even though such issues may seem specific to a small part of the task, it is not uncommon for a few sticking points to have a big effect on the overall structure of a program. For example, how are the data separated? Were they generated automatically (thus, likely to be consistent) or hand-entered by a person (and therefore likely to require error-checking)? If you don't take these issues under consideration from the very start, you may find yourself needing to rewrite parts of your program as you go, or even reversing direction and adopting a different strategy. At worst, you could introduce errors into your analysis. Once you have a good sense of all the steps that are required, you can break the problem down into smaller goals that you can work on in pieces.

The first step is to see what data you have. Open the example file `Marrus_claudanielis.txt` in your text editor and get a feel for how the file is structured:

```
Dive△          Date△       Lat△        Lon△          Depth△  Notes
Tiburon 596△   19-Jul-03△  36 36.12 N△ 122 22.48 W△  1190△   holotype
JSL II 1411△   16-Sep-86△  39 56.4 N△  70 14.3 W△    518△    paratype
JSL II 930△    18-Aug-84△  40 05.03 N△ 69 03.01 W△   686△    Youngbluth (1989)
```

The data are organized one specimen per row, beginning with a header row that describes which data can be found in each column. There is a tab between each data field (shown by the △ symbol, which you can reveal in TextWrangler using Show Invisibles), and some fields can contain spaces. The latitude and longitude are each in their own column. These coordinates are formatted in degrees and minutes (the minutes can have fractions), with a letter indicating North, South, East, or West. There is a single space between the degrees, minutes, and compass letter. The Notes field at the end of the line can also contain spaces, or it can be blank.

Now take stock of where you want to go. The intermediate step will be to split the values from the columns. The ultimate goal is to write a file that Google Earth can load, formatted in Keyhole Markup Language, or KML. There are a couple of ways to figure out file formats. If you are lucky, the file format will have a good technical definition that explains all its parts and ensures that there is no confusion across programs or among programmers. In this case, Google's KML file format has a thorough definition (see `code.google.com/apis/kml/documentation/`), but more often than not, you will be faced with file formats that have never been properly specified. In those cases you must take a look at an existing file and reverse-engineer its structure.

---

[1] This is equivalent to creating an outline before you write a paper. We may not always operate this way, but the results, especially in a more complex paper, are usually better when we do.

Here, the example file `Marrus_claudanielis.kml` is what you are hoping to produce at the end of the process, and so it will give you a sense of how a KML file is organized. To take a look at its structure, open it in a text editor rather than in Google Earth. At this point, don't worry about all the formatting characters; we will get to that stage later. For now, just identify the bits of data that will need to be generated from the input file and figure out how these bits will need to be modified, going from one file type to the other. A quick examination indicates that the position coordinates need to be in decimal degree format, rather than in degrees and minutes, with a minus sign rather than letters to indicate South and West. The depths also need to be in negative numbers, reflecting that they are below sea level.

Now you know some specifics about your starting and ending points. Rather than take on the entire transformation at once, you will first focus on extracting the data out of the input, in the process converting the geographic coordinates from degrees and minutes to decimal degrees. You will then write these converted data to a file, though not yet in KML format—that can wait until you have taken care of all other data processing that needs to be done.

Reading lines from a file

Considerations before reading a data file

As in most programming languages, there is more than one way to read files in Python. In particular, you can read the whole file into memory at once, or step through it line-by-line (Figure 10.1). Both methods involve creating a variable that represents the stream of data contained in the file.

When you are considering how to read your dataset, there are two main questions to ask: First, is each data value contained within a single line (as is the case with `Marrus_claudanielis.txt`, in which each line contains information on one specimen), or can a data value continue across lines (as with molecular sequence data, where one sequence can be split across several lines)? Second, does the output depend only on the values contained in a single line (such as when converting units, or generating a new value based on other values in that line) or does it depend on knowing values from other lines in the file (such as a sorted list or a running average)?

The simplest case is when there is a one-to-one correspondence between input and output lines, as in file type A in Figure 10.1. Here, no data are combined across lines, and the input and output lines are in the same order. In this case, file processing can generally be handled with a single `for` loop. Strategies for laying out your data are discussed in Chapter 15, but in general you want to strive from the outset—before you even begin your experiment—to organize your records so that associated information is not spread across several lines.

On the other hand, a common task in biological computing is to take a file that is in a non-standard format, and reorganize it into a simple table that can be processed by MATLAB, R, or a spreadsheet program.

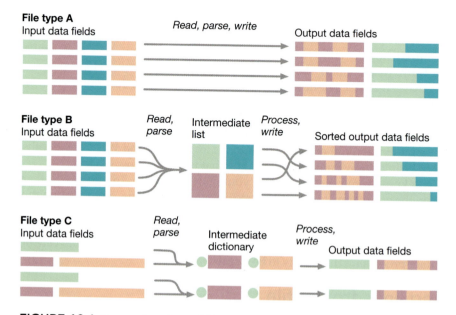

FIGURE 10.1 Processing data within and across lines of a file Parcels of similar data are represented by colored blocks in three different file types. In file type A, each line of output is generated from data within only one line of input, and the order of the lines is not changed. No data need to be stored or reorganized between lines, and each line can be processed in the order it occurs. In file type B, the order or content of data in the output may depend on all input data (such as when the output data are sorted across lines), or the values in the output may depend on a value derived from combined input (such as when all lines are normalized by the average). All input lines must be parsed into internal variables, such as lists, before any data can be processed and written to the output. Each of these input/output processes typically requires its own loop. In file type C, data from two or more lines of input are combined into each output line, such as when a specimen name occurs above a block of data and needs to be incorporated into each output line. For each line of output, data from multiple lines of input must be read, parsed, stored, and processed.

When reading and writing files where there is not a one-to-one correspondence between the lines of the "before" and "after" files, as in file types B and C in Figure 10.1, it is often useful to employ two loops: the first loop reads the data from the input file and stores it in internal variables, and the second loop processes the data and writes the new file. The file-loading loop may also be used to import data from several files into a combined data variable, with the second loop serving to output the aggregated data set; you will explore this approach in the next chapter. Once you create a program to do such operations on one type of file, it is relatively quick to modify the program to handle many other data formats.

Opening and reading a text file

To begin, this simple program will open a text file, read each line one by one, display each line to the screen as it is read, and then close the file:

```python
#!/usr/bin/env python

# Set the input file name
# (The program must be run from within the directory
# that contains this data file)
InFileName = "Marrus_claudanielis.txt"

# Open the input file for reading
InFile = open(InFileName, 'r')

# Initialize the counter used to keep track of line numbers
LineNumber = 0

# Loop through each line in the file
for Line in InFile:
    # Remove the line-ending characters
    Line = Line.strip('\n')
    # Print the line
    print LineNumber,':', Line

    # Index the counter used to keep track of line numbers
    LineNumber = LineNumber + 1

# After the loop is completed, close the file
InFile.close()
```

Enter the lines as shown into a text file, then save the file in your ~/scripts directory as latlon_1.py (or just copy latlon_1.py from the example scripts to ~/scripts). Make the file executable, then open a terminal window and cd to your examples folder, where the data file Marrus_claudanielis.txt is located. Execute the new program by typing latlon_1.py.

If the program fails, try making sure that the data file is present in your current directory:

```
ls Marrus_claudanielis.txt
```

If all of the lines of the file are displayed by your program as a single line, check the end-of-line characters in your data file to make sure they are appropriate for your system (see Chapter 1). This is a common problem when reading data files, and fortunately it is usually simple to address by opening the file in a text editor, changing the end-of-line character, and saving the file again. You can also try using 'rU' instead of 'r' in the program's open() statement. There are ways to do batch conversions of end-of-line characters for more than one file at a time; these will be discussed in Chapter 16.

The comments (the text following the # characters) explain most of what is happening, including some new things you haven't seen until now. The built-in function open() is called with two parameters, one specifying the name of the file to be opened and the other specifying the file mode; in this case, the mode is 'r' for read. The file-object returned by open() is assigned to the variable InFile. Note that the variables InFileName and InFile are different. Whereas InFileName is just a string that contains the path of the file to be opened, InFile is an object that allows us to interact with the file and its contents in a read-only mode.[2] Mixing up these two kinds of variables is a common mistake for beginning programmers.

In pseudocode, the basic process for reading a file is this:

```
InFile = open(FileName, 'r')     ← Open up a pipeline to the file
for Line in InFile:       ← Loop through the lines one by one
    pull values from each Line
    store them, do calculations, or write output
InFile.close()   ← Close the file object (not the filename)
```

To keep track of where you are in processing the file, you will keep a running count by adding 1 to the variable LineNumber each time through the loop. In this case, the count serves to give you an idea of how the program is progressing; more generally, internal markers of this sort can be useful for debugging and locating irregularities.

The for loop processes the file one line at a time. At each iteration of the loop, the for statement places the next line from the file into the string Line, after which the statements within the loop have access to the data values.

Removing line endings with .strip()

Lines are separated within files by end-of-line characters. Although these characters are normally invisible to you, they are retained at the end of each line as the line is read in. It is good practice to strip them off, to avoid having them interfere with later analyses. Here, the first step of processing each line is therefore to remove the end-of-line character with .strip('\n'). This string method returns a copy of the string it acts on (in the present program, this variable is Line), but first strips away the specified character (in this case, \n) from either end of the string. If the specified character doesn't occur at the beginning or end of the string, then the new string is identical to the old one. If you don't specify a character for .strip(), all white space, including spaces and tabs, is removed from the beginning and the end of the string. If you are going to be reading files from various sources, you can strip both kinds of line endings in succession with Line.strip('\n').strip('\r').

For the moment, the program just prints out the line number, followed by a colon and the line itself. Remember that the print function adds an end-of-line character

You also need to remove \r for Windows files.

---

[2] Within the for loop, the InFile object that you create acts like a list of lines in the file. You can create an actual list variable containing all the lines of the file using the command FileList = InFile.readlines().

to the text it displays; thus, even though you are stripping this character as each line is read, it is effectively added back when the line is displayed. The final statement in the loop increments the line counter—that is, adds 1 to its value—before returning to the `for` statement to process the next line. After the loop is done, the file is closed with the `.close()` method, which is built into the `InFile` variable created by `open()`. Remember to close `InFile`, not `InFileName`, and to place the close statement outside of the loop, rather than have it indented within the loop.

Note that the `LineNumber` variable starts out with a value of 0 and is then incremented by 1 as the last step of each cycle of the loop. Because `LineNumber` isn't incremented until the *end* of each loop, the first line is labeled as line 0, the second line as line 1, etc. We could have started the count at 1 so that the first line was 1 rather than 0, but it is conventional in computer programs to start counters like this at 0. In this case, the starting number wouldn't have made a substantial difference, but in many cases, such as when indexing lists, it is important to start with a 0. It is therefore best to always number items starting with 0, so that there isn't any confusion about which numbering system is in use.

Skipping the header line

As it stands, the program treats each line the same. However, the example file has a header line, and you will want to skip this line when you extract data from the specimen records. Since you are already keeping track of the line numbers, all you need to do now is insert an `if` statement that only considers lines greater than 0. Since the first line is numbered 0, that line is now skipped. Modify your program to add an `if` statement as follows:

```python
#!/usr/bin/env python

# Set the input file name
InFileName = 'Marrus_claudanielis.txt'
# Open the input file for reading
InFile = open(InFileName, 'r')
# Initialize the counter used to keep track of line numbers
LineNumber = 0

# Loop through each line in the file
for Line in InFile:
    if LineNumber > 0:       # ← To skip the header line; note next few lines are now further indented
        # Remove the line ending characters
        Line = Line.strip('\n')
        # Print the line
        print LineNumber,":",  Line
    LineNumber = LineNumber + 1

# After the loop is completed, close the file
InFile.close()
```

The line that increments the counter, `LineNumber = LineNumber + 1`, is not under control of the `if` statement—its indentation is the same as that of the `if` statement, so it is executed with each cycle of the loop. If it were within the block of code controlled by the `if`, it would never be executed and `LineNumber > 0` would never be `True`. This wouldn't create an infinite loop, since the `for` statement would still process each line of the file, but none of the lines would be analyzed or printed.

Keep practicing this process of file reading so that you become comfortable with it, and when you come up with some code that works well, reuse it for your other programs. The basic process of opening and looping through a file is one of the most powerful uses of a Python program. You can probably already imagine many potential applications.

Parsing data from lines

Splitting a line into data fields

Now that you have access to the data in this file on a line-by-line basis, you are ready to pluck the needed specimen data from each line. This could be done entirely with regular expressions—and in fact, in some cases regular expressions may be the easiest way to go, such as when formatting is inconsistent. However, when character-delimited text files are formatted in a consistent manner, they are usually most easily approached using `.split()`. This method takes a string and splits it according to a delimiter, thereby producing a list of strings—that is, a list of the values or fields occurring between the delimiters. The delimiters themselves are thrown away. In the list, the fields occur in the same order as they did in the original string.

 By default, `.split()` uses white space, in the form of both spaces and tabs, as separators for the columns of data. However, you can specify your own delimiter by adding it as a parameter in the parentheses. For example `Line.split('\t')` would split the line into a list of those elements originally separated only by tabs, not by spaces. (Remember from the chapters on regular expressions that `\t` is the character combination used to represent a **tab**.) It is desirable for now to keep the latitude/longitude values and the comments unsplit, so we will specify `'\t'`.

To see what such a list looks like, add this line to the program, indented before the `print` statement:

```
ElementList = Line.split('\t')
```

and then modify the `print` statement to display this list instead of the line itself:

```
print LineNumber, ":", ElementList
```

Rerun your edited script and see how the output has been subdivided, and how the values occur within square brackets, indicating they are contained in lists:

```
1 : ['Tiburon 596', '19-Jul-03', '36 36.12 N', '122 22.48 W', '1190', 'holotype']
2 : ['JSL II 930', '18-Aug-84', '40 05.03 N', '69 03.01 W', '686', 'Youngbluth (1989)']
3 : ['JSL II 1411', '16-Sep-86', '39 56.4 N', '70 14.3 W', '518', 'paratype']
4 : ['Ventana 1575', '11-Mar-99', '36 42.24 N', '122 02.52 W', '767', '']
```

You can also test what happens if you omit the '\t' parameter and use `ElementList.split()`. You can see that the dive, latitude, longitude, and comments are further subdivided into their own list elements, because they contain spaces.

A subtle but important thing to notice about the results is that there are empty single quote marks (' ') at the end of those lines where the Notes field is empty. When parsing a file, it is important to recognize that there may be empty data fields such as these, and to make sure that your program is not thrown off—for example, by the presence of two tabs in a row, indicating an empty field.

Selecting elements from a list

Now you are able to read in a file of delimited data and split the constituent columns into a list. Lists are ordered arrays of data, indicated in Python by square brackets, [ ]. The first element in a list is numbered 0, the second element is numbered 1, and so on. To access individual elements of the `ElementList` variable, you put the index or range of indices in [ ] after the name.

For example, each time through the loop, `ElementList[2]` (the third element in the list) will contain the latitude value, and `ElementList[3]` (the fourth element in the list) will contain the longitude value. You can test this by modifying or adding a print statement to display the depth, latitude, and longitude by themselves:

```
print ElementList[4], ElementList[2], ElementList[3]
```

When displayed this way, each field is separated by the space that is inserted by `print` when commas are used. To separate the fields with a tab, use the format operator to get more control over the string:

```
print "Depth: %s\tLat: %s\tLon: %s" % (ElementList[4], ElementList[2], \
   ElementList[3])
```

This portion of code also introduces a technique for splitting a long line into multiple lines by putting a backslash (\) at the very end of a line. This continuation character causes the line that follows to be interpreted as though it was on the same line as the backslash. In effect, you are escaping out the end-of-line character. The indentation of the second line does not matter when using the backslash in this way.

By substituting as we have into the `%s` string placeholders, we obtain this output:

```
Depth: 1190    Lat: 36 36.12 N    Lon: 122 22.48 W
Depth: 518     Lat: 39 56.4 N     Lon: 70 14.3 W
Depth: 686     Lat: 40 05.03 N    Lon: 69 03.01 W
```

Remember at this point that these values are strings and have not been converted to integers or floating point values. This is why we use %s to plug the strings directly into the output, rather than %d or %f as we would when formatting numerical values. This version of the program is saved as `latlon_2.py` in your example scripts folder.

Writing to files

The next major skill in working with files is to write your output to a new file. So far you have been printing all your output to the screen. You could capture this text using the shell's redirect operator (>>) as described in Chapter 5, but you will often want to create a file directly from within the program.

Getting set up to write data to a file is very much like getting set up to read from a file, in that you use the `open()` command and a filename:

```
OutFileName = "MarrusProcessed.txt"
OutFile = open(OutFileName,'w')
```

Here you are using the `'w'` parameter to indicate that you are getting ready to write data to the file named in `OutFileName`, overwriting any previously existing file by that name. There is also the `'a'` option, which will **append** (that is, add) to an existing file, or if necessary create such a file if it doesn't already exist. The `'w'` and `'a'` options are equivalent to the > and >> redirect operators in the shell, where the first overwrites files, and the second appends output to a file.

Now that the file is open, you can write text to it with the `.write()` method; this method is built into the file variable, in this case the variable called `OutFile`. For example, to write just the depth string to a file, within the loop you could say: `OutFile.write(ElementList[4]+"\n")`.

Note that you need to append a line ending character, `"\n"`, or else all of your output will be on the same line. This is because unlike `print`, the `.write()` method does not add a line ending by default.

Once you are done writing your output, you will close the file you have written, using the same method that you used to close the file you were reading. Again, be sure to put the `open()` and `.close()` statements *outside* the loop; you don't want to reopen the file each time:

```
OutFile.close()
```

Put these three components together in your script to generate a version, `latlon_3.py`, which writes the depth, latitude, and longitude to a file:

```python
#!/usr/bin/env python

# Read in each line of the example file, split it into
# separate components, and write certain output to a separate file

# Set the input file name
InFileName = 'Marrus_claudanielis.txt'

# Open the input file for reading
InFile = open(InFileName, 'r')

# Initialize the counter used to keep track of line numbers
LineNumber = 0

# Open the output file for writing
# Do this *before* the loop, not inside it
OutFileName=InFileName + ".kml"

OutFile=open(OutFileName,'w') # You can append instead with 'a'

# Loop through each line in the file
for Line in InFile:

    # Skip the header, line # 0
    if LineNumber > 0:

        # Remove the line ending characters
        Line=Line.strip('\n')

        # Separate the line into a list of its tab-delimited components
        ElementList=Line.split('\t')

        # Use the % operator to generate a string
        # We can use this for output both to the screen and to a file
        OutputString = "Depth: %s\tLat: %s\t Lon:%s" % \
            (ElementList[4], ElementList[2], ElementList[3])

        # Can still print to the screen then write to a file
        print OutputString

        # Unlike print statements, .write needs a linefeed
        OutFile.write(OutputString+"\n")

    # Index the counter used to keep track of line numbers
    LineNumber = LineNumber + 1

# After the loop is completed, close the files
InFile.close()
OutFile.close()
```

⚡ When writing to files, you need to be careful with your filenames and file-related variables. When you open a file for reading, it is not a serious problem to accidentally specify the wrong filename. You will just open a file you weren't expecting, or you will get a "No such file" error. However, when you are *writing* to files, even the act of opening a file will erase its contents! It is often good practice to check to see if a file exists before trying to open it. One way to do this is with the `os.path.exists()` function. This function is part of the `os` module, which you can access by adding `import os` to the top of your program, after the shebang line but before you invoke any of the module's functions. (You will learn more about importing modules shortly.)

Recapping basic file reading and writing

At this point, you can read lines of data from a file, extract individual fields, and write these fields back out to another file. In the present program, `latlon_3.py`, the sequence of events goes like this:

- Open the files to be used for reading and writing with the `open()` function

- Read in a line as part of a `for` loop

- Check to see if the line number is greater than 0

- Split the line into a list of strings according to tabs

- Generate a formatted output string

- Print that string to the screen (this behavior is used for testing or debugging, but otherwise stays turned off)

- Write the string to a file

- Increment the line number

- Loop back to read and process the next line

- When done with the last line, close the input and output files

This general sequence will be used repeatedly in your work. For the most part, the things that will change from one program to another are how the line values are processed and how the output string is formulated. You will now explore variations of both of these aspects—first, to convert the latitude/longitude values to decimals, and second, to eventually write the output as a Google Earth KML file.

Parsing values with regular expressions

You have split a line into individual elements, but some of these elements now require additional parsing. To convert the latitude and longitude values to their decimal equivalents, you have to separate the corresponding text fields into their three components: degrees, minutes, and hemisphere (i.e., N, S, E, or W). There are a few ways to do this, but here you will use regular expressions, the search and replace tools you learned about in Chapters 2 and 3.

For the next section, we will mainly be showing portions of the program that occur within the `for` loop after the line is read in, and not showing the full body of the code. We will show the full code with all these modifications at the end of this section.

Importing the `re` module

While there is support for regular expressions in Python, these commands aren't available by default. They are bundled into a module that you must first import into your program. Modules are collections of computer code that are packaged together, and the act of importing them into a program makes them available for your use. Modules will be discussed in greater detail in Chapter 12, where you will find information on additional built-in modules that come with the Python language, more options for importing tools from modules, and information on installing third-party modules that don't come with Python by default.

The name of the regular expressions module is `re`. To import this module into your program add the following line below the shebang:

```
import re
```

It is customary to do your imports close to the top of the file. This ensures that the appropriate modules are loaded before you use them, and makes it easier to glance at a program and get a sense of what tools it uses.

You can also import modules with this same command when working at the Python interactive prompt. Once a module is imported, you can take advantage of the `dir()` and `help()` functions, such as `help(re)`, to get a bit of information on what is available in it.

Using regular expressions with the `re` module

There are several functions in the `re` module, including `re.search()` and `re.sub()`. Regular expressions have the following general format in Python:

`Result = re.search(RegularExpressionString, StringToSearch)`

In our case, the string to search will be one of the latitude/longitude values in the list named `ElementList`. Review the output of `latlon_2.py` or `latlon_3.py` to see the format of `ElementList[2]` (the first line of which is shown here) and `ElementList[3]`:

```
"36 36.12 N"
```

Now think back to your regular expression skills. In words, you now want to match "one or more digits, a space, one or more digits or decimal points, a space, a letter." Translated to regular expressionese, this becomes:

```
\d+ [\d\.]+ \w
```

You want to capture the three elements independently, so put parentheses around them:

```
(\d+) ([\d\.]+) (\w)
```

To put this into a Python regular expression and search for it within the string `ElementList[2]`, you will create that query as a string:

```
SearchStr = '(\d+) ([\d\.]+) (\w)'
Result = re.search(SearchStr,ElementList[2])
```

For clarity, we have defined the search string as a variable, `SearchStr`, and then used the variable for the search. However, you could also put the search string directly within the `re.search()` statement where it currently says `SearchStr`.

PYTHON SEARCH STRINGS When you construct search terms in Python for regular expressions, you can generally use the same wildcards and symbols as in **TextWrangler**. However, some issues arise with backslashes. This is because Python uses backslashes in the same way regular expressions do—to escape special characters. This means that escaped characters are escaped by Python when the search string is defined, before being passed to the regular expression. If the escape sequence is one that is interpreted the same way by both Python and regular expressions, this will have no effect on the result. The two characters `\t`, for instance, will be replaced with an actual tab character by Python before being passed to the regular expression. This tab will match tabs in the string that is being searched, just as the original escape sequence `\t` would have. Likewise, no problematic issues arise when the escape sequence is specific to regular expressions. The two characters `\d`, for instance, don't mean anything to Python, and are passed as-is to regular expressions.

However, if you want to search for a backslash (\) with a regular expression, you need to escape it with a backslash. You therefore need two backslashes (\\) in the search term being passed to the regular expression. But Python also escapes backslashes, so if you put two backslashes in the string you write in your Python code, Python will pass only one of them to the regular expression! You therefore need to escape each of the two backslashes with its own backlash. As an example, searching for a single backslash requires an initial search string with four backslashes. The need to write \\\\ to get \\ to search for \ is so nefarious that it has been given its own name, the "backslash plague."

Apart from writing four backslashes, there is another solution to this problem. This is to turn off character escaping in Python by placing an `r` immediately before the quotes that mark a string (`r` stands for "raw"). For example, if you set `s="c:\\"`, Python will escape the backslash; thus the resulting string `s` will contain only one backslash and will have the value `"c:\"`. Defining a raw string with the statement `s=r"c:\\"`, on the other hand, suppresses Python's character escaping, and the string will have the value `"c:\\"`.

When this command is run it generates a variable called `Result`. But how do you access the results themselves? In fact, search results are a special type of Python object, containing information about what matched within the string, where it matched, and the captured text. These results are stored as groups within the variable `Result`. All the results together can be retrieved with `Result.groups()`; they can also be retrieved individually with the `.group()` method applied to the variable (note that there is no `s` at the end of the method name). The zeroth group (that is, the very first group) is the full match of your search expression, while subsequent groups (`1,2,3`) contain the substrings captured by each pair of parentheses. The following commands retrieve the captured groups:

```
DegreeString = Result.group(1)  # this is equivalent to \1 in TextWrangler
MinuteString = Result.group(2)
Compass      = Result.group(3)   ← The spacing here is just to make it more readable
```

Because you are going to work with the first two values as floating point numbers, you will need to convert them with the `float()` function. Since you don't need the strings themselves for anything, you might as well do this conversion at the same time that you retrieve the regular expression results:

```
Degrees = float(Result.group(1))
Minutes = float(Result.group(2))
Compass = Result.group(3).upper() # make sure it is capital too
DecimalDegree = Degrees + Minutes/60

# If the compass direction indicates the coordinate is South or
# West, make the sign of the coordinate negative
if Compass == 'S' or Compass == 'W':
    DecimalDegree = -DecimalDegree
```

This code is a lot to digest at once. You are using a regular expression to extract three substrings, then converting two of these to floating point numbers, and then using the three values in combination to do a couple of simple calculations. If you want to reinforce the ways that regular expressions are built, try creating some strings in the Python interpreter and testing some searches.

Summary of using `re.search()` and `re.sub()`

Usage of the `re` module is summarized in the following session at Python's interactive prompt. We haven't walked you through an example of the `re.sub()` function, but its usage is relatively clear from the context:

```
>>> import re
>>> OrigString = "Haeckel,Ernst"    ← The string you wish to search
>>> SubFind = r"(\w+),(\w+)"    ← The regular expression to search for
>>> Result = re.search(SubFind,OrigString)    ← Perform the search, store the results
>>> print Result.groups()    ← See what you found
('Haeckel', 'Ernst')    ← The two captured substrings
>>> SubReplace = r"\2\t\1"    ← Set up a string for replacement using raw string style
>>> NewString = re.sub(SubFind,SubReplace,OrigString)    ← Format for re.sub()
>>> print NewString    ← See the substituted string
'ErnstΔ   Haeckel'
```

Before you go on to incorporate the changes using re into our current program, we are going to explain one more important programming capability—that of creating your own functions. This will allow you to reuse this one bit of code on both the latitude and longitude values.

Creating custom Python functions with def

Some earlier programming examples in this book included nearly identical blocks of code that did the exact same calculations, but on different variables. In the first version of the dnacalc.py script, for example, you needed to perform the same operation on all four bases. You just copied and pasted the same code and modified it to act on A, G, C, and T in turn.

If you find yourself copying and pasting the same blocks of code within a program, this is a sign that you should pause and think of another way to achieve those repeated steps. There are several problems with reusing code by copying and pasting within a program: it gets confusing; it is a lot of work if the code is frequently reused; if you want to change one thing in the replicated procedure you will need to fix it in many places; and finally, it makes the program unnecessarily long.

Loops are a great way to reuse the same block of code. You used a loop to simplify the dnacalc.py program in the last chapter. Now we introduce another tool for reusing the same piece of code repeatedly: a **function**. You have been using functions since your first Python program. For example, consider the built-in float() function. You could conceivably write several lines of code that take a string and convert it to a floating point number, then re-copy those lines wherever you need that function. Not only would this approach make your program unwieldy, but if you wanted to change or fix the code involved, you would have to go through and reproduce that same change every where it occurred in your program.

Built-in functions are nice, but even nicer is building your own functions which allow you to extend the language to suit your own needs. You will now create the function decimalat() to allow you to reuse the code that you just wrote to convert a raw string of degrees, minutes, and compass direction to a decimal measurement. Without this function, you would need to copy and paste the code to convert both the latitude and longitude strings to floats. Your function will take a string as input (such as "36 36.12 N"), and return a numerical value (in this case, 36.60200).

Functions are defined using the term def (short for **def**inition), followed by the name you give the function, a pair of parentheses containing the names of any parameters (that is, values) to be sent to the function, and finally, a colon. The function's commands are indented beneath the def line, and at the conclusion of the definition, any values to be sent back are specified using the return statement. Functions must be defined prior to the point in your program where they are used. To build the decimalat() function, take the block of statements explained previously and place it below the line which indicates the name of your new function:

```
def decimalat(DegString):    ← DegString is the value sent to the function as input
    # This function requires that the re module is already imported
    # Take a string with format "34 56.78 N" and return decimal degrees
    SearchStr='(\d+) ([\d\.]+) (\w)'
    Result = re.search(SearchStr, DegString)

    # Get the captured character groups, as defined by the parentheses
    # in the regular expression, convert the numbers to floats, and
    # assign them to variables with meaningful names
    Degrees = float(Result.group(1))
    Minutes = float(Result.group(2))
    Compass = Result.group(3).upper() # make sure it is capital too

    # Calculate the decimal degrees
    DecimalDegree = Degrees + Minutes/60

    # If the compass direction indicates the coordinate is South or
    # West, make the sign of the coordinate negative
    if Compass == 'S' or Compass == 'W':
        DecimalDegree = -DecimalDegree

    return DecimalDegree    ← The value of DecimalDegree is the output sent back
# End of the function definition
```

At the end of the indented block which defines the function, the return statement (no parentheses are used here) tells what value the function should send back, regardless of whether this value is to be assigned to a variable or immediately used by the main program. Insert the function block near the top of your program, before you begin reading in the file. Then, within the for loop but after the line is split into separate fields, add these calls to your new function:

```
    LatDegrees = decimalat(ElementList[2])
    LonDegrees = decimalat(ElementList[3])
    print "Lat: %f, Lon: %f" % (LatDegrees,LonDegrees)
```

Notice how clean and compact the program becomes. Your function is called twice each time through the loop, once for longitude and once for latitude. You send it a string to operate on, and it provides the floating point value of the converted string, making this available for assignment to a variable.

Finally, adding `OutFile.write()` statements to your program, whether in place of or in addition to the `print` statements, will write this output to your `OutFile`. Here is the new version of the program, also available in the examples folder as `latlon_4.py`:

```python
#!/usr/bin/env python
#
# This program reads in a file containing several columns of data,
# and returns a file with decimal converted value and selected data fields.
# The process is: Read in each line of the example file, split it into
# separate components, and write certain output to a separate file

import re # Load regular expression module, used by decimalat()

# Functions must be defined before they are used
def decimalat(DegString):
    # This function requires that the re module is loaded
    # Take a string in the format "34 56.78 N" and return decimal degrees
    SearchStr='(\d+) ([\d\.]+) (\w)'
    Result = re.search(SearchStr, DegString)

    # Get the captured character groups, as defined by the parentheses
    # in the regular expression, convert the numbers to floats, and
    # assign them to variables with meaningful names
    Degrees = float(Result.group(1))
    Minutes = float(Result.group(2))
    Compass = Result.group(3).upper() # make sure it is capital too

    # Calculate the decimal degrees
    DecimalDegree = Degrees + Minutes/60

    # If the compass direction indicates the coordinate is South or
    # West, make the sign of the coordinate negative

    if Compass == 'S' or Compass == 'W':
        DecimalDegree = -DecimalDegree
    return DecimalDegree
# End of the function decimalat() definition

# Set the input file name
InFileName = 'Marrus_claudanielis.txt'

# Derive the output file name from the input file name
OutFileName = 'dec_' + InFileName

# Give the option to write to a file or just print to screen
WriteOutFile = True
```

```python
# Open the input file
InFile = open(InFileName, 'r')

HeaderLine = 'dive\tdepth\tlatitude\tlongitude\tdate\tcomment'
print HeaderLine

# Open the output file, if desired. Do this outside the loop
if WriteOutFile:
    # Open the output file
    OutFile = open(OutFileName, 'w')
    OutFile.write(HeaderLine + '\n')

# Initialize the counter used to keep track of line numbers
LineNumber = 0

# Loop over each line in the file
for Line in InFile:
    # Check the line number, don't consider if it is first line
    if LineNumber > 0:
        # Remove the line ending characters
        # print line  # uncomment for debugging
        Line=Line.strip('\n')

        # Split the line into a list of ElementList, using tab as a delimiter
        ElementList = Line.split('\t')

        # Returns a list in this format:
        # ['Tiburon 596', '19-Jul-03', '36 36.12 N', '122 22.48 W',
        # '1190', 'holotype']
        # print "ElementList:", ElementList  # uncomment for debugging

        Dive    = ElementList[0]
        Date    = ElementList[1]
        Depth   = ElementList[4]
        Comment = ElementList[5]

        LatDegrees = decimalat(ElementList[2])
        LonDegrees = decimalat(ElementList[3])
        # Create string to 5 decimal places, padded to 10 total characters
        # (using line continuation character \)
        OutString = "%s\t%4s\t%10.5f\t%10.5f\t%9s\t%s" % \
                        (Dive,Depth,LatDegrees,LonDegrees,Date,Comment)
        print OutString
        if WriteOutFile:
            OutFile.write(OutString + '\n') # remember the line feed!

    # another way to say LineNumber=LineNumber+1...
    LineNumber += 1 # this is outside the if, but inside the for loop

# Close the files
InFile.close()
if WriteOutFile:
    OutFile.close()
```

Scan through the program and make sure that you understand—or can figure out—what each section does. There are a few additional adjustments here and there in the example program which we haven't explicitly covered, but these should make sense in context. For instance, the line number counting is slightly different, taking advantage of a shorthand operator, `+=`, for incrementing a variable by a given value without writing the variable name twice. Many other programming languages also support this operator. Thus, `LineNumber += 1` does the same thing as `LineNumber = LineNumber + 1`, but with less typing. We will use this construct from here on.

Run the program at the command line, and use the `less` command to inspect the contents of the file it generates, `dec_Marrus_claudanielis.txt`.

If you have problems getting the program to execute properly, take this opportunity to practice your proofreading and debugging skills. Is the program running, but not doing what it should? Is your indentation consistent and correct? Is there punctuation missing—for example, colons after the `for` statements, the function `def`, and the `if` statements? If there are errors when you run it in the terminal window, Python will list the offending lines. Look in the vicinity of these lines for problems—the error is often located *above* the indicated line number. Also try commenting out portions of code to isolate the location of the problem, or printing out status variables to see whether the program enters the loop, what the list of elements looks like, and whether the regular expression is working. As always, if you get stuck or frustrated, consult Chapter 13 for debugging tips or seek help at `practicalcomputing.org`.

Here is a summary of the commands used when writing to a file:

```
OutFileName = "Outputfile.txt"
OutFile = open(OutFileName, 'w')
OutFile.write(DataLine + '\n')
OutFile.close()
```

Packaging data in a new format

Reformatting your file into a different tabular format, as you have done above, is a skill that you will use repeatedly. Hang on to the `latlon_4.py` script and modify it to suit your own purposes. Our ultimate goal in this chapter, though, isn't to write another tab-delimited file; rather, it is to write a KML file that can be understood by Google Earth. You are most of the way there, having read and converted the actual coordinate data. Now you need to embed these converted coordinates within the formatting characters of a KML file.

Examining markup language

The Google Earth file format is called Keyhole Markup Language, or KML. Keyhole is the name of the company that originally developed the format (before being acquired by Google), and markup language refers to a variety of formats for marking up or tagging information in a text file. Other markup languages include XML, a general data format, SVG, a graphics format, and HTML, the language of web sites. KML itself is actually just a variety of XML.

These markup languages have in common a style of bracketing or annotating each data element with a pair of opening and closing **tags**. These tags, which are themselves enclosed within angle brackets, indicate the nature of the enclosed data:

VIEWING YOUR KML FILE To see your data plotted visually on a globe, you will need to install the free **Google Earth** application. If you don't have it already, you can get it from `http://earth.google.com`. Once you have the program running, you can open KML files and the points will appear on your map.

```
<element_name>Element Content</element_name>
```

For example, you might create a file for your samples that records species name, depth, and location in an XML format as follows:

```
<sample>
 <species>Marrus claudanielis</species>
 <depth>-518</depth>
 <location>-70.238, 39.940</location>
</sample>
```

In this example, a sample tag delimits the data for a sample. The sample data include species, depth and location records, with similar corresponding tags. Notice that each opening tag is matched by a closing tag, and that the closing tag is the same as the opening tag except that its contents start with a slash (/). You can think of the opening and closing tags as a pair of parentheses or brackets. If you open one, you need to close it later; also, nested tags need to be closed in the reverse order they were opened in (just as [{}] makes sense, but [{]} doesn't.)

You are now going to generate some XML output from the table of values that you have been working with. You might be surprised how easy it will be to modify your existing converter program so that it outputs KML instead of a table of plain text values.

INTERNATIONAL CHARACTERS Not all words can be represented by the letters **A-Z** and ASCII standard symbols (see Chapter 1 and Appendix 6). If your program needs to access and display accènted chåracters and special s¥mbols (¢ ° £), you will need to add another special header line immediately below the shebang line, to tell it to expect Unicode (in this case the common variant UTF-8):

```
#!/usr/bin/env python
# coding: utf-8
```

In Python versions less than 3.0, strings that you want to be interpreted as Unicode should be preceded by a lowercase **u**, as in **u"Göteborg"**, in a manner similar to designating raw strings by preceding them with a lowercase **r**.

Preserving information during conversion

When you write your output file, not all of the fields from the original file will be required for the data to be read and displayed by Google Earth. However, it is a good idea to put all the data that is in the input file into the output file, if possible. If you put just what you think you need in the output file, chances are that at a future date you will want to use that file in a new way that you did not originally intend, which may require more data fields than you had decided to save. (On the other hand, sometimes it is not possible to preserve all the information from the input file, because of formatting restrictions on the output file; and at other times, the whole point of the file converter may be to skim off a small fraction of relevant data from a large and unwieldy input file.) Since KML files don't have fields corresponding to all the columns in your input file, you will work around this by putting each original line into the general description field of the KML file.

Converting to KML format

KML file format

There are three main sections in a KML file (and in XML files in general). These are a short header region, the repeated data entries, and a short footer region.

Header The header for a KML file can be just three lines:

```
<?xml version="1.0" encoding="UTF-8"?>
<kml xmlns="http://earth.google.com/kml/2.2">
<Document>
```

The first line indicates how the contents of the file are formatted, indicating that it is an XML file. This line stands on its own, and is a special type of tag that does not need to be closed later. The second line is the opening tag for the kml portion of the code, and the third line is the opening tag for the Document content. These last two lines will have closing tags, </Document> and </kml>, in the footer at the very end of the document. All of the location data will be placed between the opening and closing Document tags.

Placemark The main section of the KML file will include the data points that are to be plotted on the map:

```
<Placemark>
   <name>Tiburon 596</name>
   <description>Tiburon 596   19-Jul-03   36 36.12 N   122 22.48 W   1190
           holotype</description>
```

```
<Point>
    <altitudeMode>absolute</altitudeMode>
    <coordinates>-122.374667, 36.602000, -1190</coordinates>
</Point>
</Placemark>
```

Each multi-line Placemark corresponds to one line from the original input text file. Between each open-close pair of `<Placemark>` tags are nested tags for the name, coordinates, and description. The new version of the program will create a Placemark—consisting of the opening and closing `<Placemark>` tags and everything between them—each time through your loop.

Footer The last section of the file is just two tags which close the `<kml>` and `<Document>` tags opened in the header:

```
</Document>
</kml>
```

Generating the KML text

You will now modify your `latlon` script to generate the header, Placemark, and footer strings. The modified file is saved in the examples folder as `latlon_5.py`. The header and footer will be generated and written to the file outside of the loop, since there is only one copy of each of these in the file. Within the loop, you will write the code to generate and write each Placemark string to the file.

Constructing strings with triple quotes rather than single quotes allows text to span multiple lines. The resulting string will include the line endings, so this is different than spanning multiple lines by escaping out line endings with \. Triple quotes are useful for defining blocks of text like the header and Placemarks.

The combined statement to generate the `PlaceMarkString` is:

```
PlaceMarkString = '''
<Placemark>
    <name>Marrus - %s</name>
    <description>%s</description>
    <Point>
        <altitudeMode>absolute</altitudeMode>
        <coordinates>%f, %f, -%s</coordinates>
    </Point>
</Placemark>''' % (Dive, Line, LonDegrees, LatDegrees, Depth)
```

The triple-quoted string that contains all the tags is combined with the string-formatting operator (%) to insert all five values into the Placemark at once. The insertion of the values in parentheses involves two strings (%s), two floating point numbers (%f), and a final string (%s). Because **Google Earth** will want a negative elevation for the depth, you must manually insert a dash before the %s that corresponds to the depth value. To use this entire string, you can either say `print PlaceMarkString` or `output.write(PlaceMarkString)`, depending on whether you are writing to the screen or to a file.

The entire program `latlon_5.py` is presented here, and is also available in the examples folder:

```python
#!/usr/bin/env python
import re # Load regular expression module

# Read in each line of the example file, split it into
# separate components, and write output to a kml file
# that can be read by Google Earth

# Functions must be defined before they are used
def decimalat(DegString):
    # This function requires that the re module is loaded
    # Take a string in the format "34 56.78 N"
    # and return decimal degrees
    SearchStr='(\d+) ([\d\.]+) (\w)'
    Result = re.search(SearchStr, DegString)

    # Get the (captured) character groups from the search
    Degrees = float(Result.group(1))
    Minutes = float(Result.group(2))
    Compass = Result.group(3).upper() # make sure it is capital too

    # Calculate the decimal degrees
    DecimalDegree = Degrees + Minutes/60

    if Compass == 'S' or Compass == 'W':
        DecimalDegree = -DecimalDegree
    return DecimalDegree
# End of the function decimalat() definition

# Set the input file name
InFileName = 'Marrus_claudanielis.txt'

# Derive the output file name from the input file name
OutFileName = InFileName + ".kml"

# Give the option to write to a file or just print to screen
WriteOutFile = True
```

```python
# Open the input file
InFile = open(InFileName, 'r')

# Open the header to the output file. Do this outside the loop
Headstring='''<?xml version=\"1.0\" encoding=\"UTF-8\"?>
<kml xmlns=\"http://earth.google.com/kml/2.2\">
<Document>'''

if WriteOutFile:
    OutFile = open(OutFileName, 'w')      # Open the output file
    OutFile.write(Headstring)
else:
    print Headstring

# Initialize the counter used to keep track of line numbers
LineNumber = 0

# Loop over each line in the file
for Line in InFile:
    # Check the line number, process if you are past
    # the first line (number == 0)
    if LineNumber > 0:
        # Remove the line ending characters
        # print line  # uncomment for debugging
        Line=Line.strip('\n')

        # Split the line into ElementList, using tab as a delimiter
        ElementList = Line.split('\t')

        # Returns a list in this format:
# ['Tiburon 596', '19-Jul-03', '36 36.12 N',
# '122 22.48 W', '1190', 'holo']
        # print "ElementList:", ElementList  # uncomment for debugging

        Dive = ElementList[0]    # the whole string
        Date = ElementList[1]
        Depth= ElementList[4]    # A string, not a number
        Comment=ElementList[5]

        LatDegrees = decimalat(ElementList[2])   # using our special function
        LonDegrees = decimalat(ElementList[3])
# Indentation for triple-quoted strings does not have to
# follow normal python rules, although the variable name
# itself has to appear on the proper line
        PlacemarkString = '''
<Placemark>
    <name>Marrus - %s</name>
```

```
    <description>%s</description>
    <Point>
        <altitudeMode>absolute</altitudeMode>
        <coordinates>%f, %f, -%s</coordinates>
    </Point>
</Placemark>''' % (Dive, Line, LonDegrees, LatDegrees, Depth)

        # Write the PlacemarkString to the output file
        # This is indented to be within the for loop
        if WriteOutFile:
            OutFile.write(PlacemarkString)
        else:
            print PlacemarkString

    LineNumber += 1 # This is outside the if, but inside the for loop

# Close the files
InFile.close()
if WriteOutFile:
    print "Saved",LineNumber,"records from",InFileName, "as", \
        OutFileName    # Shown on the screen, not in the file
    # After all the records have been printed,
    # write the closing tags for the kml file
    OutFile.write('\n</Document>\n</kml>\n')
    OutFile.close()
else:
    print '\n</Document>\n</kml>\n'
```

Run the script, and open the resulting KML file in Google Earth.

This script will be another important tool in your toolkit. You should be able to modify this template to read nearly any column-organized data file and output it in another format. When doing this, it is usually a good idea to open the template file, do a Save As... with the name of your new project, and *then* begin your editing. That way you can make changes without worrying about breaking your working script.

SUMMARY

You have learned how to:

- Approach file parsing
- Read files with open(FileName,'r') and a for loop
- Remove trailing characters with .strip('\n')
- Parse strings into lists with .split()
- Access list elements with [ ]

- Write to files with `open(FileName, 'w')` and `OutFile.write()`
- Continue Python lines with `\`
- Use regular expression searches with `re.search()` and `re.sub()`
- Access regular expression search results with `.group()`
- Create custom functions with `def` *function_name*`:`
- Generate XML and KML files
- Use triple-quoted strings (`"""` `"""`) to span multiple lines

Moving forward

- Find other types of input files containing latitude/longitude data and modify your program to support them. Make sure you understand how to allow for different formats of degrees and minutes (the `re.search()` function) and different numbers of header lines, to skip lines with comments marked by #, and to account for the possibility of footer lines at the bottom, all via the `if` statement within the `for` loop.

- Look at the HTML source of a web page presenting a table of values, and determine how you would replicate this by printing header, content, and footer sections.

MERGING FILES

Many instruments and archives store data into a series of files. Conversely, when it comes time to analyze data, you will often want to combine data from several files into a single file. In some cases this is as simple as joining the files end to end, but usually it requires combining the data in more complex ways. In this chapter, you will learn to read data from multiple files, combine the data in new ways, and then write the data to a file in a different format. You will also learn how to allow the user to specify filenames, folders, and other parameters when your program is run from the command line.

Reading from more than one file

In Chapter 5 you saw how to use the shell commands `cat` and `grep` to join several files into one long file. This is great for a multifile time series or other datasets where the values can be placed end to end with minimal modification. However, it is more difficult to add values side by side. For instance, you will often have several data files that have lines in X Y format where the X values of all the files are the same. Analysis may require rearranging and combining the data to form a single file with lines in an X Y Y Y format. For example, X could be a specimen number and the Y values could be different measurements for each specimen or X could indicate different measurements for a particular specimen and Y the value of those measurements in different experiments.

In the next example, you will see how to generate a list of all the files of a certain type within a directory, then extract values from those files and combine them into a single master file. Figure 11.1, later in this chapter, will illustrate this process visually. The example will also demonstrate several new Python capabilities. These include taking user input directly from the command line with `sys.argv`, opening several files in succession, building a list, and printing status messages using `sys.stderr.write()`. Otherwise the program is very similar to the file-reading program shown in Chapter 10.

In the terminal, change directories via `cd ~/pcfb/examples/spectra/` to the example folder we will be using. The directory contains a series of emission spectra (intensity at a certain wavelength) for various colors of light-emitting diodes:

```
host:spectra lucy$ ls
LEDBlue.txt     LEDGreen.txt     LEDRed.txt     LEDYellow.txt
host:spectra lucy$ less LEDBlue.txt
x          BlueLED
350.12    4
350.48    8
350.85    3
351.21    11
351.58    13
351.94    12
...
850.06    10
850.36    6
850.67    7
850.97    10
(END)
```

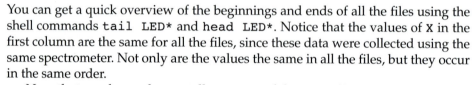

You can get a quick overview of the beginnings and ends of all the files using the shell commands `tail LED*` and `head LED*`. Notice that the values of `x` in the first column are the same for all the files, since these data were collected using the same spectrometer. Not only are the values the same in all the files, but they occur in the same order.

Now that you know the overall structure of the input files, you can start coding. Switch to a text editor, create a new text file, and save it as `filestoXYYY.py` in your `~/scripts` folder. Although it is not yet a script, you can make it executable with the usual command:

```
host:spectra lucy$ chmod u+x ~/scripts/filestoXYYY.py
```

Because this command specifies the path of the script file relative to your home directory, it will work without your needing to move out of the `examples/spectra` folder.

Getting user input with `sys.argv`

The first component of this program will be a mechanism for getting user input by passing arguments when the program is launched. You have already been using command-line arguments with other programs, such as `ls`. For example:

```
ls -l LED*
```

passes the arguments `-l` and `LED*` to the `ls` command.

In Python, command-line arguments are accessed through the `sys.argv` variable, which is provided via the `sys` module. To see how this works, enter these lines into your blank `filestoXYYY.py` script:

```python
#!/usr/bin/env python
import sys
for MyArg in sys.argv:
    print MyArg
```

Once you import the `sys` module in your script, you have access to the `sys.argv` variable. This variable is a Python list of arguments sent to a program when the user executes it. It provides a way to pass data from the command-line world to the Python world.

Save the file and run it in the terminal window:

```
host:spectra lucy$ filestoXYYY.py
/Users/lucy/scripts/filestoXYYY.py
```

Interesting! Even though you didn't pass any arguments to `filestoXYYY.py`, the `sys.argv` variable was not empty. This is because the zeroth element of the `sys.argv` list is the name of the script being run. This is important to remember when you use arguments in Python. In other words, don't assume that the first argument you have passed to the program on the command line is the first argument in the list. Try running `filestoXYYY.py` again, and add some arguments of your own:

```
host:spectra lucy$ filestoXYYY.py first second "third and fourth"
/Users/lucy/scripts/filestoXYYY.py
first
second
third and fourth
```

As you can see, each argument was separated by a space on the command line, and each was brought into the program as a string. If you want to include spaces as part of an argument, therefore, you need to either surround it with quotes, or escape each space with a backslash. It's also important to know that if you redirect the output of the program to a file using >, the command-line text from > onward is not included in the arguments.

So far this is pretty straightforward. Now try:

```
host:spectra lucy$ filestoXYYY.py LED*.txt
/Users/lucy/scripts/filestoXYYY.py
LEDBlue.txt
LEDGreen.txt
LEDRed.txt
LEDYellow.txt
```

Perhaps this was not the output you were expecting? (If you don't get similar output, make sure you are in your spectra folder.) In this case, the shell itself is expanding `LED*.txt` to generate a list of matching filenames. This expanded list of files is then sent to the Python script as individual arguments. The shell is acting as a middleman between the user and script. You can get this same list using the shell's echo command:

```
echo LED*.txt
```

Notice that the file list is returned in alphabetical order here. If you were to run this from a different directory and no files matched the string, then you would just get the string itself, with the asterisk, as a normal text argument.

Converting arguments to a file list

Here you will take advantage of the shell's ability to give us a file list, and you will operate on each of these files. First, add a bit of user interface robustness to your program by checking to make sure an argument has been given. If there is only one argument—the name of the program, which is always the first (that is, zeroth) element of the `sys.argv` list—then you will print out some text describing how the program should be used. Otherwise, use the argument list as a list of files and continue. Your program should look something like this:

```python
#!/usr/bin/env python

Usage = """
filestoXYYY.py - version 1.0
Example file for PCfB Chapter 11
Convert a series of X Y tab-delimited files
to X Y Y Y format and print them to the screen.
Usage:
   filestoXYYY.py *.txt > combinedfile.dat
"""

import sys

if len(sys.argv)<2:
    print Usage
else:
    FileList= sys.argv[1:]
    for InfileName in FileList:
        print InfileName
```

First, try running this program without any arguments. Next, run it with ~/scripts/*.py or *.txt as an argument, or even with a list of full filenames separated by spaces: LEDBlue.txt   LEDRed.txt. Remember that the additional information sent to the file begins at the second element of sys.argv. When coding, you can access the arguments from this second element to the end of the list by using the index number followed by a colon alone:

> ⚡ All arguments passed to your program by **sys.argv**, even numbers, arrive as strings. If the information entered by the user needs to be used as an integer or floating point value, it will have to be converted first using the function int() or float().

```
sys.argv[1:]
```

Providing feedback with sys.stderr.write()

This example program, like many others you will write, is designed to be used with the redirect operator > rather than by opening and writing to a destination file. The advantage of this approach is that it gives the user the ability to see the output, make sure it is what was expected, and then choose an appropriate filename in which to store the result. However, a problem with this approach is that any feedback to the user—including progress reports and other information that is useful when the program is running, but not useful when analyzing results—will also be captured in the file. This can complicate later analyses, since such information would likely need to be removed before further work with the file.[1]

Fortunately, there is a way to open up a separate pipeline of communication from the program. You have been using the system's standard input/output, known as stdio, each time you display something on the screen with a print statement. This output is what is captured during redirection with > or >>. You can also send output to the screen through a different pipe, used to report errors and warnings, known as stderr. This is writable as though you were writing to a file:

```
sys.stderr.write("Processing file %s\n" % (InfileName))
```

The sys.stderr.write() statement, like the file-related .write() statement, takes a string as a parameter and writes it as-is. If you want a line ending you have to add it yourself. You can build up a string from multiple pieces of data using the formatting operator % to convert and insert values, as shown here, or through string addition—that is, joining strings with +. Unlike the print statement, you can't mix integers and strings without converting the integers with the str() function or % operator.

---

[1] The two authors disagree as to whether this complication means you should provide more feedback or less when writing a program. One of us feels that this is an incentive to minimize feedback, while the other feels it is best to provide as much feedback as possible, using the sys.stderr.write() mechanism described in this section. We both agree that as a beginning programmer, you should select the approach which best fits your own personal preference.

Now substitute the `sys.stderr.write()` command in place of the `print` statement in your loop, where it currently prints `InfileName`. The output will still appear on the screen, but this way it will not be written to a file if you redirect the output with >. This is a good way for the program to provide feedback on its status, without causing you to go back and comment out these diagnostic statements when you are *really* ready to use it.

RETRIEVING THE CONTENTS OF A DIRECTORY
Although it is often desirable to ask the user which files they want to analyze, there will be times when you want to retrieve a directory list in your program without user input. Two convenient functions can do this. To get a full listing of a particular path, you can `import os` and use `os.listdir()`. This function takes a path in the format of `'/Users/lucy/Documents/'` as a parameter. If the directory does not exist, an error is returned. To do a wildcard search, you can `import glob`[a] and use `glob.glob()`[b] with a path in the format `'/Users/lucy/Documents/*.txt'` as the parameter. If the directory does not exist, an empty list is returned rather than an error.

---

[a] We are not making this up.
[b] Don't look at us.

Looping through the file list

Now you have a `for` loop that cycles through the filenames and prints each one. You can now insert code for file opening, reading, and output into this loop.

Now take stock of everything the program needs to do with each file (Figure 11.1). It should open the file and read in the lines, skipping the header line. If this is the first file, you will create a list variable containing the X and Y values together as a text string. For each subsequent file, you will extract the Y value from each line, then append it to the corresponding line of text from the first file, separated by a tab. The table is built column by column, value by value. Once this master data list has been constructed, you will print out the aggregated lines, one by one. Again, for this program to do something meaningful, each file must have data that correspond line by line.

The code to accomplish the first part of this task, building the master list, should be indented within the file-listing `for` loop you have created. (The beginning of

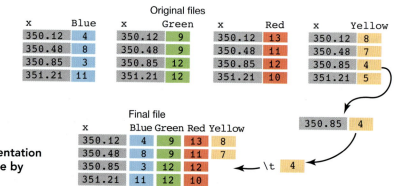

FIGURE 11.1 Graphic representation of the data reorganization done by `filestoXYYY.py`

the program is not repeated here, but the additional portion of your code should look something like this. It can also be seen in its entirety later in this chapter and as `filestoXYYY.py` in the example scripts folder.)

```
FileNum = 0
MasterList = []
    for InfileName in FileList: # statements done once per file
        Infile = open(InfileName, 'r')
        # the line number within each file, resets for each file
        LineNumber = 0
        RecordNum = 0  # the record number within the table

        for Line in Infile:
            if LineNumber > 0:  # skip first line
                Line=Line.strip('\n')
                if FileNum == 0:  # first file only, save both x & y
                    MasterList.append(Line)
                else:
                    ElementList = Line.split('\t')
                    MasterList[RecordNum] += ( '\t' + ElementList[1] )
                    RecordNum += 1
            LineNumber += 1
        Infile.close()
        FileNum += 1 # the last statement in the file loop
```

The important thing to understand here is what happens to `MasterList` while reading the first file versus what happens when reading subsequent files.

Before opening any files, `MasterList` is given the initial value of an empty list, as indicated by the brackets [ ]. This tells the program that the variable is going to be treated as a list of some sort. The first time through the file loop, each line that is read in is appended to the end of `MasterList`, creating a list of strings. After the first five data lines, the list would look like this:

```
['350.12\t4',
 '350.48\t8',
 '350.85\t3',   ← Think of it as two columns of tab-separated values
 '351.21\t11',
 '351.58\t13']
```

With later files, `FileNum` is greater than 0, so the code under the `else` statement is executed instead. First the line which has been read from the file is split at each tab, and put into the variable `ElementList`:

```
['350.12','9']
```

At this point, the X value for each row has already been stored in `MasterList`, so you only want to grab the second element, which has an index of 1. Add a tab before this item and add it onto the corresponding line of `MasterList`, which has already been defined the first time through the loop, to give:

```
'350.12\t4\t9'
```

To keep track of what record in `MasterList` corresponds to the line currently being parsed, use the `RecordNum` variable, which is reset for each file and increments each time through the loop. This is used as an index to indicate the appropriate record, based on the list that was built the first time through the loop.

Before beginning to read in values from the third file, `RecordNum` is reset to 0 so that it starts over at the first item in `MasterList`. Once the file loop is completed, the items in `MasterList` will look like this:

```
['350.12\t4\t9\t13\t8',
 '350.48\t8\t9\t11\t7',
 '350.85\t3\t12\t12\t4',
 '351.21\t11\t12\t10\t5',
 '351.58\t13\t7\t14\t8',
 ...
```

To see this (long) list, you could add a `print MasterList` statement at the end.

Printing the output and generating a header line

Once the files have been processed, generating output is simply a matter of looping through the items in `MasterList`:

```
for Item in MasterList:
    print Item
```

This code is indented to the same level as the file-reading `for` loop, so that it remains inside the first `else` statement—meaning that the user entered more than 0 arguments upon execution.

One bit of information lost during this process is which spectrum came from which file. An easy way to preserve this is to use the filenames themselves to label the top of the Y columns. This variable, which you'll call `Header`, should be pre-initialized with the name of value in the X column, in this case `'lambda'` (for the wavelength). Each time through the file-reading loop, you can add `\t` plus the filename (stored in the variable `InfileName`) to the `Header` string:

```
Header += '\t' + InfileName
```

Then, at the end of the program, just before you print the `MasterList`, you can print the `Header`. You could also consider stripping off any text after the base filename, for example "`.txt`", using the `re.sub()` function.

Avoiding hardcoded software

Scripts usually start out working well for a certain task and then get repurposed for other tasks. In our case, the script works fine if there is one header line in the files, but it will fail or generate contaminated data if there are additional lines. The number of header lines in this case is indicated only in one place in the program:

```
if LineNumber > 0:  # skip first line
```

If you wanted to reuse this program for a file that had 17 header lines (it happens) then you would have to open this program, figure out how it works again, and determine a number to put into this `if` statement ("Let's see, 18—no, 16!"). The situation is even worse when a value is used at several places in the program, meaning you need to look at every statement with a 0 throughout the program to determine if it is doing something based on the header lines.

For such reasons, it is often best to create an extra variable that is used anywhere there is a decision involving a value that may be modified later. In this case, you will define `LinesToSkip = 1` near the beginning of the script, and then change the `if` statement to read:

```
if LineNumber > (LinesToSkip - 1):
```

Now when you return to this script with data files of a different origin, you just have to change the variable definition at the beginning, and you will know that the rest of the occurrences of this variable should work without a problem.

Using the same type of construction, you could add a bit of code allowing the user to enter the number of lines to skip as the first argument, before the list of filenames:

```
filestoXYYY.py 17 *.txt
```

We won't add this option here, but take a moment to consider how you would implement this, including what other statements would need to be modified for your new program to work (not the least of which is the `Usage` string).

With a few modifications, the full code for reading files and converting data in X Y format to X Y Y Y is summarized here. It is also available as a program file in the examples folder, `filestoXYYY.py`. Use this program as a starting point when you need to read in data from multiple files, even if the exact transformation you need is not the same.

```python
#!/usr/bin/env python

Usage = """
filestoXYYY.py - version 1.0
Convert a series of X Y tab-delimited files
to X Y Y Y format and print them to the screen.
Usage:
  filestoXYYY.py *.txt > combinedfile.dat
"""
import sys

if len(sys.argv)<2:
    print Usage
else:
    FileList= sys.argv[1:]
    # for InfileName in FileList:
    #     print InfileName
    Header = 'lambda'
    LinesToSkip=1

    # change this for comma-delimited files
    Delimiter='\t'
    MasterList=[]
    FileNum=0
    for InfileName in FileList:
        # use the name of the file as the column Header
        Header += "\t" + InfileName
        Infile = open(InfileName, 'r')
        LineNumber = 0 # reset for each file
        RecordNum = 0

        for Line in Infile:
            # skip first Line and blanks
            if LineNumber > (LinesToSkip-1) and len(Line)>3:
                Line=Line.strip('\n')
                if FileNum==0:
                    MasterList.append(Line)
                else:
                  ElementList=Line.split(Delimiter)
                    if len(ElementList)>1:
                        MasterList[RecordNum] += "\t" + ElementList[1]
                        RecordNum+=1
                    else:
                        sys.stderr.write("Line %d not XY format in file %s\n" \
                        % (LineNumber,InfileName))
            LineNumber += 1 # the last statement in the Line/Infile loop

        FileNum += 1
        Infile.close() # the last statement in the file loop
```

```
# output the results
# these are indented one level to stay within the first "else:"
print Header
for Item in MasterList:
    print Item

sys.stderr.write("Converted %d file(s)\n" % FileNum)
```

If you try to apply this program to other data files, but it does not seem to read the lines properly, try modifying the statement for stripping end-of-line characters by changing it to the following, which will also strip off the alternative end-of-line character \r:

```
Line=Line.strip('\n').strip('\r')
```

Another solution is to open the file using the parameter 'rU' instead of 'r'.

```
Infile = open(InfileName, 'rU')
```

This will recognize either line-ending character (\n or \r) as it reads in the lines and will make your program compatible with Unix or Windows line endings.

Other applications of file reading

As discussed at the beginning of Chapter 10, data files are often presented with a single data element spanning several lines within the file. One such format is a molecular sequence format called FASTA. Files with this format put the sequence name of each record on a line beginning with >, with the subsequent line or lines containing the corresponding DNA or amino acid sequence:

```
>Avictoria
MSKGEELFTGVVPILVELDGDVNGQKFSVRGEGEGDATYGKLTLKFICT---TGKLPVP
WPTLVTTFSYGVQCFSRYPDHMKQHDFLKSA---M-PEGYVQERTIFY---------KD
>Pontella
MPDMKLECHISGTMNGEEFELIGSGDGNTDQGRMTNNMKSI---KGPLSFSPYLLSHIL
GYGYYHFATFPAGYE--NIYLHA---MKNGGYSNVRTERY---------EDGGIISITF
```

You have a couple of options for reading files of this type: you can read the names into one list and the sequences into another, or you can load the names as keys in a dictionary with the sequences as corresponding values. The dictionary approach makes it easy to access a particular sequence within your program, but because dictionaries cannot have repeated keys, it won't be able to handle situations where two sequences have identical names. A possible solution demonstrating both approaches is shown here and is available in the example scripts as seqread.py:

```python
#!/usr/bin/env python

Usage="""
seqread.py - version 1
Reads in a file in fasta format into a list
and a dictionary.

The resulting list is formatted
[['name1','sequence1sequence1sequence1'],
 ['name2','sequence2sequence2sequence2']]

Usage:
  seqread.py sequence.fta"""

import sys

# Expects a filename as the argument
if len(sys.argv) < 2:
    print Usage

else:
    InfileName= sys.argv[1]
    Infile = open(InfileName,'r')
    RecordNum = -1  # don't have the zeroth record yet

    # Set up a blank list and blank dictionary
    Sequences=[]
    SeqDict={}

    for Line in Infile:
        Line = Line.strip()
        if Line[0]=='>': # we have a new record name
            Name=Line[1:]  # chop off the > at the front

            # Make a two-item list with the name as the first element,
            # and an empty string as the second

            Sequences.append([Name,''])
            RecordNum += 1 # Now we have a record

            # Use the Name for the dictionary key
            SeqKey = Name
             # create a blank dictionary entry to append later
            SeqDict[SeqKey] = ''

        else:  # this means we are not on a line with a name
            if RecordNum > -1:  # are we past any header lines?
                # Add on to the end of the 2nd element of the list
                Sequences[RecordNum][1] += Line
                # Add to the dictionary value for the present key
                SeqDict[SeqKey] += Line
```

```
# when done with the loop, print the sequences:
# insert your processing and file output commands here
for Seq in Sequences:
    print Seq[0],":", Seq[1]

# could also print a list of all the names (=keys)
print SeqDict.keys()
```

If you turn back to look at Figure 10.1 you will see that this program fulfills the task shown in examples B and C of that figure: a few lines of input are used to create corresponding data series, either as parallel lists or as dictionary keys and values. Both methods are employed in this program, allowing you to choose whichever is more appropriate to your needs. In this case, we have made a two-dimensional list—basically two columns of data—where the first column is the names, and the second column is the sequences. To use the data, you would create a file-processing loop that cycled through each value in the list and output your desired subset of sequences or sequence-based calculations.

SUMMARY

You have learned how to:

- Gather user input using `sys.argv`

- Process several files in sequence using a `for` loop

- Use variables instead of fixed or hardcoded values in your programs

- Give warnings and feedback using `sys.stderr.write()`

Moving forward

- For some programs that take arguments, it is useful to have a default value which is used when no arguments are provided by the user. Add a `DefaultValue` variable to `filestoXYYY.py` along with the necessary logic to make it operate in this way.

- If you add a default value, as in the previous exercise, you lose the ability to print your `Usage` description automatically for naïve users. Think of ways to solve this. One way might be to check to see if the user has specifically asked for help; if so, you could see if `sys.argv[1]=='help'`. Another way might be check whether the arguments entered by the user seem to make sense, and to provide help if the check fails. For example, if the first argument should be a filename, you could check if such a file exists, using `os.path.isfile()` in the os module. If the argument should be a number, you could check it via the built-in string method `.isdigit()`.

- Add more error checking to `filestoXYYY.py`, to allow for blank lines and for lines which don't parse correctly with the `.split()` function. You can do this at the statement that checks the header lines, or at the point where the line is split into elements.

- Modify the program `filestoXYYY.py` to rearrange the columns of a file, by outputting the fields of `MasterList` in a different order.

## Chapter 12

MODULES AND LIBRARIES

This chapter further explores modules—pre-packaged Python resources that add functionality to the core language and simplify many programming tasks. Modules are reused from program to program, providing a framework for consolidating, recycling, and sharing programming code. We present a few more built-in modules (supplementing those you have already seen in previous chapters), and introduce some widely used third-party modules for data processing and analysis.

Modules are packages of computer code that can be imported into your programs as you need them. Modules contain a variety of data types and functions that extend the Python language to enable and simplify a wide range of tasks, from general-purpose mathematical calculations to drawing graphical objects on the screen. In the previous chapters you were briefly introduced to several built-in Python modules, including tools for regular expressions (the `re` module) and system operations (the `sys` and `os` modules).

In order to use a module, it must be both installed on your computer and imported into your program. There are many built-in modules that come with Python, including the modules you used in the previous chapters. These built-in modules are distributed as the Python standard library (see `http://docs.python.org/library/`), and all computers with Python will have them installed and ready to import. In addition to `re`, `sys`, and `os`, useful built-in modules include tools for `time` (for obtaining or converting dates and times) and `random` (for functions involving randomness).

In addition to the ubiquitous modules in the Python standard library, there are also many third-party modules. Anybody can write a third-party module, and this can then be provided to others to install and import into their own Python programs. These third-party modules typically provide specialized capabilities, including tools for graphics, scientific analyses, and the creation of user-interface elements. Even though they aren't part of the official Python standard library, a number of the more popular third-party modules are often installed with Python by default. You will need to check on a case-by-case basis to see if you already have a particular third-party module installed. You may get lucky and find that the module you want to use is on your computer and ready to go; if not, you will need to install it. Options for installing modules vary from module to module and computer to computer, and will be discussed later in this chapter.

Importing modules

Modules are actually collections of objects, including functions and custom variables. When you import a module, the code for these objects is read from the corresponding files and the objects become available for use within your program. It may seem like an undue burden to have to import a module when it is already installed on your computer, but there are good reasons for this added step. Specifying modules only as needed avoids the overhead that would be incurred if every Python program had to load every installed module. It also avoids cluttering your program with names of variables and functions you don't need. Assuming they have been installed on your computer, modules can be imported into your program in several ways. The simplest is to import the entire module with the `import` statement. This is how you imported modules in previous chapters. For example:

```
import re
import sys
```

When modules are imported in this way, all of the objects in the module are available for use. This manner of importing has another effect as well: it requires you to refer to the objects within the module using dot notation. You are already familiar with dot notation from Chapter 7, and from using it when calling object methods, such as `.count()` and `.strip()` for strings. In this case, because modules are objects that contain objects, the dot specifies which objects are being called within the module object. For instance, when you used the `search()` function from the regular expressions `re` module in Chapter 10, you called `re.search()` rather than just `search()`. The dot notation specified that you were calling the `search()` function of the `re` module, instead of some other function with the same name. Likewise, in later programs you have called the `stderr.write()` function from the `sys` module with `sys.stderr.write()`, and accessed the `argv` variable from the `sys` module with `sys.argv`.

It is also possible to import individual module objects without importing the module as a whole. This makes for more efficient programs if the modules are large. This alternative notation uses the `from` statement. For example:

```
from sys import argv
```

Note that there is still an `import` statement, but instead of importing an entire module, the statement imports only specified objects (`argv`) from within a specified module (`sys`). Because the object within the module is being imported directly, you don't name the module when using it. That is, no dot notation is required.

To see this in play, recall this simple program from the last chapter. It prints the arguments that are passed to it:

```
#!/usr/bin/env python
import sys
for MyArg in sys.argv:
    print MyArg
```

In this original formulation, the entire `sys` module is imported and the `argv` variable is then referred to with dot notation.

The following program has the exact same behavior as the original, but note the modifications to the `import` statement, and the omission of `sys.` before `argv`:

```
#!/usr/bin/env python
from sys import argv
for MyArg in argv:
    print MyArg
```

There is no `sys` object in this program, so a call to `sys.argv` would result in an error.

Finally, there is also a way to import all the objects from within a module without importing the module object itself. Simply use the `*` wildcard in conjunction with `from` instead of a particular object name. In this case, `*` acts like the `bash *` wild card rather than the regular expression quantifier:

```
#!/usr/bin/env python
from sys import *
for MyArg in argv:
    print MyArg
```

This is convenient if you would like access to all the contents of a module but don't want to use dot notation each time you use those objects. It can also be dangerous, however, as the names of some of the objects within the modules may be the same as the names of objects imported from other modules—or potentially, even some of the names you have created in your own code. Since you don't have the dot notation to wall off the module names from each other, different objects with the same name might end up overwriting each other. In practice this rarely happens—but when it does, it can lead to some very strange program behavior.

Which strategy you take to importing modules and objects from modules will depend upon the particular program and your personal preference, but you will routinely see both types of strategies in publicly available Python programs.

More built-in modules from the standard library

The Python standard library is large and diverse, and includes modules that will meet many of your primary needs. Drawing upon the Python standard library rather than upon third-party modules has the great advantage that you can be relatively certain these modules will be available on any system with Python installed.

The `urllib` module

The `urllib` module provides tools with similar functionality to the `curl` command you already used at the command line. It lets you download resources directly from the Internet for use within your Python program. For example:

```python
#!/usr/bin/env python
# coding: utf-8
import urllib
NewUrl = "http://practicalcomputing.org/aminoacid.html"
WebContent = urllib.urlopen(NewUrl)

for Line in WebContent:
    print Line.strip()
```

This module is useful for making "smart" `curl` commands to download, parse, extract, and save data from a Web page.

The `os` module

To perform shell operations from within a Python script, you can use the `os` (operating system) module (Table 12.1).

Especially useful is the ability to run non-Python command-line programs from within Python. This is achieved with the `os.popen()` command, with the `'r'` option indicating you want to read the output:

```python
DirOutput = os.popen('ls -F| grep \/','r').read()
```

Here, the shell `ls -F` command is sent through a `grep` command. The result is read, and the text is assigned to the variable `DirOutput`. Any shell program name can be substituted for `ls`, including pipes to other commands. This is a powerful way to build a data analysis tool that requires several programs, as described in Chapter 16. You will want to be careful with these commands, however, because

TABLE 12.1 Some useful functions of the os module, along with their shell equivalents

Module command	Operation	Shell equivalent
`os.chdir('/Users/lucy/pcfb')`	Change directory	`cd ~/pcfb`
`os.getcwd()`	Get current dir	`pwd`
`os.listdir('.')`	List directory	`ls`
`glob.glob("*.txt")`	Wildcard search for files (requires `import glob`)	`ls *.txt`
`os.path.isfile('data.txt')`	Does file exist?	
`os.path.exists('/Users/lucy')`	Does folder exist?	
`os.rename('test.txt','test2.txt')`	Rename file	`mv test.txt test2.txt`
`os.popen('pwd','r').read()`	Run a shell command and load the output into a variable	Any command; in this case pwd is shown

they can potentially wreak the same kinds of havoc incurred by unrestricted shell operations.

The math *module*

A number of basic mathematical functions are available upon importing Python's built-in math module. These tools and resources include `sin()`, `log()`, `sqrt()` and the constant `pi`:

```
>>> from math import *
>>> pi
3.1415926535897931
>>> log(8,2)
3.0
>>> sqrt(64)
8.0
```

Though the math module only has a limited number of functions, they do meet many common needs.

The random *module*

This isn't a section about a randomly selected module—it is about `random`, a module that can generate random numbers. We wrote the following script, using this module's `randint` function, to determine the order in which our names would be listed on the cover of this book.

```
#!/usr/bin/env python

import random

Casey = 418
Steve = 682
Num = -1
Possible = 1000
NumList = []

while (Num != Casey) and (Num != Steve):
    Num = random.randint(0,Possible)
    NumList.append(Num)

print NumList
print ['Casey','Steve'][Num == Steve] + ' wins!!'
```

Here, the `random.randint()` function returns an integer between 0 and the value of `Possible`. The `while` loop keeps going, generating random numbers and storing them in a list, until one of the random values matches a value defined at the start of the program.

Note that the last `print` statement is not cheating in favor of Steve. It uses the fact that the logical values `True` and `False` correspond to the numerical values `1` and `0`. This numerical interpretation is used in square brackets to index the list of names, returning `'Casey'`, the zeroth element, when the logical comparison is `False`, and `'Steve'` when it's `True`.

Other useful `random` functions are listed in Table 12.2. If you import the entire module, these are used in dot notation with the prefix `random`, as in our example script.

TABLE 12.2 Some useful functions of the `random` module

Function	Result
`randint(5,50)`	Return a random integer, in this case between 5 and 50, inclusive
`random.random()`	Return a random fraction between 0 and 1
`choice(['A',0,'B'])`	Return a randomly chosen item from the list passed as a parameter
`sample(MyList,10)`	Return a sample of 10 items from `MyList`, without replacement
`shuffle(MyList)`	Randomly rearrange the items in `MyList`, in place

Randomization analysis often requires "bootstrapping" a data set—that is, creating many data sets by resampling an original dataset with replacement. The `random` module offers sampling without replacement using `random.sample()`, but not sampling with replacement. To generate bootstrapped samples of an indicated size, you can use list comprehension (described in Chapter 9) along with the `random.choice` function:

```python
#!/usr/bin/env python
"""demo of bootstrapping via resample with replacement"""
import random

NumSamples = 100  # Number of resamplings to conduct
Bootstraps=[] # List to store the random lists

# for the demo, create a list of numbers 0 to 19
# normally these would be the original data
DataList = range(20)

# loop to perform repeated operations:
# Values of X and Y below are not used -- just counters
for X in range(NumSamples):
    # list comprehension in [] builds a list via sampling
    Resample = [random.choice(DataList) for Y in DataList]
    Bootstraps.append(Resample)

for Z in Bootstraps:
    print Z
```

Remember that list comprehension is like a one-line `for` loop. The program here uses it to repeat an operation once per value in the original data list. The actual value assigned to `Y` during the list comprehension doesn't have special meaning; it is equivalent to using `len(DataList)` to know how many times to choose a random sample from the original data set.

The `time` module

It is often necessary to keep track of time, convert times and dates from one format to another, and perform calculations with time. All of these needs can be addressed with the `time` module. A printable string version of the current date and time is produced by `time.asctime()`. (Remember ASCII refers to text characters.) You can also keep time in milliseconds using your computer's clock and the function `time.time()`, or get a list with the current time split into elements (year, month, day, etc.) with `time.localtime()`. This module also contains functions for converting between times and doing time math. As with all the modules, try the `help(time)` command or `dir(time)` for more options.

Third-party modules

The modules we have discussed up to now have all been part of the Python standard library. Third-party modules, on the other hand, aren't necessarily distributed with the core Python components. In many cases, you will have to download and install these modules on your computer before you can import them into a program. The specifics of the installation process can vary greatly. Some external modules consist of just a few text files containing Python code, while others include dozens of files and require additional libraries and programs be installed.

There are often multiple ways to install a particular module on a particular computer. These fall into two broad categories:

- First, you can download the module from its Web site. In addition to the code for the module itself, such downloads usually include detailed installation instructions (often in a file called README or INSTALL) and automated installer scripts. Even so, it can take some skill to get all the various components installed and operational, and to troubleshoot any problems that arise.

- Second, and usually easier, you can use a package manager to handle installation. Package managers are programs that provide semi-automated installation for a wide variety of software collected into ready-to-go bundles. There are several package managers for OS X; one of the most popular is MacPorts, which can be found at www.macports.org. Most versions of Linux come with built-in package managers, such as the graphical Synaptic Package Manager in Ubuntu and the apt-get command-line package manager developed for Debian. Package managers have the great advantage of also installing all the other programs or libraries a module requires. (You may hear these additional programs and libraries referred to as dependencies.) However, such automation can sometimes lead to unexpected consequences, such as installing an entirely new version of Python in addition to the one you already have. Be prepared for a package manager to take anywhere from a few seconds to several hours to install a module, depending on the module's size and dependencies.

This high degree of variation in installation approaches can get confusing, especially when there are additional choices to be made for installing a particular module. In general, there are a few steps to take when you would like to use a third-party module.

First, check to see if it is already installed. You may get lucky. As Python becomes more popular, more and more third-party modules are being included by default, just to make Python easier to use. And even if the module wasn't installed by default, it might have been installed alongside another piece of software that depends upon it.

Second, visit the module's Web site. In addition to downloads, there are often specific recommendations for each operating system. Some sites may even suggest that you use a package manager to facilitate the installation, despite the module being available as a direct download.

Third, investigate the package managers for your system to see if the module is available there. If installing the module directly from the Web site looks particularly difficult, or if you've encountered problems during installation that you can't resolve, a package manager may be the best option.

Software installation is discussed in greater detail in Chapter 21 and Appendix 1. Many of the issues explored there are relevant to installing modules, so flip ahead if you encounter problems with the third-party modules described here, or if you have any other installation-related problems. As discussed in Chapter 21, many installations require that the gcc compiler program be installed on your computer. This tool is not installed by default on OS X; rather, it is part of the optional Xcode developer tools package.[1] These tools are provided on the DVD for installing or upgrading OS X, which came with your computer. You will need these tools sooner or later, so it is a good idea to install them before you get in a pinch without them.

NumPy

To anyone familiar with MATLAB, Mathematica, or R, Python's [in]ability to handle numerical matrices with core objects and functions will seem rather limiting. For example, as you saw with list comprehension in Chapter 9, there is no direct way to access the second column of a two-dimensional list of lists. Fortunately, there are a variety of modules that add such functionality. The most widely used is the third-party numerical module, **NumPy**.[2]

NumPy is installed by default on OS X 10.6 (Snow Leopard), but if you have an earlier version of OS X you will need to install it yourself. General installers and documentation for NumPy can be found at numpy.scipy.org. Make sure you have gcc installed (test by typing which gcc at the command line, and then install the OS X Developer Tools if you don't), then download the Unix version (e.g., numpy-1.3.0.zip) from new.scipy.org/download.html. This archive will uncompress into a folder. Open a terminal window. Change into that directory with the cd command and then run the installer script:

```
host:~ lucy$ python setup.py build
```

---

[1] Package managers like MacPorts can also install the gcc compiler, as well as compilers for Fortran and other languages.

[2] Although there are many uses for Python in processing your data files, we nonetheless recommend that you learn to use R or MATLAB for more intensive numerical analyses. Because of these more suitable tools, we don't show you here how to carry out numerical analyses or generate graphics in Python.

This will print out several pages of feedback. Once it's done, install as follows:

```
host:~ lucy$ sudo python setup.py install
```

If these installation steps do not work, or if they did work and you just want to understand more about what was done, consult Chapter 21 for further information on installing software.

Once NumPy is installed on your computer, you can import it to your programs or at the interactive prompt with:

```
>>> from numpy import *
```

This style of importing the module will make the objects from the NumPy module available to you without prefixing them with `numpy`. Be aware of potential naming conflicts. For example, `e` (the mathematical constant) and `float` are both built-in NumPy objects. If you import the module components this way, you shouldn't use these names for anything else. If in doubt, use `dir()` to see all the names in use.

The first NumPy feature you will probably want to use is its support for creating and accessing array objects. The syntax for NumPy arrays is similar to that for regular Python lists:

```
>>> MyArray = array([[1,2,3],[4,5,6],[7,8,9]])
>>> MyArray
array([[1, 2, 3],
       [4, 5, 6],
       [7, 8, 9]])
```

The array object in the above example is a 2-D table of numbers. Note that the function `array()` and `array` objects are not part of the Python language itself, but were imported from `numpy`.

Now the fun begins. You can easily access rows in these 2-D structures, just as you can in Python lists. You can also access a column of numbers from this array, something not directly possible in Python without using loops or list comprehension. The syntax for extracting various subsets of arrays is very similar to that of MATLAB and R:

```
>>> MyRow = MyArray[1,:]        ← Second row, all columns
>>> MyRow
array([4, 5, 6])
>>> MyColumn = MyArray[:,1]     ← Second column, all rows
>>> MyColumn
array([2, 5, 8])
>>> MyArray[:2,:2]
array([[1, 2],
       [4, 5]])
```

As illustrated above, NumPy can extract square subsets of the arrays in addition to rows and columns.

NumPy has many other mathematical tools, and chances are that if you do any data analysis in Python, you will become a heavy user of this module. These additional tools include some pretty basic functions that aren't available in the core Python tools, including resources for statistics:

```
>>> V = [1,1,3,5,6]
>>> median(V)
3
>>> mean(V)
3.200000
>>> max(V)
6
>>> min(V)
1
```

Biopython

For simple molecular sequence processing, you can create your own scripts based on the `seqread.py` program presented earlier. For more complex tasks, though, the Biopython module contains a variety of tools for the analysis of molecular sequence data, including the ability to read and convert sequence files and to retrieve BLAST results. You can read about and download Biopython at `biopython.org`.[3] After uncompressing the source archive, it can be installed with a few commands:

```
host:~ lucy$ cd biopython-1.53
host:biopython-1.53 lucy$ python setup.py build
host:biopython-1.53 lucy$ sudo python setup.py install
```

The directory name in the first line here will depend upon the version available at the time of download. For more details on installing software, see Chapter 21.

Extensive tutorials for using Biopython are available at the Web site `tinyurl.com/pcfb-biopy` and elsewhere, so we won't go into detail here. Briefly, a special type of sequence variable is created when importing data files or when creating sequences from scratch with the `Seq()` function. These kinds of variables have many properties and built-in functions, which are revealed by using the `dir()` function:

---

[3] If you install Biopython using a package manager, it will attempt to install NumPy also, so if you would like both, try installing Biopython first.

```
>>> from Bio.Seq import Seq
>>> MySeq = Seq("ACGGCAACGTTTTGTTATGGAAACAGATGCTTT")
>>> dir(MySeq)
[... 'complement', 'count', 'data', 'endswith', 'find', 'lower',
 'lstrip', 'reverse_complement', 'rfind', 'rsplit', 'rstrip', 'split',
 'startswith', 'strip', 'tomutable', 'tostring', 'transcribe',
 'translate', 'ungap', 'upper']
>>> MySeq.translate()
 Seq('TATFCYGNRCF', ExtendedIUPACProtein())
>>> help(MySeq)
```

Standard applications of Biopython include loading sequences from FASTA files, searching them against sequence databases with BLAST, then parsing the BLAST results and organizing the sequence files according to their similarities to known sequences.

Other third-party modules

The third-party modules described here are only a small fraction of those that are available. You will encounter many others as you use and modify existing Python programs and write your own more specialized software.

There are many modules that add graphical capabilities to your programs. The matplotlib package, available at http://matplotlib.sourceforge.net/, makes it possible to generate data plots with MATLAB-like commands from within a Python script. This module requires many other software packages, and it is best installed with a package manager (e.g., MacPorts; see Chapter 21) that can automatically chase these down. An example script that uses matplotlib, called matplotCTD.py, is available in the scripts folder. This script was used to generate the plots in examples Figure 17.9C–F. If you have matplotlib installed, you can try running the script from within the examples/ctd folder using o_*.txt for the file list.

Modules also make it possible to interact with hardware in new ways. The pyserial module, for example, can read and write data from a serial port. You could use this to write custom control and logging software for a wide variety of instruments. The installation of pyserial is explained in Chapter 21 and more about designing and interfacing with hardware is discussed in Chapter 22.

If you are working with very large data sets, or data with complex relationships, you will probably want to look into using a relational database such as MySQL. The module MySQLdb enables direct connections to MySQL databases from within your Python programs. Chapter 15 is devoted to explaining databases and using Python in this context.

You will sometimes encounter .csv data files containing comma-separated values. These are not as simple to parse as tab-delimited files because the fields can contain commas within quotation marks, and a simple `.split()` function will not properly subdivide them. For this reason, we recommend using tabs as delimiters whenever possible. To help you parse comma-delimited files, however, Python has a built-in `csv` module, which can wrap around your file-opening statements to pull out the values as a list. This functionality is illustrated in the short snippet below:

```
#! /usr/bin/env python
import csv
AllRows = csv.reader(open('shaver_etal.csv','rU'))
# AllRows can be stepped through like a file object
# the Line variable will contain a list of parsed values
for Line in AllRows:
    print Line
```

Finally, there is a built-in module that is used to parse and generate XML files. The `xml` library contains a couple of alternatives for working with XML. The simpler one is probably the `xml.dom.minidom` module. The use of this module is more involved than we can cover, even in later chapters, but if you are interested in working more with Python and XML files, there are many tutorials online which walk through `minidom` usage.

Making your own modules

There is nothing magical about the libraries and modules that you have been importing in this last section. In essence, they just contain more Python code which you are adding to your script. You can easily make your own modules, containing functions and features which you have defined yourself.

For example, in Chapter 10, you created a `decimalat()` function by placing a block of code at the beginning of your program. If you want to use this code in other programs that you write, you can simply copy the `def:` block into a file, and save it in your `scripts` folder as `mymodules.py`. Be sure to include in this file whatever `import` commands are necessary for the module to run. (In this case, it would need the `re` module.)

Now, in another script, you can import everything from your `mymodules` file using:

```
from mymodules import *
```

Note that you don't need to include the `.py` file extension at the end of `mymodules`. The `decimalat()` function should now be available for use, just as if you had typed its function definition into the top of your current script.

Remember that functions shouldn't rely upon variables in the outside world—only the ones that you send in via the function parentheses—and the outside world can't use the function's internal variables either, except those sent back with the `return` command. This means that when you define a function, you must also define any variables you wish to be sent to it. You can even set default values if it makes sense to do so, for those cases where the function is called by the user without specifying variables. For example:

```
def bootstrap(DataList,Samples=10):
```

This creates a function that can be passed two parameters in a script, for example using the line `bootstrap(MyData,100)` to generate 100 bootstrap datasets. If called just as `bootstrap(MyData)`, it will default to generating 10 bootstraps, as indicated in the original definition.

> **A PRACTICAL NOTE** When studying examples of Python programs found on the Web, you may begin to notice some strange-looking `if __main__` lines near the beginnings of these programs. These lines are for more advanced applications of a program, where it has to know whether it is being run as a stand-alone program, or being imported as a module by a master program. In this book, we recommend a simpler approach, in which your program is always the master program or simple module—so for now, you can safely ignore such esoteric considerations. To use the example code that has such lines, you can either remove them and rearrange the program, or leave it in and work around them.

Going further with Python

In the last several chapters, we have only been able to skim the surface of the capabilities of Python, in hopes that you will become comfortable writing and adapting short programs. A quick reference of common Python commands is presented in Appendix 4, and debugging and program troubleshooting tips are discussed in the next chapter. Chapter 14 provides an overview to general categories of data analysis problems, and makes recommendations of which tools to use, drawing on the techniques we have discussed in Chapters 1–12.

In general, our goal is to empower you to branch out on your own. Think of the question you want to tackle ("How do I ...?") and you should be able to find the beginnings of a solution, whether in the previous example files or online. When faced with a challenge, don't hesitate to take advantage of any relevant example code. You can also consult with colleagues at `practicalcomputing.org`.

To reinforce and expand your skills, you will want to refer to any of the excellent Python resources online or in print. If you search the Web for "python tutorial," you'll retrieve something like 1.5 million hits. With that in mind, here are some of the best references, both in print and online:

- Lutz, Mark. *Learning Python*. Farnham: O'Reilly, 2007.
  This book provides a good walk-through of starting out in Python, and includes more advanced topics as well.

- Pilgrim, Mark. *Dive into Python*. Berkeley, CA: Apress, 2004.
  This is both a book and a Web site, `http://diveintopython.org`. It operates at a fairly high level, but has clear explanations of string manipulations, lists, and other essential Python components.

- Python Quick Reference, `http://rgruet.free.fr/PQR26/PQR2.6.html`.
  An excellent reference table for Python syntax, with a very clear presentation of what has changed between versions.

Python's next revision, Python 3, is available for installation but hasn't been widely adopted yet. You may begin to come across systems where it is in use. Most programs can be converted to Python 3 with a few modifications, or by using the 2to3 utility. One difference that you will notice immediately, though, is that the print statement has been changed to become a function, `print()`. In practical terms, this means it is used more like the `.write()` method you have used to write to files and to the `sys.stderr.write()` function. You can read about other differences at `docs.python.org/3.1/whatsnew/3.0.html`.

SUMMARY

In this chapter you have learned how to:

- Add specialized capabilities to Python programs using external modules, including `urllib`, `os`, `math`, `random`, `time`, `numpy`, Biopython, and `matplotlib`

- Create your own modules using function definitions

Moving forward

- Recreate the results of the `curl` example from the end of Chapter 5, but in Python with the `urllib` module.

- Modify `seqread.py` to print out sequence sizes, sorted from the smallest to the longest sequences.

- Modify some of the example programs so they use function definitions stored in a custom module.

Chapter **13**

DEBUGGING STRATEGIES

Writing a program is just one step in programming. Next comes getting your new program to work properly—in other words, to run without bugs. The most obvious bugs are errors which prevent the program from running at all. However, bugs also include problems which don't prevent a program from running, but lead to incorrect results. These sorts of bugs can be more serious because they are harder to identify in the first place. In this chapter, we describe some general strategies for avoiding, identifying, and fixing bugs. We also provide a table listing common Python error messages you may encounter, along with potential solutions.

Learning by debugging

Debugging is the process of locating and removing errors in your programs. Bugs come in two main types. The most obvious are those which cause your program to crash or not to run to completion. While frustrating, these errors can usually be located and fixed using the strategies we will outline shortly. The second type of bug is more insidious. These occur when your program runs to completion and appears to work, but is actually generating incorrect or incomplete results. This type of bug is difficult to detect and is more dangerous to your science than getting no result at all.

Addressing bugs that create incorrect output is a twofold process: first, designing your programs to avoid such bugs to begin with, and second, validating your results. The validation process can include hand-checking results for a subset of your data, comparing the results to those of another program, and building redundancies into your program to flag potential problems. You can also anticipate errors by writing your program to give you progress reports as it goes. For example, did the program process all the expected files? Did it analyze the expected number of data rows? You can even write a separate shell script to test your program under a variety of conditions and then check the results.[1]

Although debugging can be frustrating, it is also one of the best ways to improve your programming skills. This is especially true if you make notes on what you did to solve a problem—you will find yourself dealing with fewer and fewer errors as you learn to anticipate mistakes. In addition, chances are that most of the problems you will encounter have been found by others first, so searching for solutions online can lead you to new ways of approaching a particular programming challenge. This will expand your programming repertoire.

Debugging often forces you to dig a bit deeper to understand how tools you regularly use work under the hood. Stepping through the source code of modules or reading about esoteric orders of operation can be frustrating when your main goal is to fix a program so you can complete an analysis, but over time, debugging will build your depth of knowledge and make you a better programmer. It is very satisfying to write a large block of code and have it run perfectly the first time, and to use skills you learned when debugging one program to write another from scratch.

In this chapter we'll first discuss some general ways to identify and avoid bugs, and then present some specific error messages that you might encounter, along with probable causes and suggested solutions.

General strategies

Build upon working elements

When you begin to put together a program, begin by thinking through your general approach to the problem, then start building towards incremental successes. Get each element of your program working before moving too far ahead. If you are planning to read in a series of files, for example, you could get the program reading from a single file first, and then tuck that into a loop that does the same operation for several files. Or you could start by gathering and printing out the list of filenames you hope to read from. This strategy will let you identify errors before they are buried deep within many lines of code. Since you can't check the final result until all elements of the program have been written, this will often require writing test code that prints out the program status at various points. Once you are satisfied by the output of this test code, you can comment it out—functionally deleting it from your program, but leaving it available for future debugging.

 If you are writing to a file, always set up a sandbox folder—that is, a place where you and your program can play safely—and work on a copy of your data rather than on the original. This will avoid erasing or modifying critical data while you develop, test, and debug. Remember, even an act so simple as opening the wrong file for writing can erase the data that file contains. Another safety measure

---

[1] Some professional programming tool kits actually come with an "infinite monkeys" feature, which randomly pushes buttons and enters text into your program to check whether some combination of user actions will make it crash.

is to print the output filename and file contents to the screen in preliminary tests, before writing the code that actually opens the output file and writes data to it. This will help head off problems before you clobber something by accident.

Think about your assumptions

Several types of frustrating errors are due to mistaken assumptions of the most basic sort. Here are some things to be cautious of:

1. **Make sure you are editing the version of the program that you are actually using.** If at any point you have copied the program file to a data directory, saved a different version, or made a copy to send to a colleague,[2] you may end up editing a different version of the program than the one that actually executes when you type its name at the command line. One of the most frustrating debugging experiences is to identify and correct a problem, yet continue to get the same error when you run the program—simply because the copy you edited is not the copy you're executing. You can use the `which` command to get the absolute path of the program being called from the command line (for example, `which myprogram.py`). Make sure the name reported is the same file you are editing. As we have explained elsewhere, it is best to not have more than one copy of the same program on your computer. You can use a program from anywhere as long as the folder containing it is part of your PATH.

2. **Make sure you save your changes to the program file before re-executing.** The shell will execute the version of the program saved on your disk, not what you happen to have in your text editor. If you fix your program but don't save the changes, it will still fail.

3. **Check line endings and compensate for them if needed.** If your program reads text files, make sure it is really reading one line at a time. If you have incorrect end-of-line characters it may assume that all the contents of the file are in a single line. One way around this problem is to open text files with universal newline support by adding a U flag to the open command:

```
InFile = open(InFileName, 'rU')
```

When files are opened this way, all line endings are converted to newline (\n) characters.

---

[2] Incidentally, sending programs by email is more difficult than it sounds. Many virus checkers will strip off attachments that contain a shebang line, and some will even remove script text pasted into the body of a message. If you try to compress the program into a ZIP file, your attachment will probably still not go through, due to virus scanning activity. You might need to copy the script without the shebang line, or else post it on an FTP site or in a shared network folder on your computer.

4. **Check the contents of your data file.** A bug-free program can crash or generate incorrect results if there is a problem with the input data files. These problems can come in many forms. Are there really only AGCTs in your sequence files, or might there be other symbols, like –, ?, or an asterisk at the end? Are the numbers that you read in always positive, or are there negative symbols in front of some? In many cases you will find yourself working with files whose formats have not been explicitly defined; in such cases, the input file may be correct in the eyes of another program, but not what you anticipated when you wrote your own program. For instance, you may be expecting to see –1 for a missing value, but instead discover data that use NaN for this purpose instead. This can cause an error if you try to convert all values to integers.

You should anticipate formatting issues by checking that the data are as you expect before you parse or use them. If you get errors that you suspect are due to problems with the input file, add code to print the line number of the data files as the program processes the input. Then take a look at the data lines in the neighborhood of the last line that was processed before the crash. This is another example of the value of writing test code that provides feedback as the program runs.

Specific debugging techniques

Isolate the problem

When your program fails and prints out an error message, the solution can be obvious or it can be obscure. Often the problem is not on the line that is reported as generating the error. If you don't see a problem on the line that fails, start looking for errors in the preceding lines. If there is a missing parenthesis or unclosed quotation mark, for example, the program may continue beyond the line containing the error before encountering something that does not compute.

Commenting out sections of your program can help isolate the problem. Comments in effect turn these sections off, so that you can examine how the program runs without them. Once the problem has been found and fixed, it is a simple matter to remove the comment marks. You can turn a line of code into a comment in many languages just by adding a # to the beginning of the line. This works well for one or several lines of code, but gets cumbersome if you want to turn off a large block of the program; in the latter case, multiline comments are easier to use. In Python, these are demarcated by triple quotes: three quote marks in a row both before and after the code to be skipped.[3]

Comments can also be used to label the end of loops. With nested statements (several levels of indentation), it can be difficult to identify which if statement or for loop is designated by a particular indentation level. Putting a short comment,

---

[3] As you may recall from Chapter 8, triple quotes define a multiline string that includes line endings. Putting triple quotes around lines of code creates a string out of them. However, because this string isn't assigned to a variable, the end result is that the contents of the string are invisible to the rest of the program.

such as #end if negative or #end for each file, at the end of that indented block, using the proper indentation level, can help you avoid inadvertent
indentation (or outdentation) of commands into the wrong block.

Write verbose software

As you work through your program, give yourself plenty of diagnostic print
statements. Report the name of the file that has been opened, the value of the first
line or the number of lines processed, and a summary view of lists that you parse
or build from input files. Each time you insert such a statement, preface it with an
if Debug: statement. This allows you to define the variable Debug at the top of
your program as either True or False, so as to quickly turn on and off the status
reports during alternating bouts of testing and actual use. You can even make this
extra output an option that can be toggled on and off with an argument when the
program is run at the command line.

```
import sys
Debug = True

# (insert program statements here)

# wherever you want to give feedback, insert these lines
if Debug:
    print MyList
    # or you can use
    sys.stderr.write(MyList)
```

You don't have to create a fancy print statement to output your results during
debugging. Usually something simple and quick will suffice. Python will print the
entire contents of a list or variable if you just use the name, even if the variable is
not a string.

If part of a program is generating errors or you suspect that it is functioning
incorrectly, it can be very informative to print out the values of lists or dictionaries that are being used at that point in the code. For example, you might be trying
to use an index for an element that isn't there. (Remember that indexing starts at
0 for the first element, and that the last element has the index of the length of the
list minus 1). Or you might be trying to append to a list which hasn't been initial-
ized (MyList = []). Many of these errors will become obvious if you add a
print MyList statement before carrying out more operations.

You can also use the Python interpreter as a debugging device. Paste portions
of your program into the terminal, and then explore the values of variables by typing their names. This approach works best for testing parsing procedures, regular
expressions, and other string-based operations; it doesn't work as well for operations involving reading and writing files, but you can simulate the first line that
is read from a file by just defining that value manually. For example, instead of
for MyLine in MyFile: at the prompt, you can just say MyLine="" and then
paste the value of an example line between the quotes.

TABLE 13.1 Common errors running Python programs and some potential solutions

Example of reported error	Probable causes and solutions
`-bash: myscript.py: command not found`	If this error begins with `-bash` or with the name of your shell, then the program you are trying to run (`myscript.py` in this example) is either not in a folder listed in your `PATH`, or else its permissions are not set to executable. See Chapter 6 for setting the `PATH` variable, and try `chmod u+x myscript.py` at the command line to designate the file as executable.
`/Users/lucy/scripts/ myprogram.py: line 3: import: command not found`	If the `command not found` error is reported from within your Python program, rather than from `bash` (you can tell based on the text that precedes it), then there may be a problem with your shebang line. Specifically, it may not be sending the contents of your script to the Python program or other suitable interpreter. Double-check the `#!` line in the file, or copy one out of a working program. This type of error can also be generated if you misspell a built-in Python function within your program. Carefully check lines that call such functions for typos.
`bad interpreter not a directory`	The shebang line has a slash after `/usr/bin/env/`, as if `env` were a directory. Remove the slash so the interpreter understands that `env` is a program.
`usr/bin/env: bad interpreter: No such file or directory`	The part of the statement in your shebang line after `env` wasn't found. Copy the contents of the shebang line after the `#!` characters and paste into a terminal window to see if this launches Python.
`permission denied`	Permissions not executable. Fix with `chmod u+x myscript.py`
`name 'x' is not defined`	There are several possibilities here: • You misspelled a variable name in the program. • You are trying to modify a variable that was not originally defined. For example, you may be adding to a list or string that was not first defined as empty. Fix by initializing the variable, e.g., `MyList=[ ]` or `MyString=""`. • You are trying to use a function that needs to be imported from a module first. • You are using a function without the required module name in dot notation, for example `randint(5)` instead of `random.randint(5)`. Either add the module name to the function name, or else rephrase your import using `from` (as in `from random import *`) to access all the module's functions directly.

Error messages and their meanings

Common Python errors

When you run your Python script, in addition to the status messages that it prints out, you will also see error messages generated by the Python interpreter if your program crashes. In some cases, these will give a clear indication of what needs to be fixed, but in many situations they can be difficult to understand. Table 13.1 presents a list of the most common Python error messages, along with their possible causes, and suggestions on how to locate and fix the problem.

TABLE 13.1 (continued)

Example of reported error	Probable causes and solutions
`Indentation error:`	Take a guess! If the text has been pasted from a Web source, use Show Invisibles in your text editor to make sure the indentations are all just tabs or all just spaces. Also remember that invisible characters are sometimes introduced when copying and pasting.
`Attribute error`	Misspelling of a built-in function. For example, you would get this error if you have a string `MyString`, and use `MyString.lowercase()` instead of `MyString.lower()`. To fix, double-check the function or variable name that comes after the dot in your dot notation.
`type error 'xx' object is not callable`	Trying to retrieve values from a list using parentheses instead of square brackets, causing the list to be interpreted as a function name.
`traceback...zero division error`	Division by zero. This often happens when a function returns an unexpected zero, or when there is an unexpected zero in the input data. Check user and variable input to make sure that strings exist and are not blank. Test to make sure that a number is nonzero before using it as a denominator.
`Non-ASCII character '\xe2' in file`	One possibility is that the file contains one or more "curly quotes." Search for these and replace any found with "straight quotes." Another possibility is that some other symbol such as a bullet or degree sign has been used without specifying Unicode UTF-8. Add `# coding: utf-8` below the `#!` line.
`invalid syntax`	This could be many things, either in the line indicated or in preceding lines: • Missing colon after `if`, `else`, or `for` statement • Missing close parenthesis or close bracket • Using = instead of == in a logical test. • Mixing spaces and tabs when indenting

Shell errors

A common shell script error is `Illegal byte sequence`, which is sometimes reported by shell commands upon reading a file. This error originates from the presence of Unicode characters (atypical punctuation such as • ° ≠ or curly quotes) in a file being read. Some command-line programs, including `cut` and `tr`, can't process Unicode characters. If you get such an error, open the file in a text editor and remove the problematic characters using search and replace.

Other common errors stem from the use of symbols—for example, \ > * < ; / and also spaces—which have powerful meanings in the shell (and not just as emoticons). If used improperly, these symbols can result in the loss of files, so it is a good idea to test your shell commands and scripts thoroughly within a sandbox folder before using them on actual data. If function arguments include strings with punctuation, be sure there are quotes around the string and that symbols are escaped with \ where necessary. Otherwise, the shell will interpret the punctuation according to its own convention. For instance, `grep ">" test.fasta` finds all lines in the file `test.fasta` that contain >, while `grep > test.fasta` redirects the output of `grep` to `test.fasta`, overwriting all of that file's contents.

Making your program more efficient

Optimization

In some cases a program does what is expected, but is so slow or requires so much memory that it is essentially unusable. Given the speed of modern computers, the efficiency of a program may not be a significant issue for typical data processing tasks. However, as the size of your datasets grow and your analyses become more complex, improving efficiency may become essential—or at the least, a way to make your analysis experience more serene.

In most cases, there are several ways to write any given program; therefore, as you plan a project, it is good to think about which approach would be the most efficient. Most of a typical program's time will be spent doing the same operations over and over in a loop. Identify which loops will be reiterated the most, and focus on optimizing the code inside them before you move on to other, lower priorities. Make sure that variables aren't being created if they aren't being used. If a function is being called with the same arguments many times, see if you can call it once and then store the result in a variable that can be reused.

You can get a sense of the relative efficiency of different approaches by printing or saving the time when you start an operation (one function in the `time` module is `time.localtime()`) and then printing the time again, or else the elapsed time, as soon as the program completes. (If you are going to print elapsed time, you can use `time.time()`, which returns the number of seconds since a fixed date.) These tools make it possible to time operations precisely, and thus optimize speed even on a small dataset that takes no more than a few seconds to process:

```
import time
StartTime = time.time()
# perform your commands here
print "Elapsed: %.5f" % (time.time() - StartTime)
```

However, if you are trying to highly optimize a very small piece of code on a very small dataset (presumably as a test before moving to a larger dataset), the elapsed time may indeed be too short to measure precisely. To get around this, you can nest your program within a loop to repeat it a thousand times or so. This will reveal the effects of your optimizations more clearly.

Look for calculations that are done repeatedly inside a loop, even though the value does not change between loop iterations. Move these statements so they are assigned to a variable *before* entering the loop, and only recalculate when there is a chance the value will change. A similar rule applies to opening and closing files: you should not close and re-open a file each time you write a value to it. Open the file, then loop through to read or write, then close the file after exiting the loop.

try *and* except *to handle errors*

Several good tips for optimizing Python programs are discussed at `tinyurl.com/pcfb-speed`. Among them are some counterintuitive suggestions. For example, in some cases your program may be written to check if a value is present in a list or dictionary before accessing it. This takes time, regardless of whether the value is there. A quicker method is to try to access the value directly without checking, and if you get an error, assume that it wasn't there. Python has an interesting pair of commands, `try` and `except`, which can be used in combination for this strategy. A block of code indented under `try:` executes as usual up until an error occurs, in which case the program immediately exits the block and proceeds to the next unindented statement. Immediately after the `try` statement, a block of code indented under `except:` will be executed only if an error has been triggered by the preceding `try` statement. If there is no error, the `except` block is not executed, and in this situation, the program essentially "gets away with" not performing a time-consuming check. The `except` statement can further specify the type of error that is encountered. If you specify `except KeyError:` that block will be executed when trying to access a key that is not present in a dictionary and `except IOError:` is invoked when files cannot be opened because they cannot be found.

Figure 13.1 is an illustration of the same fragment of a sequence-reading program fashioned two ways: the first uses a traditional `if` statement to test if a key is in a dictionary, and the second uses `try` and `except`. In this example, the program loops through strings and tries to append them to the end of existing dictionary values. If no value is present, it defines a new key–value pair with the string in question. For a data file containing 77,000 sequences and 500,000 lines, the first

(A) Traditional

```
for Line in File:
  if Line[0]==">":
    Name=Line.strip()[1:]
  # lines with > are Names
  else:
    # check for a pre-existing key
    if Name in Dict.keys():
      Dict[Name] +=  Line.strip()
    # not a key so define
    else:
      Dict[Name] = Line.strip()
```

(B) Faster

```
for Line in File:
  if Line[0]==">":
    Name=Line.strip()[1:]
  else:
    try:
      # try to append with +=
      # assumes Name is a key
      Dict[Name] +=  Line.strip()
      # oops, not a key so define
    except KeyError:
      Dict[Name] = Line.strip()
```

FIGURE 13.1 The same fragment of a sequence-reading program fashioned two ways (A) Traditional `if` statement tests if a key is in a dictionary. (B) Attempts to append the dictionary without checking, using `try` and `except`. Version A takes 2000 times as long to complete.

approach finished in 33 minutes while the second approach took 1 second—2000 times faster!

When you're really stuck

Some errors may just prove too elusive to solve on your own in a reasonable amount of time. A fresh pair of eyes can help spot problems you have become blind to. It also helps when you have to explain to someone else how your program operates. As you step through the code, explaining it to this person line by line, you may suddenly realize the nature of the flaw. True, this other person may not have the same experience with your program that you do—but on the other hand, they may not be blinded by the assumptions you have been making, as explained in the first part of this chapter.

It is tempting to blame the system when things aren't working, and sometimes it really is "the computer's fault." Try your program on a different computer, or under a different operating system; this may reveal that a key library or piece of software is missing from your own computer.

Before you get too frustrated, reach out to the online communities of programmers, including the forums at `practicalcomputing.org`. Usually they will be eager to help. Be sure to provide all the context needed for understanding your problem, including data files, your system configuration, and the text of the programs involved. You need to give enough information that someone else could replicate the error on their own system, if they chose. By contrast, if you post only a few isolated lines with a question like "Why doesn't this work???" you are unlikely to get help.

SUMMARY

You have learned how to:

- Question your assumptions when tackling bugs

- Write a program in incremental steps, checking that each part works before proceeding to the next

- Give yourself feedback with `print` statements or `sys.stderr.write()` within your code

- Use `"""comments"""` to turn on and off portions of your program

- Interpret and solve Python error messages

- Learn from your mistakes

Moving forward

- As you encounter and troubleshoot errors, keep track of them in a file with a name like `pythonhints.txt` that you keep in your `~/scripts` directory. This is also a good place to store handy Python commands, code examples, or tricky syntax (for example, list comprehension) that you find useful.

- Try to recognize when an error is due just to syntax (e.g., an indentation error) and when it is a flaw in the design of the analysis (e.g., your script fails to process the last line in a file). Syntax errors will become less common as you write more scripts, but you always have to be vigilant against analysis errors.

- For more in-depth debugging of Python programs, you can explore some of the Python debuggers available, including iPython, IDLE, and the PyDev plug-in for Eclipse.

PART IV

COMBINING METHODS

Chapter 14

SELECTING AND
COMBINING TOOLS

You have gained experience with a range of tools in the preceding
chapters, but that is not all that is needed before you can apply them to
new problems. You also have to be able to pick the right tool for each job.
In some respects this is the more difficult skill to acquire. Here you will step
back and consider how to decide which tool to use for which problems. This
bird's-eye perspective provides an opportunity to review some of the skills
covered earlier before moving on to other specialized topics.

Your toolkit

In the preceding chapters, you have become familiar with a variety of flexible and
powerful tools for handling data. These tools fall into four broad categories:

- Regular expressions, to search and replace

- Shell commands, to interact with your computer at the command line

- Shell scripts, to combine and automate command-line operations

- Python programs, for more advanced processing

Given this range of options, there is almost always more than one possible approach to a given computing problem. Many tasks that could be addressed with
a shell script, for example, could also be accomplished with a Python program.
Sometimes a challenge is best addressed with a series of tools integrated into a
combined workflow, rather than by one tool exclusively. While it is relatively easy
to follow along with an example that walks you through how to solve a given
problem with a given tool, it can be confusing to decide which tools to use when
you approach a problem in the first place. Mapping out your analysis strategy and
making these first important decisions is the focus of the present chapter.

We will present this information along with three decision charts. The charts begin with a general problem or task and proceed to possible solutions. Find the chart that most closely applies to your task, follow the path that matches your requirements, and then consider the indicated approach. These brief charts list methods drawn from previous chapters, and will also direct you to portions of later chapters that might be useful.

Categories of data processing tasks

Getting digital data

Text files are the universal currency of data analysis, so many of your tasks will involve gathering your data into a simple text format. This might involve getting input from the user (usually yourself), exporting data from another program, or extracting data from an online source.

Data from user input   User input at the command line is a convenient way to accept input when the data are short and easily typed or pasted. The dnacalc.py program is an example where the user enters a value that is fed to the subsequent calculation. Other situations where you might consider user input are a script to reverse-complement a DNA sequence, convert among oxygen-saturation units, or change decimal latitude/longitude values to degrees, minutes, and seconds. User input can also tell the program what set of files you want it to act upon, as with the sys.argv[] variable used in your program filestoXYYY.py. In Chapter 16 you will see how to accomplish the same thing in the bash shell, using $1 to represent user input within your shell scripts.

Data from the Internet   How you interact with Internet resources depends on your needs—in particular, the number of files you anticipate processing and how often you will need to access them (Figure 14.1). If you anticipate needing to grab only a few files and process them into a usable format just once, then the simplest approach is to call up the Web page, view the source, and copy and paste the source into a new text document. From there you can use a series of regular expressions to reformat the data. The process of accessing Web sources is described in Chapter 9, and regular expressions are described in Chapters 2 and 3 and summarized in Appendix 2.

If you need to extract many files from the Web—whether a data series, some number of images, or a set of Web pages—it will probably be best to automatically download these files with curl (or alternatively with lynx, or in Linux, wget) in a shell window. This is much more convenient than clicking through the source code of a large number of pages in a web browser. The curl command is described at the end of Chapter 5. You can either use curl's ability to gather many files at once, or you can save many curl commands together in a shell script, as shown in Chapter 6.

If you are regularly retrieving data from the Web—for example, grabbing a daily record of temperatures—then it may be worth writing a Python script using

FIGURE 14.1 Decision chart for gathering data from the Internet
Gray boxes indicate the general problem. Blue boxes are progressively more specific descriptions of the task. Maroon boxes are appropriate tools. Green boxes are some specific example scripts that demonstrate the solution.

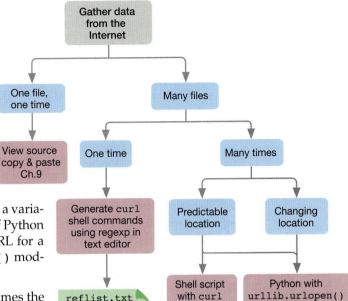

`urllib.urlopen()`. This approach can also accommodate more complex Web addresses that cannot be generated by `curl`. For instance, many URLs for time series data are built using a variation of the date, so you can write a bit of Python code that automatically generates a URL for a particular day. The `urllib.urlopen()` module is briefly illustrated in Chapter 12.

Data from other programs Sometimes the stream of data that you wish to capture comes as the output of another program. Working at the command line, you can either capture output to a file using the redirection operator (> or >>), or else send the output to another shell command using the pipe operator (|). For example:

```
history | grep "pcfb"
```

Shell operations like these can sift through a long stream of output for items of interest. For more examples of combining shell commands using pipes and functions, see Chapter 16 or consult Appendix 3.

Redirects and pipes don't work in every case to connect command-line programs, however. For example, if the first program you wish to use outputs data to the screen, but the second program requires the name of a data file as input, a pipe isn't suitable for combining them into a single command. The second program would interpret the output of the first program as a filename rather than as the data itself.

Data directly from hardware Many instruments and sensors interface with a computer through a serial port. The data stream from this equipment is often recorded with special software that is supplied by the equipment manufacturer. From there, it can usually be exported in plain text format. Often the software provided with such equipment is quite limited and can't be combined with other tools into automated workflows. When you want to use the instrument in a way that the supplied software won't allow, it may be possible to intercept the data stream directly from the serial port and then parse it any way you like. Chapter 22 gives some background on how to tap directly into these devices and interact with the physical world with custom-built electronics.

Data from spreadsheets Many people think of spreadsheet software as fundamental to managing datasets. In fact, spreadsheets are more suited to small one-off projects and simple record-keeping than they are to large analyses and complex datasets. They don't make a good central data repository because so few programs can open and manipulate the complex file formats they use. In addition they are inefficient and sometimes incapable of handling large datasets.

The first step to handling spreadsheet data in the context of a larger analysis is usually to resave the data in a basic text format. The brute force method for resaving the data is to open that file in its native program (e.g., Excel, OpenOffice, Numbers) and export the data as a text file delimited by either tabs or by commas (that is, a CSV file). While this works fine for one or two files, repeatedly clicking through the menus of a spreadsheet program can be quite tedious for even a small number of files. If you have MATLAB installed, you can write a file conversion script with MATLAB's `xlsread` command to convert Excel files to raw text. Another option is to install the `xlrd` Python package to access spreadsheet file contents. (General instructions for software installation are in Chapter 21, but briefly, you can download the source code from `http://pypi.python.org/pypi/xlrd`, uncompress the archive by double-clicking, `cd` into that directory, and type `sudo python setup.py install`.) Other spreadsheet conversion packages include the Perl-based `xls2csv` function and the PHP-based `phpexcelreader`. Alternatively, you could write a macro for OpenOffice or Excel to read in all your files and export them back out.[1]

Newer spreadsheet files with the `.xlsx` file extension are compressed archives containing XML files, which you encountered in Chapter 10. Buried among the documents in that archive is a text file containing your data. If the Python `xlrd` function does not work for your files, it might be possible to open these XML files in your text editor or with a Python script and extract the data of interest. From the command line, try uncompressing an `.xlsx` file and looking in the `xl/worksheet` directory:

```
host:~ lucy$ unzip SpreadsheetDataA.xlsx
Archive:   SpreadsheetDataA.xlsx
  inflating: [Content_Types].xml
  inflating: _rels/.rels
  inflating: xl/_rels/workbook.xml.rels
  inflating: xl/workbook.xml
...etc...
host:~ lucy$ cd xl/worksheets
host:worksheets lucy$ ls
sheet1.xml   sheet3.xml   sheet5.xml   sheet7.xml   sheet9.xml
sheet2.xml   sheet4.xml   sheet6.xml   sheet8.xml
```

---

[1]A macro is a script built within another program's own internal scripting system. OpenOffice, ImageJ, AppleScript, and Automator can create macros.

It will probably look like a mess, but `sheet1.xml` will contain the first sheet of data in your file. A shell script could fairly easily perform this unzip command on all your files, and copy the file `xl/worksheets/sheet1.xml` to `sheetA.xml`. There is a built-in Python module called `xml` which is not covered here but which gives some capability to work with XML files.

Of course, an easy way to avoid dealing with these conversions is to avoid spreadsheet data files throughout your data analysis.

Reformatting text files

The most common starting point for data processing is a text file in the "wrong" format—the data are there, but they aren't arranged in the right way for the intended use. You can use any number of general-purpose tools to reformat data in text files, and you don't need specialized software to open a text file and examine it. Whenever possible, you should save your data and request data in a plain text format instead of other specialized formats.

There are so many options for reformatting data files that it is impossible to capture all the possibilities in a single chart. We have summarized some of the primary approaches in Figures 14.2 and 14.3. The first thing to consider is whether you will just be doing a given reformatting job once on a few files, in which case regular expressions in a text editor will suffice, or whether you expect to perform the reformatting on many files or on many occasions, in which case a script is the way to go. Keep in mind that you often need to go back and reanalyze a data set—perhaps some values need to be calibrated, or you must add some important last-minute measurements that weren't initially available. Even if it will take you twice as long to create a script as to perform the same operation manually in a text editor, the script is often a worthwhile investment of time. In addition to enabling quick reanalyses, the script itself serves as documentation of how the data were modified, which can be critical for writing up your results.

Starting with regular expressions If you have only one file to process, you can use regular expressions (also known as regexp) in a text editor to rearrange and extract data elements, as explained in Chapters 2 and 3. When considering your search and replace operation, a single search expression is often not the easiest way to do a file conversion—so don't think that you have to solve it all in one brilliant regexp. You can use several expressions in a row to take care of different aspects of the reformatting. You should document the changes you make by keeping a record of the replacement patterns in your notebook and in a notebook file. For example, if you perform several replacements in succession, you can copy the search and replace terms into a separate document. This will be useful if you have to repeat the transformation or if you end up writing a Python script to recapitulate those replacements automatically. In TextWrangler you can also save searches for later, or pick from a list of recent search and replace combinations.

If you have already gone through the introduction to regular expressions in Chapters 2 and 3, then the reference table in Appendix 2 is a good way to remind yourself of the syntax and strategies.

FIGURE 14.2 Decision chart for parsing one set of text files

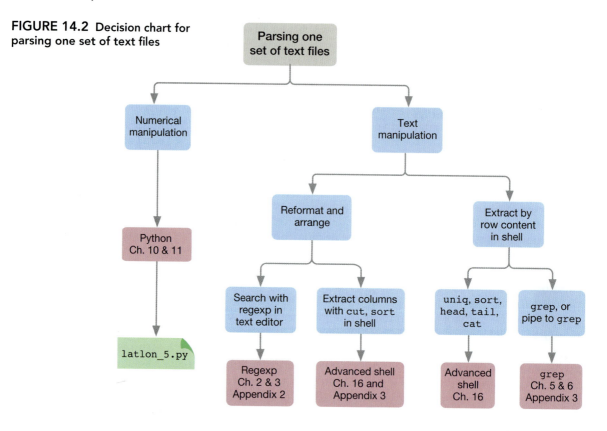

Command-line operations Another way to work interactively with your files is through the command line. This is useful when several files are involved—for example, when joining files with the `cat` command or extracting certain lines with `grep`, as described in Chapters 5 and 6. Working with text at the command line is also an important skill to have if you are manipulating files on a remote computer, as described in Chapter 20.

Some advanced shell functions can be surprisingly powerful for text manipulation, in particular those described in Chapter 16. These include the `cut` command to extract certain columns of data, the `sort` command, which lets you arrange lines either alphabetically or numerically, and the `uniq` command, which reduces a data stream to a list of non-repeated entries, and which can also count the number of entries in the process. These commands can be chained together using the pipe operator (`|`), introduced in Chapter 5 and elaborated upon in Chapter 16. Shell commands can be gathered together into a text file to serve as a script, as described in Chapter 6. Use the `history` command to review your recent shell operations so you can edit them together into a script file. Shell operations can also be combined into frequently used aliases (shortcut commands) or functions (multiline shortcuts, including the possibility for user input). These approaches

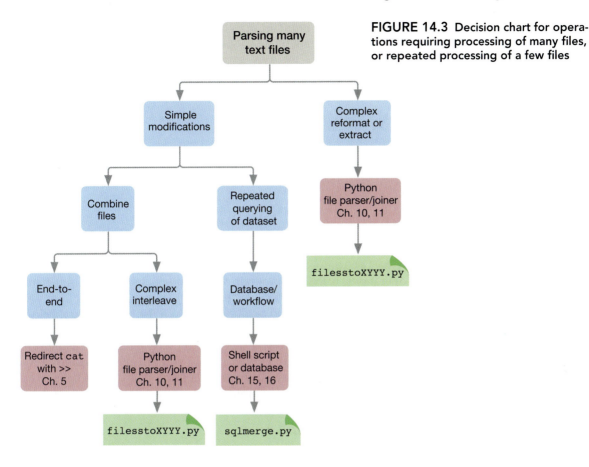

FIGURE 14.3 Decision chart for operations requiring processing of many files, or repeated processing of a few files

are most useful for relatively simple extractions on large numbers of similar files. Renaming or moving files is also an ideal role for a shell script or function.

You might think that graphics files need to be handled through a graphical user interface. However, shell scripts can readily operate on graphics files, using many of the command-line tools presented in Chapter 19. If your workflow involves images, be sure to investigate the capabilities of sips, ImageMagick (via convert and mogrify), and exiftool.

Python scripts

Python scripts are powerful tools for implementing and automating text manipulations like parsing and formatting, as well as for more sophisticated analyses. Many modules are available for specialized operations, and Python can even be used to control other programs. The example programs and scripts from this book, summarized in Table 14.1, are intended to serve as templates from which you can develop programs suited to your own research. They include three categories of interaction: user input, reading single files, and reading multiple files.

For interacting with user input in a Python script, either to create an advanced "calculator" or to operate on a set of user-specified files, you can use the `raw_input()` function or else the `sys.argv[]` variable described earlier. For more involved operations, you will typically be reading from and writing to data files using the `open` command:

```
Infile = open(InfileName, 'rU')
```

This will typically be followed by a `for` loop to cycle through the lines in the file:

```
for Line in Infile:
    # process each line here
```

Some examples of file reading and writing in Python are presented in `latlon_5.py`, `mylatlon_4.py`, and in `filestoXYYY.py`.

For many operations, a Python program with a single `for` loop is sufficient. However, for more in-depth calculations or for synthesizing data across files, you will probably need to use one loop to read the file into a list or dictionary, before generating output based on a composite of the values. A second `for` loop can cycle through to print the output lines or to save them to a file as explained in Chapter 10.

To combine shell commands with Python scripts, you can either include a program name in a `bash` script or call shell commands using Python's `os.popen()` functions, as in the `exifparse.py` example script mentioned in Chapter 16.

General considerations

One of the biggest pitfalls encountered as people take on more complicated analyses is to not keep track of what they have done. A scientific result is little more than a rumor if you can't tell someone how you got it, and you don't want to waste time second-guessing what you did if you need to do it again. As you work through analyses, keep a careful record of your scripts, searches, and discoveries. Annotate this information as much as possible. Keep these records centralized. The simplest way to do this is to keep a single plain text notebook file for analyses, just as you would keep a written lab or field notebook. Paste in your commands and results, and explain why you did things the way you did.

The best program is often the one you are most familiar with; by scavenging and repurposing from scripts you understand—including those presented in this book and other scripts you find online—you will build up a set of tools that should serve you well for a variety of tasks. A `HandyShellCommands.txt` file is a good place to jot down useful commands. You can quickly `grep` through this file to remind yourself of relevant commands for various operations. Without regular use, your shell proficiency and scripting skills can get rusty, so a combination of good note-taking and what amounts to regular exercise can keep you in good shape.

TABLE 14.1 Example programs and scripts discussed

Shell examples

`shellfunctions.sh`	A list of shell functions to use as starting points for adding tools to your own `.bash_profile`
`shellscripts.sh`	Examples of long shell commands discussed in the text, including use of the `curl` function and editing your `PATH`

Python examples

`asciihexbin.py`	Print a table showing ASCII or Unicode, hexadecimal, and binary values
`bootstrap.py`	Use the `random` module to create bootstrapped datasets (resampled from original data with replacement); could be converted to a module for use in other programs
`compositioncalc2.py`	Loop through the unique elements of a string (e.g., protein or DNA sequence) and calculate the percent composition of each element
`dnacalc2.py`	As above, but calculate additional properties of the sequence using different formulas depending on the sequence length
`exifparse.py`	Use Python to run shell commands, including the `exiftool` program; extract quantitative metadata from a series of images and print in tabular format
`filestoXYYY.py`	Take a series of files as input, extract columns of data, and join the columns together in one file
`latlon_5.py`	Convert a delimited file of locations to a Google Earth KML file
`matplotCTD.py`	Within Python, use the `matplotlib` module to plot data from a file
`mylatlon_4.py`	Read in a delimited file, convert format, and insert entries into a `mysql` database, using the `MySQLdb` module
`proteincalc.py`	For each character in a string, look up a corresponding numeric value from a dictionary and add that value to a sum
`seqread.py`	Load data from a file into a list and dictionary, using the first character of the line to determine which values become keys
`serialtest.py`	Demonstration of reading from a serial port using Python's `pyserial` module
`sqlmerge.py`	Extract values from one table in a `MySql` database, and use them to extract values from another table; merge the values into a delimited print-out

SUMMARY

You have:

- Reviewed how to gather data from users, Web sites, and other programs
- Learned some of the options for extracting data from a spreadsheet
- Reviewed the options for reformatting text files

Moving forward

- Create a text file containing some of your favorite commands and update it when you discover something new and useful.
- Make copies of the programs and scripts that are most relevant to your research, rename them, add heavy annotations, and adapt them to suit your purposes.

## Chapter 15

RELATIONAL DATABASES

There has been a strong emphasis throughout this book on storing data in easy-to-understand and portable plain text files. In general, plain text files are well-suited for many scientific needs. There are other options, though, that have many of the same advantages of text (open standards, readability, and wide support), but which add capabilities for extracting and synthesizing information from large or disparate data. This chapter illustrates these relational database management systems, with a focus on MySQL. Before introducing databases, we present some general considerations on deciding how to store data.

Spreadsheets and data organization

Developing an effective strategy for the storage and organization of data and data files is a critical part of nearly every scientific endeavor. Many data can be stored efficiently in two-dimensional grids, and we start here with some general advice on these 2-D data files. These grids usually have columns with different types of measurements, and rows with different samples or observations. Character-delimited text files represent two-dimensional data by placing each row of data on a line and separating data from different columns by a delimiter, usually either a tab or a comma (in the case of .csv or "comma-separated value" files). Although we have focused on text files for much of this book, we do realize that for many people, a spreadsheet is the primary way they enter, interact with, and analyze their results. A spreadsheet is just a graphical representation of this two-dimensional grid, with tools for editing and calculating values.

There is a tendency to organize character-delimited text files and spreadsheets the way you would arrange the table of a publication, with separate headers and sub-tables for different experiments (Figure 15.1A), or with alternating rows of treatment and control (Figure 15.1C). This patchwork approach to grids is rarely suitable for subsequent work like analyzing the data, importing them into other programs, or performing numerical or statistical analyses. Problems usually arise when information for a particular observation is spread across multiple rows. In contrast, the most general and flexible way to store data in tables is to make sure that each row has all the data needed to interpret that row.

In suboptimal approaches of the type shown in Figure 15.1A, the row shown in green contains descriptors of the data in rows that follow. In essence, the data are separated into multiple tables stacked on top of each other, and interpreting a row with data requires knowing something about the descriptor row somewhere above it. A preferable approach, shown in Figure 15.1B, would be to give the descriptors their own column, and repeat the descriptor within each row to which it applies. This way all the information about a row is found right within a row, and a program can read the data directly without parsing descriptors separately from other pertinent rows.

Another suboptimal approach, shown in Figure 15.1C, is to put the pairs of control and treatment values (or background and signal) on alternating lines, as represented by the blue and green boxes. If possible, these values should be put in the same row, as in Figure 15.1D. These modifications turn the file into one large grid with a single header.

The reasons for organizing character-delimited text files and spreadsheets as one large grid are numerous. Nearly all databases, statistics programs, and analysis programs such as MATLAB and R import and organize data in this format. Even in a spreadsheet, one formula added into a new column can draw information from other values located relative to it in the same row, instead of from miscellaneous places in the table. This allows you to use a single formula down the entire length of a column, making your analyses and graphing operations as efficient as possible. If you were to draw arrows to the cells that

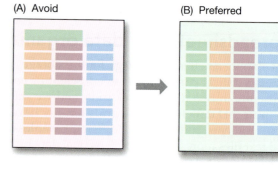

(A) Avoid   (B) Preferred

(C) Avoid   (D) Preferred

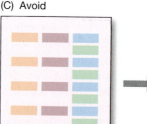

FIGURE 15.1 Approaches to organizing data in spreadsheets and character-delimited text files Each colored block represents a different type of data or a different comparable group of measurements such as date, temperature, category of treatment, or control and response variables. (A) and (C) show common but difficult-to-analyze approaches to organizing data. (B) and (D) are possible ways to reorganize the data so that each column contains a single type of information, and each row contains all the information relevant to a record.

are being used to calculate your results, would you get a simple network, or something that looks like a subway map? *In general, all data in a column should hold the same type of values, and each row should hold information that corresponds to a particular measurement.* The type of measurement, rather than being indicated by its position in the table, can be designated by a separate column devoted to that purpose.

When you are using a spreadsheet program, there are a few things you can do to improve clarity and facilitate the analysis of the data. Raw values should be entered into the cells; don't do any calculations or conversions before entering the data. Let the spreadsheet do the work of making conversions, and keep your data as unmodified as possible. This will save time and also provide a better record of the processing that the data have undergone during analysis. Likewise, don't type values directly into a formula; keep the values in their own cells and refer to those cell locations with your formulas. Again, this will make it simpler to update data and calculations, and easier to keep track of processing steps.

Some data cannot readily be coerced into a two-dimensional table with uniform columns, or at least not without duplicating excessive amounts of information within the file. Data with a nested structure, such as phylogenies, are not well suited to tabular organization. In other cases, it may be necessary to repeat most information from row to row, rather than just one or two columns. If you find yourself in such a situation, where a simple two-dimensional grid is insufficient or inefficient for your needs, you should think about moving beyond spreadsheets and character-delimited text files, and instead investigate keeping your data in a relational database, as we will discuss in the remainder of this chapter. (Of course, grids and databases are not the only way to store data in text files; many general file formats have been developed for complex datasets. The most widely used of these formats is XML, which was introduced briefly in Chapter 10 but isn't covered further in this book.)

Data management systems

The pervasiveness of spreadsheets can give the false expectation that all data are best stored in a single two-dimensional grid. This is not the case. For example, imagine that you want to track information about field sites and about multiple specimens observed at each field site. If there are many specimens per field site, shoehorning all the data into a single grid would require repeating the site information many times (once for each specimen found at a site). If you want to update the field site information, you will need to change it in many places. This approach has other problems besides redundancy. If you have complex data for several different but related elements of a study, these different data may not correspond in such a way that each record can be placed into the row of a single table. If you create individual files or grids within files to store different types of associated data (such as specimen data that include geographic coordinates for collection sites, molecular sequences for multiple genes, and one or more photographs), you cannot easily retrieve information from one of these tables based on data stored in

This chapter stands on its own, so if relational databases are not applicable to your analysis needs at this time, you can skip ahead without affecting your ability to use other sections of the book.

another table. Nonetheless this is often exactly what you would want to do.

This is where relational databases come in. A relational database management system, or RDBMS, is a server program that runs continuously in the background and manages one or more databases. These databases are collections of structured information. Although databases are stored as files, the user doesn't interact with the files directly—the management system acts as a middleman. It takes care of the creation, organization, and optimization of the files, as well as all direct interaction with them. It also listens for requests to add, edit, or look up data. These requests can come from other software on the same computer, or the computer can be configured to accept requests over network connections, so that the database and program using it don't even need to be in the same location. Commercial database management systems include FileMaker, Microsoft Office Access, Microsoft SQL Server, and the Oracle software suites. Open source options include MySQL (maintained by Oracle), PostgreSQL, and SQLite. These management systems are designed to work with a wide range of database sizes, from dozens of entries to billions. Complex tricks are used behind the scenes to process requests quickly and to optimize file organization for speed and memory efficiency. It is generally much faster to find a particular piece of information in a database than it would be to scan through a large text file. The files used to store the data are different from system to system, but because the user never directly interacts with these files, the differences don't usually matter.

Interactions with all modern database management systems are performed in a database language known as **Structured Query Language**, or SQL.[1] This language includes commands, functions and variables, and it follows a formal syntax. Since nearly all systems use a closely related variation of SQL, if you learn the basics of SQL once, you'll be well prepared to use most database software. Database management systems come with command-line and graphical interfaces for creating and interacting with databases by submitting SQL commands, but you can communicate with them in other ways as well. In addition to their direct interfaces, such systems also have back-door interfaces that allow for interaction with their databases from within other software packages. R, MATLAB, Python, web servers, and many other tools can be configured to interact directly with a database. This obviates the need to import and export data to and from files. In Python, for example, the ability to interact with a database is added with modules, one of which will be introduced later in this chapter. SQL queries can then be constructed as strings and sent to the database management system. Any data that are returned can be accessed from within Python.

In addition to accessing data, there are several important logistical advantages to using a database management system. Database files are centralized, so it is easy to back them up, and this avoids nightmare situations where there are redundant

---

[1] The proper pronunciation of SQL is open to debate, but here we pronounce each letter.

versions of data files and different ones are updated independently. The RDBMS takes care of a lot of the overhead of database management that you would otherwise need to build into your software. It also makes a good centralized warehouse. If you are working a project that involves several analysis programs, getting them to all talk with each other through multiple data file intermediates can be one of the biggest challenges. If they can all talk via the central database server, then it is much easier to pass data between them. More than one program can even access the database at the same time.

Relational databases don't only change the way that information is stored and retrieved on the computer, they also allow for important flexibility in the way that the data are organized and how you interact with them. The driving concept is that each piece of information is stored only once, and then linked through relations to other pieces of data rather than copied. This makes updates easier, reduces the chance of inconsistencies, and makes the database more efficient in term of both memory and computation. Most of these advantages become more apparent for larger and complex analysis projects. For smaller projects, database systems may be overkill.

Anatomy of a database

One RDBMS server can host any number of databases (Figure 15.2). Each database in turn contains two-dimensional tables that hold the actual data. Each column in a table has a different type of data, and each row contains a record. In this respect, database tables are recognizable to anyone who has used a spreadsheet—although as you will see later, databases provide much more powerful tools for interacting with the data and linking data across tables. There can be one table or hundreds of tables in a database, each containing a collection of records that pertain to a particular category of information. A database for a library might, for instance, have one table for books, one table for patrons, and another table describing which patrons have checked out which books.

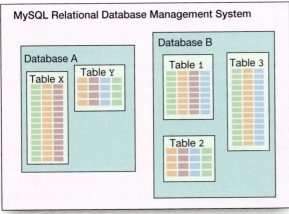
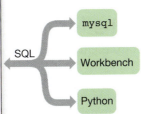

FIGURE 15.2 The anatomy of a database system Each server (pink box) can store databases, each with their own tables. Queries and commands come from the outside world by means of SQL commands, delivered by a variety of means.

TABLE 15.1 Common RDBMS data types	
Data type	Description
INTEGER	An integer ranging in value from -2147483648 to 2147483647; INT can be used as an abbreviation for INTEGER
FLOAT	A floating point number, including scientific notation: 3.14159 or 6.022e+23
DATE	A date in 'YYYY-MM-DD' format
DATETIME	A date and time in 'YYYY-MM-DD HH:MM:SS' format
TEXT	A string containing up to 65535 characters
TINYTEXT	A string containing up to 255 characters
BLOB	A piece of information encoded in binary, including images or other non-text data; there are four sizes of blob data types, with different storage capacities

The type of data in each column of a database table must be specified, and an error will result if you try to add data that don't conform to the specified type. Many of the data types used in databases are the same as those you have already encountered in Python (see Chapter 7), such as integers, floats, and strings. The naming of the types is a little bit different, and there are some types that are available in one context but not another. The most frequently used data types are listed in Table 15.1; additional types can be found at dev.mysql.com/doc/refman/5.1/en/data-types.html.

When you create a new table, one of the columns must be specified as the **primary key**, and each row of the table must have a unique primary key value which distinguishes that record or row. Using the primary key, you can unambiguously identify or extract any particular row of the table, in the same way that the key of a Python dictionary is uniquely associated with a particular value. The primary key is usually an integer value that is automatically computed by the database management system and stored when a new row is added.

Installing MySQL

Several excellent open-source relational database management systems are available, each optimized for different types of uses, but all perfectly adequate for many scientific tasks. Here we will provide specifics for getting started with MySQL, which is freely available and widely used in biology and science in general.

If you are using OS X or Windows, it is easiest to download and install MySQL directly from the project download page at www.mysql.com/downloads. There are different versions of the MySQL Server (the database management system itself) and a variety of ancillary files. To get started on your own computer, the two downloads you will need are MySQL Community Server, which is the actual RD-

BMS, and the MySQL Workbench, a graphical interface that facilities database maintenance and data visualization (Figure 15.3). If you have a computer running Ubuntu Linux, you can install the server through the Synaptic Package Manager, but you will still need to install the Workbench via the MySQL Web site.

There are a number of options available for the OS X download, including source and binary files built for different versions of OS X and for different computer hardware. The DMG archive is the simplest package format to install. Make sure that you get the correct file for your operating system version (e.g., OS X 10.6, 10.5, or 10.4; you can check the version of your operating system by clicking on About This Mac in the Apple menu). You have the option of choosing between 32-bit and 64-bit versions of the software. All Apple computers built since 2007 are 64 bit. If you have an older computer and aren't sure if it is 32 or 64 bit, the 32-bit version should work fine for most needs.

The OS X MySQL Community Server installation requires installing several different components, as described in the `ReadMe.txt` file provided with the installers. Run the two package installers; mysql–xxx.pkg installs the RDBMS, while MySQLStartupItem.pkg installs the launcher that starts the database software when the computer boots up. Drag the MySQL.prefPane to your /Library/PreferencePanes folder. Be sure to use the Library folder at the root level of your computer (this will require password authentication), not the folder of the same name in your home directory. This panel will allow you to start and stop MySQL within the System Preferences GUI.

There are a variety of command-line programs that come with MySQL. The installer will place these in `/usr/local/mysql/bin`, which is not in your default `PATH`. Edit your `.bash_profile` to add this directory to your path. (See Chapter 6 if you are rusty on how to do this. Briefly, you will add `:/usr/local/mysql/bin` to the end of the existing `export PATH` command, inside the quote mark.)

Restart your computer, open System Preferences, and click on the MySQL icon. In the MySQL preference pane it should say The MySQL Server Instance is running. If it is not running, click the Start MySQL Server button. If this results in an error window, consult the online MySQL documentation to troubleshoot the problem. If it wasn't initially running but works fine when you start it manually, make sure that you installed the MySQLStartupItem.pkg, that the Automatically Start MySQL Server on Startup box is checked, and then restart the computer again.

Consult the ReadMe file for your configuration information.

Once the MySQL server is running, install the MySQL Workbench. The installation is simpler than that for the server software: the program just needs to be dragged to the Applications folder. Launch the Workbench. The most important Workbench tools for getting started are the SQL Development tools on the left side of the window. It also has many advanced features that won't immediately be of use. Before you can interact with the database server you installed, you have to connect to it. Remember that database management systems don't automatically assume that a database lives on the same computer where it is being used, since it may also be accessed on a remote computer over the network. In fact, connect-

OTHER OPTIONS FOR SQL GUIs Another free GUI for interacting with SQL database systems is the Java-based, cross-platform **SQuirrel SQL Client**. You can download it for free from the links at `squirrelsql.org`. Be sure to install the Data Import and MySQL plug-ins as part of the installation process. If you choose this route, follow along with SQuirrel SQL wherever the **Workbench** is used. An inexpensive commercial option is **Navicat**, which also runs on nearly all platforms, but which has different program files for different databases.

ing to the local database is done as if the database server were on a different computer, but by using the special `localhost` address `127.0.0.1`, the connection is redirected back to the local computer. More detail on network connections and addresses, including `localhost`, is provided in Chapter 20.

Click on the New Connection icon in the SQL Development portion of the MySQL Workbench. All of the default values are for connecting to the local database. All you need to do is give the connection a name. Type `localhost` into the Connection Name box. Click Test Connection to make sure all the settings and installations are set up correctly. You will be asked for a password, which is blank by default. Click OK again. You should get a window that says Connection to MySQL at 127.0.0.1:3306 with user root, Connection parameters are correct. Click OK twice more, and you will get back to the main Workbench screen and will see the `localhost` connection you just created in the connections box.

Getting started with MySQL and SQL

There are many ways to connect to and interact with the MySQL server. In the previous section, you connected using the supplied MySQL Workbench graphical application. In this chapter we will use this graphical interface mostly to observe the changes to the database that are made through other types of connections. In many respects, the most convenient interface to the database is—yes, yet again—the command line. Later in the chapter you will also learn to connect to the database from within a Python program. The commands discussed in this chapter are summarized in Appendix 7 for quick reference.

Connecting to the MySQL server at the command line

Open a Terminal window, and enter the following commands (shown in bold):

```
lucy$ mysql -u root
Welcome to the MySQL monitor. Commands end with ; or \g.
Your MySQL connection id is 1426
Server version: 5.1.48 MySQL Community Server (GPL)

Copyright (c) 2000, 2010, Oracle and/or its affiliates. All rights reserved.
This software comes with ABSOLUTELY NO WARRANTY. This is free software,
and you are welcome to modify and redistribute it under the GPL v2 license

Type 'help;' or '\h' for help. Type '\c' to clear the current input.
mysql> SHOW DATABASES;
```

```
+--------------------+
| Database           |
+--------------------+
| information_schema |
| mysql              |
| test               |
+--------------------+
3 rows in set (0.00 sec)
mysql> EXIT;
Bye
lucy$
```

This launches the command-line program mysql, which provides an interface for interacting with the database server. It is not the server itself, as the name might imply, but, like the MySQL Workbench, a stand-alone program that can issue commands to the server and present the results. The -u root argument specifies that you want to connect to the database as the MySQL user root.[2] No password is needed since one hasn't been configured yet. MySQL users are different than the system users who have accounts on your operating systems, and they are configured completely separately. No network address is specified, so mysql assumes that you want to connect to the local MySQL server. Like the Workbench, mysql can connect to servers on other computers, but we won't get into the specifics of how to do that here.

Linux users should add –p to the command if they added a password as recommended by the installer:
mysql –u root –p

If the connection is successful, you will get an introduction that describes a bit about the server, your connection to the server, and some other status information. If the connection isn't successful or the program won't run, make sure that the MySQL server is running and that mysql is in your PATH. You will then get a mysql> prompt. Like the shell prompt that takes bash commands or the interactive python prompt that takes Python commands, the mysql> prompt is waiting for database commands in SQL. As was the case with the other languages we have covered, SQL has far more features than we'll be able to address here. The goal is simply to provide you with enough information to get started, implement some simple projects, and help you determine whether a relational database is something that would be of use for your work.

The first thing to notice about SQL is that commands end with a semicolon. By pressing ⌶return⌶ without a semicolon, you can split a long command across multiple lines, since the command won't be executed until it is ended with a ; followed by ⌶return⌶. This multi-line format is often used for displaying and entering SQL commands, making them more readable than if they were all on a single long line. Another convention is that special words like functions and built-in variables are usually presented in ALL CAPS in SQL, though it is not strictly required for all database systems.

---

[2] Just as the shell root user is the most powerful superuser, with privileges to modify any file, the MySQL root user has permission to change any aspect of any MySQL database on the computer.

When entering commands into **mysql**, if you make a mistake, you may be tempted to type [ctrl]C to terminate that command and start over, like you can in the **bash** shell. Don't do it. At the **mysql>** prompt, [ctrl]C will terminate the whole **mysql** program. Instead, type \c at the end of the text you have already entered, and press [return]. This will end that line of input, or that continuing command, without causing it to be operated.

In the example above, SHOW DATABASES; is the first SQL command issued. The results of the command are presented in a simple table below the command, along with information on how long it took to carry out the operation. SQL uses plain English words for many statements, so this command is easy to understand: you are asking for a list of the databases managed by the server. One server can have many user-created databases for different projects; for example, different lab groups can have separate databases on the same server (see Figure 15.2).

The output of this command shows that there are three databases on the server: information_schema, mysql, and test. The first two of these databases are used by the server itself to store configurations, and should not be modified. The last database is an empty test database for checking to make sure the system is functioning, as you just did. Finally, the EXIT; command closes the connection between the mysql program and the server and returns control to the shell prompt.

During your mysql session, you can type HELP; by itself to get a list of general commands and topics, or HELP followed by a command name to get more specific information on that operation.

To get another view of the process as you work, within MySQL Workbench, double-click the localhost connection that you created earlier. This will open a new SQL Editor tab, and on the left side of the window you will see the test database (Figure 15.3). The other configuration databases are hidden from view. As you work with databases, you can enter SQL commands at the command line and monitor your progress with this graphical view. It is much like navigating your filesystem at the command line and using the Finder as a graphical interface to the same files.

Creating a database and tables

So far you have connected to your local database server and looked around a bit. (Not much is there right out of the box.) Over the course of the following pages you will create a database and load in the data from several different files. These data are of a couple different types, but all were collected as part of the same project. You have already worked with one of the files, the geographical coordinates where several specimens of the deep-sea siphonophore *Marrus claudanielis* were collected with remotely operated underwater vehicles. In addition to loading these specimen data into the database, you will also load data from the CTD (conductivity, temperature, depth) instrument which collected environmental information during the dives. These two tables together will allow you to extract environmental data for the locations where the specimens were collected. This is an example of a project that includes multiple types of data that would be difficult to store efficiently in a single text file or spreadsheet.

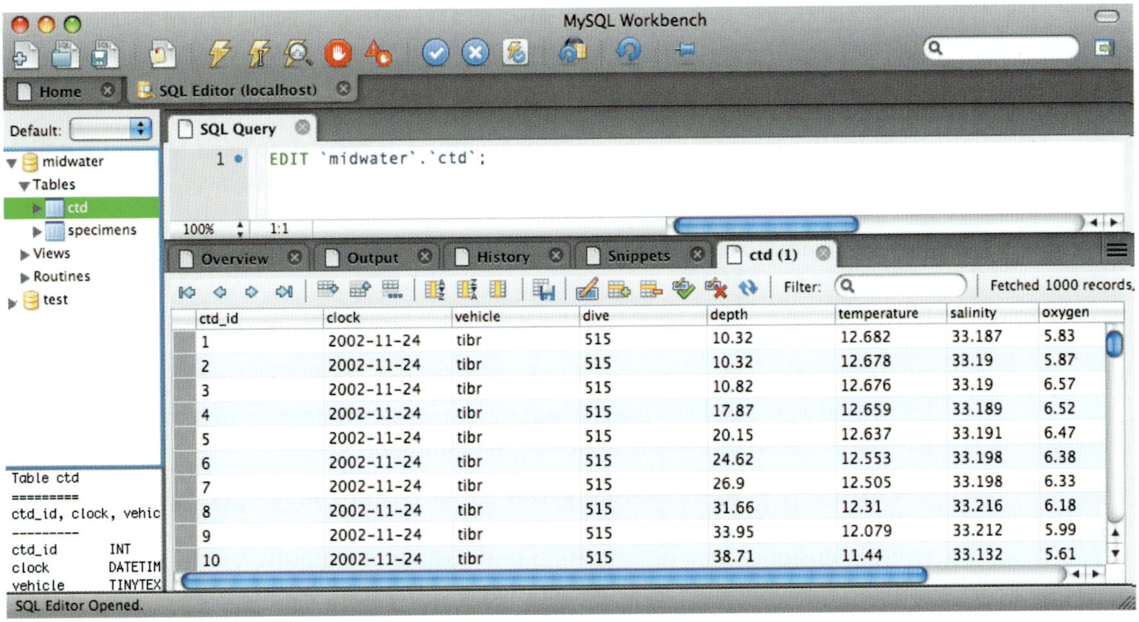

ctd_id	clock	vehicle	dive	depth	temperature	salinity	oxygen
1	2002-11-24	tibr	515	10.32	12.682	33.187	5.83
2	2002-11-24	tibr	515	10.32	12.678	33.19	5.87
3	2002-11-24	tibr	515	10.82	12.676	33.19	6.57
4	2002-11-24	tibr	515	17.87	12.659	33.189	6.52
5	2002-11-24	tibr	515	20.15	12.637	33.191	6.47
6	2002-11-24	tibr	515	24.62	12.553	33.198	6.38
7	2002-11-24	tibr	515	26.9	12.505	33.198	6.33
8	2002-11-24	tibr	515	31.66	12.31	33.216	6.18
9	2002-11-24	tibr	515	33.95	12.079	33.212	5.99
10	2002-11-24	tibr	515	38.71	11.44	33.132	5.61

FIGURE 15.3 The MySQL Workbench GUI, viewing the contents of a table that you will create in the course of this chapter Alternative SQL database GUIs will provide similar views and options for managing and querying your database tables.

Creating and selecting the database When starting out with a project, the first step is to create an empty database using the `CREATE DATABASE` command. In the terminal window, type the following:

```
lucy$ mysql -u root

mysql> CREATE DATABASE midwater;
Query OK, 1 row affected (0.09 sec)

mysql> SHOW DATABASES;
+--------------------+
| Database           |
+--------------------+
| information_schema |
| midwater           |
| mysql              |
| test               |
+--------------------+
4 rows in set (0.00 sec)

mysql>
```

You can see that there are now four databases, including the newly created `midwater` database. If you have the `localhost` connection open in MySQL Workbench, click the refresh button, the circular arrow, in the Overview pane. You will also see the database appear there too.

Since the server is responsible for several databases, you need to specify which one you want to work with. The `USE` command selects a database:

```
mysql> USE midwater;
Database changed
mysql>
```

When you issue further commands during this session, the MySQL server will know that they apply to the `midwater` database. You can switch to another database at any time by issuing another `USE` command.

Creating the specimens table After creating an empty database and selecting it with `USE`, the next step is to start to create tables. In a database, tables are two-dimensional data organization units, somewhat equivalent to a spreadsheet. You can create a table that has no data records in it, but there has to be at least one column (also called a field) to begin with. You can always add and remove columns later, but it is best to anticipate the needs of the table and create them all at the start. The first table you will create is the `specimens` table, into which you will load the data from the `Marrus_claudanielis.txt` file.

The first field to consider when designing a table is the primary key, the special column that contains a value which uniquely identifies each row. By convention it is best to make the first column the primary key and to use a series of unique integers as its entries. It is common practice to let the database generate a value for the primary key automatically when each row is created, so the value doesn't correspond to a value already present in the input file. After the primary key, the remaining table fields will roughly correspond to the columns of the file `Marrus_claudanielis.txt`. In your text editor, open this file from your `~/pcfb/examples` folder as you did at the start of Chapter 10. Select the Show Invisibles option in TextWrangler so that you can also inspect the white space characters:

```
Dive△          Date△       Lat△         Lon△          Depth△   Notes
Tiburon 596△   19-Jul-03△  36 36.12 N△  122 22.48 W△  1190△    holotype
JSL II 1411△   16-Sep-86△  39 56.4 N△   70 14.3 W△    518△     paratype
JSL II 930△    18-Aug-84△  40 05.03 N△  69 03.01 W△   686△     Youngbluth (1989)
```

The fields of each column are delimited with a tab (which in TextWrangler is displayed as the △ symbol). This will be important to know when parsing the data. There is a header line that describes what each column of data contains; this is helpful for understanding the structure of the file, but will need to be skipped when parsing it.

There are six columns of data. The first column, `Dive`, contains a string that specifies both the name of the submersible vehicle (e.g., `Tiburon`) and the number of the dive (e.g., `596`). You will split this into a string and an integer, and then store each value separately in the database. `Date` is presented as a string, but we will parse it into a special date type.[3] `Lat` and `Lon` contain the latitude and longitude as strings with degrees, minutes, and compass direction. As discussed in Chapter 10, this isn't a very convenient format for storing and analyzing position information, and again we will convert it to decimal degrees so that it can be stored as a float. After `Lat` and `Lon` comes `Depth`, which in the examined part of the file contains integers. The depth, however, is a continuous measurement and it is possible that some files might be floating point numbers. You will therefore convert it to a float and store it that way. Finally, there is a `Notes` column with a string that can contain spaces and punctuation characters.

You now know that the table needs to contain the following columns:

Column name	Type
specimen_id	INTEGER ← This will be the primary key
vehicle	TINYTEXT
dive	INTEGER
date	DATE
lat	FLOAT
lon	FLOAT
depth	FLOAT
notes	TEXT

There are a few things to note above. The column names are descriptive; it is just as important to label the parts of your database with informative names as it is to give variables meaningful names when programming. By convention, the primary key for a table is often given a name similar to that of the table itself, combined with the suffix `_id`. The `vehicle` column has type `TINYTEXT` rather than type `TEXT`. This will make the database more memory-efficient, and it won't cause problems so long as you know that the names of the vehicles will never be longer than 255 characters. `TEXT` is used for the `notes` column since there is a reasonable chance a note will exceed 255 characters.

Now that you have gathered this information together, you can begin to construct the SQL command that will create a table with these columns. The com-

---

[3] It is probably not good practice to use the name `date`, because that is the name of a data type in SQL, but it will not cause problems in this case.

mand for creating a table is, unsurprisingly, CREATE TABLE. The following is the full command for constructing the table as designed above. (Note that you don't type ->. This is part of the prompt when a command is continued across multiple lines.) You can also find the commands from this chapter in the file mysql_commands.txt for copying and pasting:

```
mysql> CREATE TABLE specimens (
    -> specimen_id INTEGER NOT NULL AUTO_INCREMENT PRIMARY KEY,
    -> vehicle TINYTEXT,
    -> dive INTEGER,
    -> date DATE,
    -> lat FLOAT,
    -> lon FLOAT,
    -> depth FLOAT,
    -> notes TEXT
    -> );
Query OK, 0 rows affected (0.10 sec)

mysql> SHOW TABLES;
+--------------------+
| Tables_in_midwater |
+--------------------+
| specimens          |
+--------------------+
1 row in set (0.00 sec)
```

Here the CREATE TABLE command is spread across multiple lines and isn't executed until [return] is pressed after typing ;. This command could also be entered in a single line. A one-line approach is good for scripts and automating data entry, but it makes the commands less readable and more error-prone when operating at the prompt.

CREATE TABLE is followed by the name of the table you want to create, and then, in parentheses, information about each column, separated by commas. That information includes the name of the field, its type, and optionally some additional information. The only column with additional parameters in this example is the primary key specimen_id. The statement NOT NULL indicates that no row can be missing a value for this column, so creating a row without a value for this field would result in an error. The database would work without this option, but including it makes the table more robust. Since NOT NULL isn't specified for the other fields, they can have missing data. AUTO_INCREMENT specifies that a new unique integer will be automatically placed in this column each time a row is added, the value of which will be one higher than the previous row that was added. This ensures that the rows have unique primary key values, and takes care of creating this value in the background so you don't have to worry about it when adding data.

PRIMARY KEY specifies that this column is the primary key. Some variant of this first field definition will probably be used in most of your table creation commands.
After the CREATE TABLE command there are a couple ways to take stock of the changes that were made to the database. SHOW TABLES simply gives a list of the tables available in the active database. The DESCRIBE command, followed by a table name, gives a summary of the structure of a table. An inspection of the results of this command confirms that the table was created as expected. Just as the shell command ls is indispensable for navigating your filesystem, SHOW and DESCRIBE are quick tools to get your bearings as you navigate a database:

```
mysql> DESCRIBE specimens;
+-------------+----------+------+-----+---------+----------------+
| Field       | Type     | Null | Key | Default | Extra          |
+-------------+----------+------+-----+---------+----------------+
| specimen_id | int(11)  | NO   | PRI | NULL    | auto_increment |
| vehicle     | tinytext | YES  |     | NULL    |                |
| dive        | int(11)  | YES  |     | NULL    |                |
| date        | date     | YES  |     | NULL    |                |
| lat         | float    | YES  |     | NULL    |                |
| lon         | float    | YES  |     | NULL    |                |
| depth       | float    | YES  |     | NULL    |                |
| notes       | text     | YES  |     | NULL    |                |
+-------------+----------+------+-----+---------+----------------+
8 rows in set (0.01 sec)

mysql>
```

Adding rows of data to tables and displaying table contents

You have now created a database and an empty table and are ready to add data. Before we explore tools for importing datasets from a file, it is important to understand how to add data one row at a time with SQL commands. A new row of data is added with the INSERT command. Here is a test command to try out; you can either copy it from the mysql_commands.txt file or type it in:

```
mysql> INSERT INTO specimens SET
    -> vehicle='Tiburon',
    -> dive=596,
    -> date='2003-07-19',
    -> lat=36.602,
    -> lon=-122.375,
    -> depth=1190,
    -> notes='holotype';
Query OK, 1 row affected (0.01 sec)
```

The latitude and longitude have been converted to float values by hand. The date string was reformatted to be consistent with the DATE type. The INSERT command itself is straightforward. At a minimum you need to specify the table that you are inserting the data into (here, specimens). SET is then followed by a list of column names and the values to be stored in each of them. As with the CREATE command, these are separated by commas, and they can all come on one line or in sequential lines. There are several ways to specify new data to be added to a table with the INSERT command, including some shorter formats that don't require you to enter all the column names. Specifying the column names helps avoid errors in which values get offset into another field, and it also makes your commands more readable.

To examine the contents of your table thus far, use the SELECT command with the following syntax:

```
mysql> SELECT * FROM specimens;
+-------------+---------+------+------------+--------+----------+-------+----------+
| specimen_id | vehicle | dive | date       | lat    | lon      | depth | notes    |
+-------------+---------+------+------------+--------+----------+-------+----------+
|           1 | Tiburon |  596 | 2003-07-19 | 36.602 | -122.375 |  1190 | holotype |
+-------------+---------+------+------------+--------+----------+-------+----------+
1 row in set (0.00 sec)
```

SELECT is a very powerful command, and this simple use of it gives little indication of its potential. It is the SELECT command that will later allow us to combine data across tables and to look at specific subsets of data. Here, though, the * indicates that it should show all columns of data in the table, and the lack of other clauses for refining the search leads it to display all the rows in the table. (So far you have just inserted the one row.)

You can also view your new table and data from the MySQL Workbench. After connecting to the localhost MySQL server, select the midwater database from the Overview tab. There you will see a list of available tables. Double click on specimens to view its contents. A two-dimensional grid will appear, displaying each row and column of the table (see Figure 15.3). There are buttons for adding and deleting rows, and you can click on cells to directly edit the data. (If you edit any cells, you must click the Apply changes to data button, the one with a green check mark, for them to take effect.) This can be convenient for getting an overview of the database and making minor modifications, but use caution during database interactions, because there often is not an Undo button.

Go ahead and delete the test row you created. Either highlight it in MySQL Workbench and click the Delete row button, or issue the command DELETE FROM specimens; from within the mysql command-line interface. Be very careful with the DELETE command. As you can see it is very easy to delete all the data in a table with just a few words, leaving only the empty table behind. Like the SELECT command, DELETE assumes you want to operate on all the rows unless you specify details on which subset of rows it should consider.

Interacting with MySQL from Python

The command-line program `mysql` and the graphical MySQL Workbench are only two of many possible ways of interacting with the MySQL database server; you can also automate interactions with your database in a variety of ways. In the present example, the data in the `Marrus_claudanielis.txt` file require some parsing and calculation before they are loaded into the database. Python is a convenient tool for such manipulations, and it can interact directly with the database once the conversions are done. In fact, you already wrote a program, `latlon.py`, in Chapter 10 that made some of these manipulations on the exact same file. Here you will repurpose that program to write data to the MySQL database rather than to an output file. The first step is to remove unneeded parts from the script and take care of the other text reformatting issues, so that each field of data is in the correct format for loading into the database. Next, an SQL statement is built that will insert these data into the database. This is simply a string of text formatted exactly as a command that would be entered at the `mysql>` prompt. Once the command is properly generated, you will use a module that connects to the MySQL server from within Python to execute this statement and add the data.

Parsing the input text

The final version of the previous script, `latlon_5.py`, will serve as the starting point for our new script, `mylatlon.py`. The first step is to strip the old script down, removing the now-unneeded code for writing the KML file. This stripped-down version of the script, along with a line for printing the parsed data to the screen, is saved as `mylatlon_1.py`. (We don't show the program here since it is so similar to what was presented in Chapter 10, but you can look at the code for the program in the `scripts` folder.) Here is the output produced when `mylatlon_1.py` is run; note that the path to the `Marrus_claudanielis.txt` file isn't specified in the program, so you need to be in the same directory as that file to run it:

```
host:ctd lucy$ mylatlon_1.py
Tiburon 596 19-Jul-03 36.602 -122.3746667 1190 holotype
JSL II 1411 16-Sep-86 39.94 -70.23833333 518 paratype
JSL II 930 18-Aug-84 40.08383333 -69.05016667 686 Youngbluth (1989)
Ventana 1575 11-Mar-99 36.704 -122.042 767
Ventana 1777 16-Jun-00 36.71 -122.045 934
Ventana 2243 9-Sep-02 36.708 -122.064 1001
Tiburon 515 24-Nov-02 36.7 -122.033 1156
Tiburon 531 13-Mar-03 24.317 -109.203 1144
Tiburon 547 31-Mar-03 24.234 -109.667 1126
JSL II 3457 26-Sep-03 40.29617 -68.1113333 862 Francesc Pages (pers.comm)
```

The initial `mylatlon_1.py` script reads the input file, skips the header line, converts the latitude and longitude to decimal degrees, and writes all the data to the screen. This takes care of most of the file parsing, but there are still two text reformatting issues that must be addressed. First, the `Dive` variable needs to be divided into the vehicle name and dive number. This will be done with a regular expression (keeping in mind that the vehicle name may have a space in it). Second, the date needs to be converted from the format found in the file to that expected by the MySQL `DATE` type (`11-Mar-99` becomes `1999-03-11`).

The date conversion could be done from scratch, but it would require a dictionary to convert the abbreviated month names to month numbers and implement rules for when to add 19 to the start of an abbreviated year (e.g., for 1999) and when to add 20 (e.g., for 2003). Fortunately, the built-in Python `datetime` module can handle all these conversions already. The `datetime` module has a `datetime` class for storing dates and times. This `datetime` class has a method called `.strptime()` that can parse `datetime` data from a string according to a specified format. It also has another method called `.strftime()` that can create a string in a specified format from a `datetime` variable. There are many formatting options, which are described at `docs.python.org/library/datetime.html`.

> **SOLVING A PROBLEM IN MORE THAN ONE WAY** As it happens, SQL has its own `STR_TO_DATE()` function that operates almost identically to the Python `.strptime()` method, so you could also do a conversion as part of the SQL command when it is entered. There are usually several ways to solve a given problem, and although we chose to do the conversion in Python this time, you should investigate the range of data-handling options available to you in SQL.

The formatting characters used here are `%d` for day, `%b` for abbreviated month name, `%y` for two-digit year (without the century), `%Y` for the full four-digit year, and `%m` for the numeric representation of the month. In addition to these changes to how the program parses text, you will also change the type of the `Depth` record from a string to a float.

To start implementing these changes, add the following line below the other `import` commands, near the top of the script:

```
from datetime import datetime
```

Note that you are importing a `datetime` class from a module called `datetime`. These are two different objects, one nested within the other. Using the same name for nested objects like this is confusing, and it should be avoided in your own code. The loop for parsing the date should be reorganized as follows:

```
# Loop over each line in the file
for Line in InFile:
    # Check line number, process if past the first line (number == 0)
    if LineNumber > 0:
        # Remove the line ending characters
```

```
# print line   # uncomment for debugging
Line = Line.strip('\n')
# Split the line into a list of ElementList, using tab as a delimiter
ElementList = Line.split('\t')
# Returns a list in this format:
# ['Tiburon 596', '19-Jul-03', '36 36.12 N', \
#   '122 22.48 W', '1190', 'holotype']

Dive     = ElementList[0] # includes vehicle and dive number
Date     = ElementList[1]
Depth    = float(ElementList[4])
Comment  = ElementList[5]

LatDegrees = decimalat(ElementList[2])
LonDegrees = decimalat(ElementList[3])

# NEW CODE ADDED BELOW HERE
#Isolate the vehicle and dive number from the Dive field
SearchStr='(.+?) (\d+)'
Result = re.search(SearchStr, Dive)
Vehicle = Result.group(1)
DiveNum = int(Result.group(2))

# Reformat date
# Create a datetime object from a string
DateParsed = datetime.strptime(Date, "%d-%b-%y")
DateOut = DateParsed.strftime("%Y-%m-%d") # string from datetime object

print Vehicle, DiveNum, DateOut, LatDegrees, LonDegrees, Depth, Comment

LineNumber += 1 # This is outside the if, but inside the for loop
```

Run the script to confirm that you get a reformatted date beginning with the four-digit year, and that the dive number is parsed correctly.

Formulating SQL from the data

Now that each data field has been parsed from the file and all fields formatted appropriately, these data can be packaged for insertion into the database. Each line of data will be added to the database with an INSERT INTO command, just like the one from the SQL example. This SQL command is just a string that describes what you want to do with some data. Before the final code for connecting to the database is added to the program, you will print the SQL command to the screen. This is a good step to take with any program that can modify a database. It allows you to catch potential problems before difficult-to-fix database mistakes are made.

A simple way to build up a large string with many fields is with the % string formatting operator, introduced in Chapter 8. Triple quotes are used so that the string can be split across multiple lines for readability. Comment out the existing `print` line and add the code for creating the SQL statement immediately below it:

```
# print Vehicle, DiveNum, DateOut, LatDegrees, LonDegrees, Depth, Comment
SQL = """INSERT INTO specimens SET
vehicle='%s',
dive=%d,
date='%s',
lat=%.4f,
lon=%.4f,
depth=%.1f,
notes='%s' ;
""" % (Vehicle, DiveNum, DateOut, LatDegrees, LonDegrees, Depth, Comment)

print SQL
```

Notice that the SQL command we are generating requires quotation marks around strings, and therefore single quotes are used within the triple-quoted string. This version of the program is saved as `mylatlon_3.py`. The output of the program is now a series of SQL commands, the first two of which are shown here:

```
host:ctd lucy$ mylatlon_3.py
INSERT INTO specimens SET
        vehicle='Tiburon',
        dive=596,
        date='2003-07-19',
        lat=36.6020,
        lon=-122.3747,
        depth=1190.0,
        notes='holotype' ;

INSERT INTO specimens SET
        vehicle='JSL II',
        dive=1411,
        date='1986-09-16',
        lat=39.9400,
        lon=-70.2383,
        depth=518.0,
        notes='paratype' ;
```

These SQL commands will enter the specimen data into the database row by row. You could even copy and paste one of these commands right into an open `mysql` session to enter the data. (The semicolons are optional when commands are submitted through Python, but we have included them here so you can paste the output directly at a `mysql>` prompt.)

Executing SQL commands from Python

All that remains is to connect to the MySQL server from within Python and execute the commands. There are a few steps to this.

Installing the MySQLdb Python module There are quite a few things that have to happen behind the scenes to make a connection to the MySQL database. Fortunately, there is a Python module called MySQLdb which takes care of all the work under the hood. The most complicated part of connecting to MySQL from Python is installing this module, but you only have to do it once. MySQLdb has some known installation issues with older versions of OS X, MySQL, and Python. If you encounter any errors or something doesn't seem consistent with the instructions provided here, consult Chapter 21, search the Web for the error message you get, or visit practicalcomputing.org for additional guidance. If you are running Ubuntu Linux, MySQLdb can be installed with the Synaptic Package Manager. Note that you must install MySQL first before installing the module.

If you are running OS X, download the MySQLdb module from sourceforge.net/projects/mysql-python/. Double-click the archive to uncompress and expand it.[4] Open a terminal window, type cd space and drag the icon for the folder you just expanded to the terminal window, and press return to move into that directory. Read the installation instructions in the README file, and then proceed with the installation:

```
host:ctd lucy$ cd ~/Downloads/MySQL-python-1.2.3/
host:MySQL lucy$ cat README
host:MySQL lucy$ python setup.py build
host:MySQL lucy$ sudo python setup.py install
```

The names of the files and folders may differ slightly if you download a later version of MySQLdb. The cat README command will display information about the installation process in the supplied README file. If the suggested installation commands differ from the python and sudo commands shown here, then follow the instructions in the README file instead.

It is a good idea to test that the module loads correctly using the interactive Python prompt before giving it a try in a program. Move to the examples/ctd folder, and try to load the module in Python:

```
host:MySQL lucy$ cd ~/pcfb/examples/ctd
host:ctd lucy$ python
Python 2.6.1 (r261:67515, Feb 11 2010, 00:51:29)
[GCC 4.2.1 (Apple Inc. build 5646)] on darwin
Type "help", "copyright", "credits" or "license" for more information.
>>> import MySQLdb
>>>
```

---

[4] For more information about special cases of unarchiving and installing software, see Chapter 21.

In this example, there were no warnings or errors. Even if you get some warnings, the module may still work fine. Restarting your computer may help resolve some errors.

Establishing the database connection Once you have installed `MySQLdb` on your computer, you still need to import it into your Python program to use it. Add the following line below the `import` statements that are already at the top of your script:

```
import MySQLdb
```

Next, you need to create a connection to the MySQL database. You can create a single connection to the database near the top of your program, and then use and reuse it wherever you like. Place the following lines right before the `for` loop:

```
MyConnection = MySQLdb.connect( host = "localhost", \
user= "root", passwd = "", db = "midwater")
MyCursor = MyConnection.cursor()
```

The first of these lines (shown here split into two with a \ to escape the line ending) creates the actual connection object, `MyConnection`. It needs information about the network address of the MySQL server (`localhost` since you are running the server right on your computer), username, password (left blank here since we haven't created one), and the database on the server that you want to connect to. Once you have the connection object, you use it to create a cursor object. You can think of this database cursor like the cursor at the command line: it is the point at which you interact with the MySQL program. This cursor is used to submit commands and retrieve results.

At the very end of the program, add two lines to close the cursor and database connection:

```
MyCursor.close()
MyConnection.close()
```

Executing SQL commands At this point, you have generated SQL command strings and have a connection to the database. All you need to do is execute the SQL commands so that the program adds the data to the database. Now that you have taken care of all the housekeeping, the actual execution command is only a single line. Add it right after the line that prints the SQL variable to the screen:

```
MyCursor.execute(SQL)
```

This line executes the SQL string you created earlier, using the database cursor you opened before the loop. Each time through the loop, it executes a new SQL command and adds another row of data to the database.

The final program, saved as `mylatlon_4.py`, is below:

```python
#! /usr/bin/env python
"""

mylatlon_4.py
import latitude longitude records from a text file,
format them into a SQL command, and enter the records into a database
"""


import re # Load regular expression module
from datetime import datetime # Load datetime class from the datetime module
import MySQLdb

# Functions must be defined before they are used
def decimalat(DegString):
    # This function requires that the re module is loaded
    # Take a string in the format "34 56.78 N" and return decimal degrees
    SearchStr='(\d+) ([\d\.]+) (\w)'
    Result = re.search(SearchStr, DegString)

    # Get the (captured) character groups from the search
    Degrees = float(Result.group(1))
    Minutes = float(Result.group(2))
    Compass = Result.group(3).upper() # make sure it is capital too
    # Calculate the decimal degrees
    DecimalDegree = Degrees + Minutes/60

    if Compass == 'S' or Compass == 'W':
        DecimalDegree = -DecimalDegree
    return DecimalDegree
# End of the function definition

# Set the input file name
InFileName = 'Marrus_claudanielis.txt'

# Open the input file
InFile = open(InFileName, 'r')

# Initialize the counter used to keep track of line numbers
LineNumber = 0

# Create the database connection
# Often you will want to use a variable instead of a fixed string
# for the database name

MyConnection = MySQLdb.connect( host = "localhost", user = "root", \
     passwd = "", db = "midwater")
MyCursor = MyConnection.cursor()
```

```python
# Loop over each line in the file
for Line in InFile:
    # Check the line number, process if past the first line (number == 0)
    if LineNumber > 0:
        # Remove the line ending characters
        # print line  # uncomment for debugging
        Line = Line.strip('\n')
        # Split the line into a list of ElementList, using tab as a delimiter
        ElementList = Line.split('\t')
        # Returns a list in this format:
        # ['Tiburon 596', '19-Jul-03', '36 36.12 N', '122 22.48 W',
        # '1190', 'holotype']

        Dive      = ElementList[0] # includes vehicle and dive number
        Date      = ElementList[1]
        Depth     = float(ElementList[4])
        Comment = ElementList[5]
        LatDegrees = decimalat(ElementList[2])
        LonDegrees = decimalat(ElementList[3])

        #Isolate the vehicle and dive number from the Dive field

        SearchStr='(.+?) (\d+)'
        Result = re.search(SearchStr, Dive)
        Vehicle = Result.group(1)
        DiveNum = int(Result.group(2))

        #Reformat date
        # Create a datetime object from a string
        DateParsed = datetime.strptime(Date, "%d-%b-%y")
        # Create a string from a datetime object
        DateOut = DateParsed.strftime("%Y-%m-%d")
        #print Vehicle, DiveNum, DateOut, LatDegrees, LonDegrees, Depth, Comment
        SQL = """INSERT INTO specimens SET
    vehicle='%s',
    dive=%d,
    date='%s',
    lat=%.4f,
    lon=%.4f,
    depth=%.1f,
    notes='%s' ;
""" % (Vehicle, DiveNum, DateOut, LatDegrees, LonDegrees, Depth, Comment)
        print SQL
        MyCursor.execute(SQL)
    LineNumber += 1 # This is outside the if, but inside the for loop

# Close the files
InFile.close()
MyCursor.close()
MyConnection.commit()
MyConnection.close()
```

From the folder that contains the `Marrus_claudanielis.txt` file, execute `mylatlon_4.py`. The output will look the same as the output of `mylatlon_3.py`, but behind the scenes the data are being added to the database! (Avoid re-running this command during testing or it will add duplicate records to your database.) From within the `mysql` command-line interface, use the `SELECT` command to take a look again at the contents of the `specimens` table. If you are starting a new `mysql` session, remember to first select the `midwater` database with the `USE midwater;` command. The output below has been edited slightly to fit on the page:

```
mysql> SELECT * FROM specimens;
+-------------+---------+------+------------+--------+----------+-------+--------------------+
|specimen_id  | vehicle | dive | date       | lat    | lon      |depth  | notes              |
+-------------+---------+------+------------+--------+----------+-------+--------------------+
|           4 | Tiburon |  596 | 2003-07-19 | 36.602 | -122.375 | 1190  | holotype           |
|           5 | JSL II  | 1411 | 1986-09-16 | 39.94  | -70.2383 |  518  | paratype           |
|           6 | JSL II  |  930 | 1984-08-18 | 40.084 | -69.0502 |  686  | Youngbluth (1989)  |
|           7 | Ventana | 1575 | 1999-03-11 | 36.704 | -122.042 |  767  |                    |
|           8 | Ventana | 1777 | 2000-06-16 | 36.71  | -122.045 |  934  |                    |
|           9 | Ventana | 2243 | 2002-09-09 | 36.708 | -122.064 | 1001  |                    |
|          10 | Tiburon |  515 | 2002-11-24 | 36.7   | -122.033 | 1156  |                    |
|          11 | Tiburon |  531 | 2003-03-13 | 24.317 | -109.203 | 1144  |                    |
|          12 | Tiburon |  547 | 2003-03-31 | 24.234 | -109.667 | 1126  |                    |
|          13 | JSL II  | 3457 | 2003-09-26 | 40.296 | -68.1113 |  862  | Pages (pers.comm)  |
+-------------+---------+------+------------+--------+----------+-------+--------------------+
10 rows in set (0.00 sec)
```

You can also inspect the modified table from the MySQL Workbench graphical interface. Your values for `specimen_id` might vary from those shown here, because the `AUTO_INCREMENT` counter keeps track of all the rows that have ever been added even if they have subsequently been removed.

Bulk-importing text files into a table

Typically, when starting to work with a database, you will already have your data stored in text files or spreadsheets, which you want to import into database tables. In the next component of this example you will load a new data table named `ctd` with environmental data measured from the CTD sensors on the submarines that collected six of these specimens. Unlike the `specimens` data, the columns of the files correspond exactly to the columns of the table that you create, and no conversion is needed. While you could import these data into the database with another custom Python program, because they are already formatted, you can add the data with simple SQL commands. Before you can add any data, though, you will need to create the table.

Creating the `ctd` table

Open a new terminal window to get a shell prompt, and change into the `~/pcfb/examples/ctd/` directory. Then, generate a list of the files that start with the word `Marrus` and view the header of the first file using the `head` command, which is described fully in the next chapter:

```
host:~ lucy$ cd ~/pcfb/examples/ctd
host:ctd lucy$ ls Marrus*
Marrus_ctdTib515.txt     Marrus_ctdTib596.txt     Marrus_ctdVen2243.txt
Marrus_ctdTib531.txt     Marrus_ctdVen1575.txt
Marrus_ctdTib547.txt     Marrus_ctdVen1777.txt
host:ctd lucy$ head Marrus_ctdTib515.txt
rovCtdDtg,vehicle,depth,temper,salin,oxyg,lat,lon
2002-11-24 14:24:15,tibr,10.32,12.682,33.187,5.83,36.571183,-122.52263
2002-11-24 14:24:45,tibr,10.32,12.678,33.19,5.87,36.70004157,-122.03345157
2002-11-24 14:25:15,tibr,10.82,12.676,33.19,6.57,36.70000171,-122.03336828
2002-11-24 14:25:45,tibr,17.87,12.659,33.189,6.52,36.700005,-122.03334412
2002-11-24 14:26:15,tibr,20.15,12.637,33.191,6.47,36.69998512,-122.03335
...
host:ctd lucy$
```

The first things that stand out are that the values are separated by commas, and that there is a header row that describes what each of the values are. From this information you can generate and execute a CREATE TABLE command that has all the needed fields:

```
mysql> CREATE TABLE ctd (
    -> ctd_id INTEGER NOT NULL AUTO_INCREMENT PRIMARY KEY,
    -> clock DATETIME,
    -> vehicle TINYTEXT,
    -> dive INTEGER,
    -> depth FLOAT,
    -> temperature FLOAT,
    -> salinity FLOAT,
    -> oxygen FLOAT,
    -> lat FLOAT,
    -> lon FLOAT
    -> );
```

Some of these fields correspond to the fields of the `specimens` table, and this will ultimately help you link corresponding pieces of information. The `date` field in these files happens to be properly formatted for MySQL to import directly.[5] Otherwise, you might have to convert them with regular expressions, a separate program, or one of the SQL date functions like `STR_TO_DATE`.

---

[5] What a fortunate coincidence...

Importing data files with the LOAD DATA command

The command for loading data into a table from a text file is long but for the most part self-explanatory. Here is an example for the first CTD file:

```
LOAD DATA LOCAL INFILE '~/pcfb/examples/ctd/Marrus_ctdTib515.txt'
    INTO TABLE ctd
    FIELDS TERMINATED BY ','
    IGNORE 1 LINES
    (clock,vehicle,depth,temperature,salinity,oxygen,lat,lon)
    SET dive=515;
```

In this example, there are only two parts of this command that will change from file to file: the name of the file and the value placed into the dive variable, which is derived from the filename. The first line of the command specifies that it is a LOAD DATA statement and that the file is on the local computer, and provides the path to the file. The INTO TABLE portion of the statement specifies which table to load the data into. The next two lines indicate that the fields are separated by commas, and that the first line is a header line that should be skipped. The names of the table columns that the fields should be loaded into are then specified, within parentheses, in the order that they occur in the file. The last line sets the value of the dive field in the table to 515 for all the added rows. This dive number is in the filename, but isn't located within the file itself.

To generate the commands for loading data from all the files, list the ctd directory using the command ls -1 Marrus*. (The flag is the number 1). This will show a column listing just the CTD file names. Note that there are no files from the JSL II submarine, so specimens collected with that vehicle will not have corresponding temperature information. Copy the file list into a text editor and use regular expressions to modify this list into a series of one-line commands as described here. Search for the following:

```
(\w+?(\d+)\.txt)
```

Then replace all with the text below,[6] which you can copy from the file mysql_commands.txt:

```
LOAD DATA LOCAL INFILE '~/pcfb/examples/ctd/\1'
INTO TABLE ctd
FIELDS TERMINATED BY ',' IGNORE 1 LINES
(clock,vehicle,depth,temperature, salinity,oxygen,lat,lon)
SET dive=\2;
```

---

[6] This search is a bit tricky because it uses nested parentheses to capture replacement text. The text in the outermost pair, including the dive number, is saved as \1. The inner parentheses save the dive number alone as \2.

The result of this replacement will be a series of commands, shown below, which will load each file into your database.

The full set of additional commands is also saved in the `mysql_commands.txt` example file. You could also type the command once, and then recycle it while replacing the file name on the first line and dive number on the last line. As with the `bash` shell, you can use the ⤒ key to move back through your command history to edit the names. Another very useful aspect of the `mysql>` prompt is that `tab` will also auto-complete `mysql` commands and variable names that are known to the database. Note that if you paste all of these lines at once, it may overwhelm the terminal program's buffer, and some of them may get garbled. On our computers, pasting four commands at a time worked without any problems.

```
mysql> LOAD DATA LOCAL INFILE '~/pcfb/examples/ctd/Marrus_ctdTib515.txt'
   INTO TABLE ctd  FIELDS TERMINATED BY ','  IGNORE 1 LINES
   (clock,vehicle,depth,temperature,salinity,oxygen,lat,lon)  SET dive=515;

mysql> LOAD DATA LOCAL INFILE '~/pcfb/examples/ctd/Marrus_ctdTib531.txt'
   INTO TABLE ctd  FIELDS TERMINATED BY ','  IGNORE 1 LINES
   (clock,vehicle,depth,temperature,salinity,oxygen,lat,lon)  SET dive=531;

mysql> LOAD DATA LOCAL INFILE '~/pcfb/examples/ctd/Marrus_ctdTib547.txt'
   INTO TABLE ctd  FIELDS TERMINATED BY ','  IGNORE 1 LINES
   (clock,vehicle,depth,temperature,salinity,oxygen,lat,lon)  SET dive=547;

mysql> LOAD DATA LOCAL INFILE '~/pcfb/examples/ctd/Marrus_ctdTib596.txt'
   INTO TABLE ctd  FIELDS TERMINATED BY ','  IGNORE 1 LINES
   (clock,vehicle,depth,temperature,salinity,oxygen,lat,lon)  SET dive=596;

mysql> LOAD DATA LOCAL INFILE '~/pcfb/examples/ctd/Marrus_ctdVen1575.txt'
   INTO TABLE ctd  FIELDS TERMINATED BY ','  IGNORE 1 LINES
   (clock,vehicle,depth,temperature,salinity,oxygen,lat,lon)  SET dive=1575;

mysql> LOAD DATA LOCAL INFILE '~/pcfb/examples/ctd/Marrus_ctdVen1777.txt'
   INTO TABLE ctd  FIELDS TERMINATED BY ','  IGNORE 1 LINES
   (clock,vehicle,depth,temperature,salinity,oxygen,lat,lon)  SET dive=1777;

mysql> LOAD DATA LOCAL INFILE '~/pcfb/examples/ctd/Marrus_ctdVen2243.txt'
   INTO TABLE ctd  FIELDS TERMINATED BY ','  IGNORE 1 LINES
   (clock,vehicle,depth,temperature,salinity,oxygen,lat,lon)  SET dive=2243;

mysql>
```

After these commands have run, all of the CTD data have been loaded from files into the `midwater` database's `ctd` table. We will explore this table in a later section.

Other approaches to automating the import of files into a database are postulated at the end of the chapter.

Exporting and importing databases as SQL files

It is common to distribute databases as SQL files. These are text files with all the commands needed to create tables (and sometimes the database itself as well), and to add all the rows of data to the tables. This is achieved with the `mysqldump` command, which is also a convenient way to backup a database. Not only are the data themselves preserved, so is the structure of the database.

We have provided both tables of the database created above as a SQL file called `midwater.sql`, also available in the `ctd` folder. It was created with the shell command:

```
host:ctd lucy$ mysqldump -u root midwater > midwater.sql
```

(Note that this is a `bash` command, not a `mysql` command.) Open up the `midwater.sql` file in a text editor. Some of the commands in the file will be familiar, whereas some use statements you have seen but in different formulations, and some aren't covered in this book.

If you are unable to create and load the database as described in previous sections but would still like to follow along with the data-mining examples below, you can load the database from the `midwater.sql` file. First create the empty `midwater` database in `mysql`, and then at the shell enter the following command to execute the SQL commands in the file:

```
host:~ lucy$ mysql -u root midwater < ~/pcfb/examples/ctd/midwater.sql
```

This general strategy works for executing any set of SQL statements; they don't have to be commands for creating and filling tables.[7] Because SQL commands remain largely the same among various implementations of relational database systems, you might be able to import this file or a slightly modified version into another database management program.

Exploring data with SQL

Now that all the data for this project are in the database, you can use SQL commands to summarize, update, and extract information.

Summarizing tables with SELECT *and* COUNT

With the small `specimens` table, we already examined all the rows using the `SELECT * FROM specimens` command. Here are some more uses of `SELECT`, and

---

[7] This use of the < operator has not been covered in this book, but it is a variation of the redirection operator > that you have been using to save output to a file. When used in the other direction, pointing left, it causes a file to be used as input to a program or command.

ways to refine the output. A basic question about a database is, "How many rows does my table contain?" You can use a slightly modified SELECT statement to get an answer:

```
mysql> SELECT COUNT(*) FROM specimens;
+----------+
| COUNT(*) |
+----------+
|       10 |
+----------+
1 row in set (0.00 sec)

mysql> SELECT COUNT(*) FROM ctd;
+----------+
| COUNT(*) |
+----------+
|     3738 |
+----------+
1 row in set (0.00 sec)
```

By replacing * with COUNT(*), you now retrieve a single row that contains the count of the number of rows retrieved by SELECT, rather than all the data rows themselves. The command COUNT is one of SQL's statistical functions, and it accepts parameters passed to it within parentheses. Other math and statistical operators are described below and summarized in Table 15.2.

It is also possible to extract data from only particular columns with SELECT. Instead of using *, which is a wildcard for all columns, you can specify the columns you want, separating them by commas:

```
mysql> SELECT vehicle,date FROM specimens;
+----------+------------+
| vehicle  | date       |
+----------+------------+
| Tiburon  | 2003-07-19 |
| JSL II   | 1986-09-16 |
| JSL II   | 1984-08-18 |
| Ventana  | 1999-03-11 |
| Ventana  | 2000-06-16 |
| Ventana  | 2002-09-09 |
| Tiburon  | 2002-11-24 |
| Tiburon  | 2003-03-13 |
| Tiburon  | 2003-03-31 |
| JSL II   | 2003-09-26 |
+----------+------------+
10 rows in set (0.00 sec)
```

Collating data with GROUP BY

A common task is to see how many distinct values a given column has. It would, for instance, be informative to know how many vehicles have records. This can be done in a couple of different ways:

```
mysql> SELECT DISTINCT vehicle FROM specimens;
+----------+
| vehicle  |
+----------+
| Tiburon  |
| JSL II   |
| Ventana  |
+----------+
3 rows in set (0.03 sec)

mysql> SELECT vehicle,COUNT(*) FROM specimens GROUP BY vehicle;
+----------+----------+
| vehicle  | COUNT(*) |
+----------+----------+
| JSL II   |        3 |
| Tiburon  |        4 |
| Ventana  |        3 |
+----------+----------+
3 rows in set (0.21 sec)

mysql> SELECT vehicle,dive,COUNT(*) FROM ctd GROUP BY vehicle, dive;
+----------+------+----------+
| vehicle  | dive | COUNT(*) |
+----------+------+----------+
| tibr     |  515 |      491 |
| tibr     |  531 |     1348 |
| tibr     |  547 |      486 |
| tibr     |  596 |      760 |
| vnta     | 1575 |      100 |
| vnta     | 1777 |      210 |
| vnta     | 2243 |      343 |
+----------+------+----------+
7 rows in set (0.00 sec)

mysql>
```

The SELECT DISTINCT command is the simplest to type, but it doesn't tell you how many rows there are for each vehicle type. To get the count for each variable, use the alternative command that includes the GROUP BY clause. It groups the rows by shared vehicle values, and then counts the number of rows in each of these groups with COUNT(*). Note that both vehicle and COUNT(*) are selected;

TABLE 15.2 Selected SQL math and statistical operators and functions	
Function or operator	**Meaning**
+, −, *, /	Basic math operators
AVG	Average of the values
COUNT	Count of the values
MAX	Maximum value
MIN	Minimum value
STD	Standard deviation
SUM	Sum of the values

if only COUNT(*) is specified then you will get counts without knowing which vehicles they are associated with. You can use multiple columns for GROUP BY, so that each row in the result will be for a unique observed combination of these columns.

Mathematical operations in SQL

In addition to returning values from a table, SQL can perform mathematical and statistical operations on the data that are retrieved (see Table 15.2). To use these operators, construct a formula within parentheses, connecting field names with math symbols, as with (depth * 3.3). You can also use the statistical functions by placing a field name in parentheses after the parameter name. For example, to get the average depth for dives, grouped by each vehicle, you can use the following command:

```
mysql> SELECT vehicle, AVG(depth) FROM specimens GROUP BY vehicle;
+---------+-------------------+
| vehicle | avg(depth)        |
+---------+-------------------+
| JSL II  | 688.666666666667  |
| Tiburon |              1154 |
| Ventana | 900.666666666667  |
+---------+-------------------+
3 rows in set (0.06 sec)
```

Refining selections by row with WHERE

In addition to isolating particular columns, you can also isolate particular rows from a table. This is done with the WHERE clause (Figure 15.4). To see only the rows that pertain to the vehicle Tiburon, use the following command:

```
mysql> SELECT * FROM specimens WHERE vehicle='Tiburon';
+-------------+---------+------+------------+--------+----------+-------+----------+
| specimen_id | vehicle | dive | date       | lat    | lon      | depth | notes    |
+-------------+---------+------+------------+--------+----------+-------+----------+
|           4 | Tiburon |  596 | 2003-07-19 | 36.602 | -122.375 |  1190 | holotype |
|          10 | Tiburon |  515 | 2002-11-24 |   36.7 | -122.033 |  1156 |          |
|          11 | Tiburon |  531 | 2003-03-13 | 24.317 | -109.203 |  1144 |          |
|          12 | Tiburon |  547 | 2003-03-31 | 24.234 | -109.667 |  1126 |          |
+-------------+---------+------+------------+--------+----------+-------+----------+
4 rows in set (0.00 sec)
```

Note that the equality operator in SQL is a single = sign, not == as in Python and many other languages. You can slice data by combined criteria that pertain to columns (dive, data) and rows (WHERE vehicle = 'Tiburon') together:

```
mysql> SELECT dive,date FROM specimens WHERE vehicle='Tiburon';
+------+------------+
| dive | date       |
+------+------------+
|  596 | 2003-07-19 |
|  515 | 2002-11-24 |
|  531 | 2003-03-13 |
|  547 | 2003-03-31 |
+------+------------+
```

The WHERE phrase is very adaptable, and you can use it with approximate matches using LIKE or even regular expressions with REGEXP. Instead of testing for an exact match to a string with =, you can use LIKE to retrieve matches to a portion of the string. LIKE uses % as a wildcard the same way that * is used in the bash shell and how .* is used in regular expressions:

```
mysql> SELECT vehicle,dive FROM specimens WHERE vehicle LIKE 'TIB%';
+---------+------+
| vehicle | dive |
+---------+------+
| Tiburon |  596 |
| Tiburon |  515 |
| Tiburon |  531 |
| Tiburon |  547 |
+---------+------+
```

To perform a regular expression search, you use WHERE *field* REGEXP *query*, and *query* is a string containing the search term. Any field values which match the regular expression will be returned. The SQL regular expression syntax matches most closely with the terms used in the bash shell (see Appendix 2), and it doesn't include all of the wildcards like \w and \d. You are able to specify the beginning and ending of strings with ^ and $, any character with a period, and range of characters with square brackets [A-Z]:

```
mysql> SELECT vehicle,dive FROM specimens WHERE vehicle REGEXP '^V';
+---------+------+
| vehicle | dive |
+---------+------+
| Ventana | 1575 |
| Ventana | 1777 |
| Ventana | 2243 |
+---------+------+
```

Expressions built with WHERE also commonly use numerical comparison operators and logical statements, just like `if` statements in Python. This will probably be one of your most common uses of a SELECT statement. For example:

```
mysql> SELECT vehicle,dive FROM specimens WHERE dive < 1000;
+---------+------+
| vehicle | dive |
+---------+------+
| Tiburon |  596 |
| Tiburon |  515 |
| Tiburon |  531 |
| Tiburon |  547 |
| JSL II  |  930 |
+---------+------+
```

If you build up a logical sequence of comparisons, be sure to think out the use of AND and OR. Two comparisons linked by an OR will return the merged set of values where those tests are true. For example, to return the combined set of dive numbers for Tiburon and JSL II, you could use:

```
SELECT vehicle, dive FROM specimens
    WHERE vehicle LIKE "Tib%" OR vehicle LIKE "JSL%";
```

If you wanted records from Tiburon and JSL, but tried using an AND statement, you would get no results. We will use WHERE with AND to return a desired subset of the CTD records in a later example.

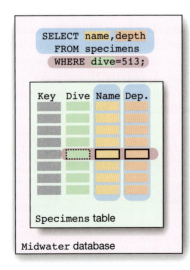

FIGURE 15.4 A graphical view of extracting data from a database The SELECT command starts by defining the columns (blue fields) to retrieve. The WHERE command (maroon stripe) refines which rows of these columns should be extracted, based on values in those or other columns.

Modifying rows with UPDATE

Once data are loaded into a database you will often want to modify them. In this database, different names have been used for the vehicles in the `specimen` and `ctd` files:

```
mysql> SELECT vehicle,COUNT(*) FROM specimens GROUP BY vehicle;
+---------+----------+
| vehicle | COUNT(*) |
+---------+----------+
| JSL II  |        3 |
| Tiburon |        4 |
| Ventana |        3 |
+---------+----------+
3 rows in set (0.01 sec)

mysql> SELECT vehicle,COUNT(*) FROM ctd GROUP BY vehicle;
+---------+----------+
| vehicle | COUNT(*) |
+---------+----------+
| tibr    |     3085 |
| vnta    |      653 |
+---------+----------+
2 rows in set (0.09 sec)
```

This is less than desirable. Below are UPDATE commands that change the vehicle abbreviations in the `ctd` table to match the full vehicle names used in the `specimens` table:

```
mysql> UPDATE ctd SET vehicle='Tiburon' WHERE vehicle='tibr';
Query OK, 3085 rows affected (0.07 sec)
Rows matched: 3085  Changed: 3085  Warnings: 0

mysql> UPDATE ctd SET vehicle='Ventana' WHERE vehicle='vnta';
Query OK, 653 rows affected (0.05 sec)
Rows matched: 653  Changed: 653  Warnings: 0
```

Here the WHERE clause is acting just as it did for the SELECT command: it is restricting the command to a subset of rows where the specified criteria are true. The SET clause is acting as it did in the INSERT INTO and LOAD DATA from commands: it is assigning a particular value to a particular field.

The previous examples were exploring data tables, but this command is altering the data table in a way that is important for the rest of the operations in this chapter. Execute these UPDATE commands (also available in example file) before proceeding.

Selecting data across tables

So far there has been no interaction between the tables in databases. They have been loaded with data and analyzed independently. Combining data across tables is one of the most powerful abilities of relational databases. It is also where relational databases get their name—relationships can be defined between data across tables. This allows for complex database structures that would be very inefficient to represent with two-dimensional grids, even though each table in the database is two-dimensional.

Combining data across tables is not as complicated as you might think. In most cases it just requires modifying the SELECT statement so it is pulling data from multiple tables according to particular relationships. To access more than one table, you put the names of the tables after the FROM statement, separated by commas. Because you are now querying against two tables with different dimensions and different fields, some of which may have the same name, you need to specify which table you are talking about when you indicate a field name. This is done with a dot notation similar to some of the methods you used in Python. A particular field is specified with the name of the table, a dot, and then the name of the field. To indicate the vehicle field of the specimen table, for instance, you would use specimen.vehicle.

In the following series of SELECT commands you will extract environmental data from the ctd table corresponding to the collection depth of particular organisms from the specimens table. As a first step, just select the vehicle, dive, and depth fields for the holotype specimen (the specimen that was chosen as the representative of the entire species):

```
mysql> SELECT specimens.vehicle, specimens.dive, specimens.depth
    -> FROM specimens
    -> WHERE specimens.notes="holotype";
+----------+------+-------+
| vehicle  | dive | depth |
+----------+------+-------+
| Tiburon  |  596 |  1190 |
+----------+------+-------+
```

To list all the CTD data for this dive by this vehicle, try the following command:

```
mysql> SELECT ctd.* FROM ctd, specimens
    -> WHERE specimens.notes="holotype"
    -> AND ctd.vehicle=specimens.vehicle AND ctd.dive=specimens.dive;
```

This will return hundreds of rows of data taken during dive 596 by the remotely operated underwater vehicle Tiburon.[8] Neither the dive number or vehicle were stated explicitly, though. The portion of the CTD data displayed was restricted by

---

[8] To only print the first 10 rows, add LIMIT 10 to the end of the command.

specifying that the notes field for specimens had to be "`holotype`", and then that the dive and vehicle fields in the `ctd` table had to match the corresponding values from the selected row from the specimens table.

At this point you could scroll through these data, look for the depth that is closest to that of the specimen, and extract the desired CTD data—or you could let the database system do the work for you. Depth can't be specified across tables in the same way that vehicle and dive were since the CTD measurements are taken at intervals, and the chances of getting measurements at the exact depth where the specimen was collected are slim.

The following command shows selected CTD data for the row that has the minimal depth, as specified with `MIN(ctd.depth)`, out of the depths that are greater than or equal to the collection depth, as specified by `ctd.depth >= specimens.depth`. This is one method to find the closest value to a known value. This command also displays all the vehicle and dive values from the `specimens` table for this record, showing how each row of the result can contain data that are gathered from across multiple tables:

```
mysql> SELECT specimens.vehicle, specimens.dive, MIN(ctd.depth),
    -> ctd.temperature, ctd.salinity, ctd.oxygen
    -> FROM ctd, specimens
    -> WHERE specimens.notes="holotype"
    -> AND ctd.vehicle=specimens.vehicle AND ctd.dive=specimens.dive
    -> AND ctd.depth >= specimens.depth;
+---------+------+----------------+--------+----------+--------+
| vehicle | dive | MIN(ctd.depth) | temp.  | salinity | oxygen |
+---------+------+----------------+--------+----------+--------+
| Tiburon |  596 | 1190.170043945 | 3.405  |  34.284  |  0.05  |
+---------+------+----------------+--------+----------+--------+
```

Next you will add a command like this into a Python program, to retrieve a combined set of values for each specimen, not just for the holotype.

Generating output using Python

For most data extraction procedures there are ways to achieve everything with a series of SQL commands. As you saw at the beginning of this chapter, though, sometimes it is easier to process files using the familiar Python environment, generating customized SQL queries and sending them to MySQL using the `MySQLdb` module. In the next example, you will take the command derived above to extract CTD data and use a Python script to apply it to each row in the `specimen` table. The resulting output is a tab-delimited table that could quickly be formatted for use in a publication. For this script to work, you need to have generated or imported the `specimens` and `ctd` tables in the `midwater` database, and performed the `UPDATE` command to standardize vehicle names, as described in the body of this chapter.

```python
#! /usr/bin/env python
"""

sqlmerge.py
using the mysql database 'midwater', with its tables 'ctd' and 'specimens',
look up the dive and depth for each specimen, and extract the corresponding
temperature, salinity, and oxygen from the ctd table

output the combined results as a tab-delimited table
"""

import re         # Load regular expression module
import MySQLdb    # must be installed separately

# Create the database connection. Often you will want to use a
# a variable to hold the database name, instead of a fixed string
MyConnection = MySQLdb.connect( host = "localhost", user = "root", \
                                passwd = "", db = "midwater")
MyCursor = MyConnection.cursor()
SQL = """SELECT specimen_id,vehicle,dive,date,depth,lat,lon from specimens;"""
SQLLen = MyCursor.execute(SQL)  # returns the number of records retrieved

# MyCursor is now "loaded" with the results of the SQL command
# AllOut will become a list of all the records selected
AllOut = MyCursor.fetchall()
# print AllOut ## Debugging
# Print the header line
print "Vehicle\tDive\tDate\tDepth\tLat.\tLong.\tTemperature\tSalinity\tOxygen"

# Step through each record and create a new SQL command to retrieve
# the corresponding values from the other DB
for Index in range(SQLLen):
    # two dimensional indexing:
    # from the Indexed record, take the first item (the primary_key)
    Spec_id = AllOut[Index][0]

    # Other ways to print debugging information
#   vehicle,dive,date,depth,lat,lon =    AllOut[Index][1:]
#   print "%s\t%d\t%s\t%.1f\t%.4f\t%.4f\t" % AllOut[Index][1:]
#   vehicle,dive,date, depth, lat,lon,

# insert spec_id (the primary key) into each command
    SQL = """SELECT MIN(ctd.depth),ctd.temperature,ctd.salinity,ctd.oxygen
    from ctd, specimens where
    specimens.specimen_id=%d and specimens.vehicle=ctd.vehicle and
    specimens.dive=ctd.dive and ctd.depth>=specimens.depth ; """ % Spec_id

    # print SQL    ## Uncomment to test the command structure before running

    SQLLen = MyCursor.execute(SQL)
    NewOut = MyCursor.fetchall()
```

```
    if SQLLen < 1 or NewOut[0][0]==None:   # Some records don't have CTD data
        print  "%s\t%d\t%s\t%.1f\t%.4f\t%.4f\t" % AllOut[Index][1:] \
            + "NaN\tNaN\tNaN"
    else:
        print  "%s\t%d\t%s\t%.1f\t%.4f\t%.4f\t" % AllOut[Index][1:] \
            + "%.2f\t%.3f\t%.2f" % NewOut[0][1:]
# Close the files
MyCursor.close()
MyConnection.close()
```

In this example script, saved as `sqlmerge.py` in the `~/pcfb/scripts` folder, the results fetched from the SQL command are loaded into the variable `AllOut` using the `.fetchall()` function. This is equivalent to the file operator `.readlines()` which loads all lines of a file into a variable.

The basic approach for one of these retrievals is to execute the query and then to load the data from the `MyCursor` object:

```
MyCursor.execute(SQL)
AllOut = MyCursor.fetchall()
```

Instead of `.fetchall()` you could also use a loop and the `.fetchone()` method to retrieve one record at time from the `MyCursor` results. This would be a better approach for large datasets, as all the results aren't stored in the memory at once.

To print or otherwise access the contents of each line stored in `AllOut`, you use two indices in square brackets: the first shows which line to use and the second tells which field within that line you want to print. The order of the fields in `AllOut` correspond to their order in the first SQL command that we ran.

For instance, the second line of the `AllOut` variable would be `AllOut[1]`, corresponding to the specimen collected during Tiburon dive 515. Within this line, the fourth value `AllOut[1][3]` is the depth, so to print it to one decimal place, you could use:

```
print "%.1f" % AllOut[1][3]
```

From this first query, the script then uses the specimen information define a second query. This query is used to isolate a row of the CTD data that corresponds to the same dive and depth at which that specimen was collected. This approach lets you use information from one table to guide the extraction of information from another table.

Looking ahead

This is just the briefest view of data management with relational databases. There are many other ways to combine data across tables, and important best practices to follow to keep your database from growing unwieldy as it gets larger. If you would like to use databases in your research, we strongly suggest that you continue learning about them by following up this chapter with other resources, including the resources indicated at the end of the chapter.

Database users and security

All of the MySQL examples in this chapter have been executed as the `root` user on the `localhost` database system without password protection. This was done only to simplify the introduction to database systems. If you use databases in your research, and certainly if you are using a database with network access or on a shared computer, you should add password protection and create additional users with restricted privileges to access and modify the data. There are a couple of reasons for this. First, if you are logged in as `root` it is very simple to make a mistake that wipes out your entire database or makes widespread changes that you may not initially notice but that have a large effect on your data. If you log in as a user with restricted privileges, you create a line of defense against these mistakes. Second, if you provide network access to your database or it is installed on a shared machine, you will want to provide secure restricted access to the data. Even if someone isn't after your super-secret science data, it is an invitation to hackers to compromise your machine, whether to turn it to their own purposes or just because they enjoy wasting other people's time—as well as their own—with computer vandalism.

Creating a `root` password

SQL isn't only used for interacting with your data; it is also used to modify MySQL user and password settings. This is because the information about database system configuration and users is stored right in the database itself. The command for changing the password of an existing user is `SET PASSWORD`. If you don't specify a particular user when you issue the command, it changes the password of the user currently logged in. The following example changes the password for the root MySQL user to `mypass`, and then shows how to log in with the new password:

```
$ mysql -u root
mysql> SET PASSWORD = PASSWORD('mypass');
Query OK, 0 rows affected (0.83 sec)
mysql> EXIT;
Bye
$ mysql -u root -p
Enter password:
mysql>
```

Go ahead and change the password of the root MySQL user on your computer, but pick a more secure password than `mypass`. When you login the next time, add the `-p` argument to the `mysql` command. This will lead the `mysql` program to ask for your password, which will not be displayed as you type it (just press ⌐return⌐ when you are done). To login as the root user through the MySQL Workbench interface, you will also need to add the new password to the `localhost` connection you created earlier. This is done through the Manage Connections link at the lower left of the Workbench home screen. You would also have to add the password to your `sqlmerge.py` script, although that modification is better performed after setting up a restricted-access user, as described next.

Adding a new MySQL user

Here we will create a user account specifically for accessing the database from Python programs. This user can select, update, and insert rows into tables in the `midwater` database, but can't delete elements or alter the underlying structure of the tables or database. We will call this user `python_user`, and give them the password `ventana`.

The SQL command for adding a new user is `CREATE USER`, which specifies the user's name, password, and where they can connect to the database system from. By default, a new user generated with the `CREATE USER` command isn't allowed to do anything with any part of the database. Each privilege must be granted explicitly with the `GRANT` command.

Here are the commands to create the user `python_user` and allow this new user basic access to the database:

```
mysql> CREATE USER 'python_user'@'localhost' IDENTIFIED BY 'ventana';
Query OK, 0 rows affected (0.28 sec)

mysql> GRANT SELECT, INSERT, UPDATE ON midwater.*
    -> TO 'python_user'@'localhost';
Query OK, 0 rows affected (0.00 sec)
```

Within the `CREATE USER` command, the @ symbol separates the new user name from the network address it can connect from. In this chapter we only cover connecting to the local database, so the address is `localhost`. The password for the new user is specified with the `IDENTIFIED BY` clause.

The `GRANT` command requires a list of the commands you want the user to be able to execute. Note that `CREATE` and `DELETE` are not among the commands listed in this example, so the user can't create new databases or tables even though they can perform a variety of other functions. `ON` specifies which databases and tables these privileges apply to. The dot notation is used to specify the database and table, and the wildcard * can act is a stand-in for all databases or all tables within a database. To apply this command to all tables of the `midwater` database, you specify `midwater.*`. A series of `GRANT` commands can be used to apply different permissions to different databases and tables. TO specifies the user the GRANT command applies to, and, like CREATE USER, requires that the network location of the user is specified. This last point is worth noting, because if you allow a user access from one network address, they might not be able to connect when logged in from another, even as root or with a VNC connection. (See Chapter 20 for more on addresses and VPN.)

Now that you have configured a special user for Python programs to access the database, you can modify your scripts to connect with this user. Replace the existing MySQLdb.connect line in your script with the following:

```
MyConnection = MySQLdb.connect( host = "localhost", \
    user = "python_user", passwd = "ventana", db = "midwater")
```

Nothing else in the program needs to be changed. For a script that is just designed to retrieve data, it would probably be safest to define a basic user who only has SELECT privileges, so the database couldn't accidentally be altered or deleted by an error in your program.

It is a common practice to create special users for particular programs and languages to interact with the database, rather than for particular people. Typically for a web service, you would have a user with limited privileges who must be connecting from a specified location. It is important to keep in mind that both the username and password for this user appear in your programs. If you share your programs, remove or change this connection information. The password might occur in other scripts that could be opened by anyone with access to the computer. Don't recycle passwords that you have used for other purposes.

SUMMARY

You have learned:

- To organize files that contain two dimensional data, such as a spreadsheet or a character delimited text file, as a single grid for optimal analysis.

- That not all data are best represented in two dimensions, and that relational databases provide a good alternative for storing and analyzing these more complex data sets.

- How to:

    Install the relational database management system MySQL

    Create databases and tables

    Execute MySQL commands from within Python

    Add data to a database directly from a file

    Extract data from tables with the SELECT command

Extract and combine data across tables

Manage users and passwords

Moving forward

- Review the SQL commands summarized in Appendix 7 for quick reference.

- Investigate the many options for using the WHERE command at the MySQL help pages on the Web site.

- There are many ways to tackle bulk import of data files into a data table. In this chapter, the CTD data files were imported using a regular expression search to convert file names to SQL commands.

    Write a shell script to load the data into the table, invoked with a command such as `loadctd Marrus*.txt`.

    Write a Python program, perhaps derived from `filestoXYYY.py`, to perform the same import, taking advantage of functions in the `MySQLdb` library.

Recommended reading

The official MySQL developer page at `http://dev.mysql.com` is a great resource, with documentation and tutorials.

There are a number of excellent books on databases, MySQL, and SQL, including the ever-reliable O'Reilly reference books. We also suggest *PHP and MySQL Web Development* (Upper Saddle River, NJ: Addison-Wesley, 2008) by Luke Welling and Laura Thompson as a jumping-off point for learning more about the tools explored in this chapter. This book covers many web development topics beyond the scope of this book, but the introduction to MySQL and databases is excellent.

## Chapter 16

ADVANCED SHELL AND PIPELINES

In Chapters 4 through 6 you became familiar with basic shell commands and with collecting them together in a file to create brief shell scripts. Here we return to the command-line environment and present new programs and operations, along with other ways to join commands together to form time-saving functions. We also show how to use shell operations to link together programs into data processing pipelines. These semi-autonomous scripts will help automate your analyses and keep a record of processing operations for future use.

Additional useful shell commands

Extract lines with head and tail

You have already seen how the `cat` and `less` shell commands will show the contents of a file. Two related commands are `head`, which shows the first lines, and `tail`, which shows the last lines of a file. These are used to get a quick glimpse into the contents of a long file, or, when added to the end of a command following the pipe symbol, to show just a portion of the printout of a shell command. Both `head` and `tail` can be modified to show X lines by adding `-n  X` after the command name. For example, to see the first 5 lines of a data file:

```
head -n 5 ~/pcfb/examples/ctd.txt
```

Or, to see the 15 most recent items in your command history:

```
history | tail -n 15
```

Normally, the –n option used with `tail` will show you the last X lines. However, when a plus sign is added in front of X, it will show you the lines in the file starting with line X and continuing thereafter. For example, to skip a single header line:

```
tail -n +2 ~/pcfb/examples/ThalassocalyceData.txt
```

TABLE 16.1 Options for the `cut` command	
`-f 1,3`	Return columns 1 and 3, delimited by tabs
`-d ","`	Use commas as the delimiters, instead of tabs; this flag is used in conjunction with the –f option
`-c 3-8`	Return characters 3 through 8 from the file or stream of data

Both `head` and `tail` can accept more than one filename. The filenames can be specified either one after another, or via a wildcard—for example, `*.txt`. This makes for a convenient way to peek into many files at once.

Extract columns with `cut`

The `grep`, `head`, and `tail` commands slice out lines of interest, extracting horizontal selections from a file. To pull vertical columns from a file, on the other hand, you can use the `cut` command (Table 16.1). This operates based either on character position within the column when using the –c flag, or on delimited fields when using the –f flag. With the –c flag, numbers are given to indicate which characters to extract; with the –f flag, the numbers indicate which columns to extract.

To pull out the first few characters of lines containing the > character without displaying the > itself, you can use:

```
grep ">" ~/pcfb/examples/FPexamples.fta | cut -c 2-11
CAA58790.1
AAZ67342.1
ACX47247.1
ABC68474.1
AAQ01183.1
```

The `cut` command is most often used to extract columns of data delimited by spaces, tabs, or some other character, using the –f option. The number following –f indicates which field, list of fields, or range of fields to retrieve:

```
cut -f 2-4 ~/pcfb/examples/ThalassocalyceData.txt
Depth     Latitude   Longitude
348.7     36.71804   -122.0574
520.3     36.749134  -122.03682
118.36    36.83848   -121.96761
200.2     36.723267  -122.05352
100.85    36.726974  -122.04878
1509.6    36.584644  -122.52111
```

By default, `cut` expects tabs as the delimiter; it will not split on spaces without the use of the `-d` flag followed by a space between quote marks. Bear in mind that a consecutive sequence of spaces will not be treated as a single separator, but as multiple separators. Multiple spaces are often used, for instance, to justify columns of text that have entries with variable numbers of characters. You may need to reformat data to accommodate this behavior. You could, for example, open the file in your text editor and use a regular expression to replace one or more spaces, as designated with " +", with a single space, " ".

To indicate a comma as the delimiter, use `-d` followed by a comma, again with quotes around it, as is safest for shell operations whenever you want a punctuation mark interpreted as a string:

```
head -n 20 ~/pcfb/examples/ctd.txt | cut -f 5,7,9 -d ","
depth,temper,salin
2.78,15.299,33.132
3.47,15.3,33.133
8.64,15.298,33.134
15.29,15.295,33.134
21.84,15.003,33.155
...
88.93,10.614,33.439
95.28,10.403,33.464
101.53,10.188,33.486
107.68,10.03,33.539
```

Sorting lines with `sort`

The output stream produced by any of these commands, as well as the lines of a file or of a series of files, can be sorted into alphabetical order with the built-in `sort` command (Table 16.2). By default, sorting starts with the first character of the line and the first column of data:

```
grep ">" ~/pcfb/examples/FPexcerpt.fta | sort
>Anthomed
>Avictoria
>BfloGFP
>Pontella
>ccalRFP1
>ccalYFP1
>ceriOFP
>discRFP
>ptilGFP
>rfloGFP
>rfloRFP
>rrenGFP
```

TABLE 16.2 Options for the `sort` command

`-n`	Sort by numeric value rather than alphabetically
`-r`	Sort in reverse order, z to a or high numbers to low numbers
`-k 3`	Sort lines based on column 3, with columns delimited by spaces or tabs
`-t ","`	Use commas for delimiters, instead of the default of tabs or white space
`-u`	Return only a single unique representative of repeated items

To sort based on other columns (whether separated by tabs or spaces), use the `-k` flag (possibly standing for "kolumn"?) followed by the number of the column you wish to use for sorting. Note that because values are sorted in ASCII order (see Appendix 6), blanks come alphabetically before the letter A, so if there are empty values in a column of a tab-delimited file (in other words, two tabs in a row), those fields will be sorted to the top of the list. Another behavior is that all capital letters come before lowercase letters, so capital z is alphabetically before lowercase a.

Sorting can proceed numerically instead of alphabetically when the `-n` flag (**n**umeric) is used. Compare the output of commands sorting by the second column of data, first without and then with `-n` :

```
tail -n +2 ~/pcfb/examples/ThalassocalyceData.txt | sort -k 2
Thalassocalyce  100.85  36.726974  -122.04878  2.30  1999-08-09
Thalassocalyce  118.36  36.83848   -121.96761  1.52  1999-05-14
Thalassocalyce  1509.6  36.584644  -122.52111  0.95  2000-04-17
Thalassocalyce  200.2   36.723267  -122.05352  1.63  1999-08-09
Thalassocalyce  348.7   36.71804   -122.0574   1.48  1992-03-02
Thalassocalyce  520.3   36.749134  -122.03682  0.52  1992-05-05

tail -n +2 ~/pcfb/examples/ThalassocalyceData.txt | sort -k 2 -n
Thalassocalyce  100.85  36.726974  -122.04878  2.30  1999-08-09
Thalassocalyce  118.36  36.83848   -121.96761  1.52  1999-05-14
Thalassocalyce  200.2   36.723267  -122.05352  1.63  1999-08-09
Thalassocalyce  348.7   36.71804   -122.0574   1.48  1992-03-02
Thalassocalyce  520.3   36.749134  -122.03682  0.52  1992-05-05
Thalassocalyce  1509.6  36.584644  -122.52111  0.95  2000-04-17
```

When sorting is done alphabetically, the value `1509.6` falls between `118` and `200`; however, when it is done numerically, this same value is placed at the end of the list where it belongs.[1]

Isolating unique lines with `uniq`

Another powerful and frequently used command for extracting a subset of values from a file, or summarizing a stream of text, is `uniq` (Table 16.3). This command removes consecutive identical lines from a file, leaving one unique representative.

---

[1] Graphical interfaces for different filesystems don't all sort numbers in filenames the same way: some sort alphabetically, while others sort numerically. Now and then it can pay to know which of these two methods your particular operating system favors.

In order to be removed, the matching lines have to occur in immediate succession, without any intervening different lines. To get a single representative of each unique line from the entire file, in most cases you would need to first sort the lines with the `sort` command to group matching lines together. (The `sort` command actually includes a flag to return only unique records (–u), but `uniq` has other capabilities that make it useful in its own right.)

TABLE 16.3 Options for the `uniq` command

-c	Count the number of occurrences of each unique line
-f 4	Ignore the first 4 fields (columns delimited by any number of spaces) in determining uniqueness
-i	Ignore case when determining uniqueness

The `uniq` command can be used with the –c flag to count the number of occurrences of a line or value as it consolidates them. This gives a quick way, for example, to assess the number of occurrences of each taxon in a data file, or the most common annotations in a table.

Combining advanced shell functions

The output from a shell command can be sent to the screen or captured to a file. It can also be piped into another shell command, forming a chain of operations that can distill heavily processed values from a complex file. In this example, the starting point is a PDB file that describes the three-dimensional position of each atom and amino acid of a protein. We will build a combined shell command using `cut`, `sort`, and `uniq` to extract a table from this complex file, showing the frequency of occurrence of each amino acid.

To begin, move into the `examples` directory and take a look at the first few lines of each of the `*.pdb` files, using the `head` command:

```
host: lucy$ cd ~/pcfb/examples
host:examples lucy$ head -n 2 *.pdb
==> structure_1ema.pdb <==
HEADER    FLUORESCENT PROTEIN                       01-AUG-96   1EMA
TITLE     GREEN FLUORESCENT PROTEIN FROM AEQUOREA VICTORIA

==> structure_1g7k.pdb <==
HEADER    LUMINESCENT PROTEIN                       10-NOV-00   1G7K
TITLE     CRYSTAL STRUCTURE OF DSRED, A RED FLUORESCENT PROTEIN FROM

==> structure_1gfl.pdb <==
HEADER    FLUORESCENT PROTEIN                       23-AUG-96   1GFL
TITLE     STRUCTURE OF GREEN FLUORESCENT PROTEIN

==> structure_1s36.pdb <==
HEADER    LUMINESCENT PROTEIN                       12-JAN-04   1S36
TITLE     CRYSTAL STRUCTURE OF A CA2+-DISCHARGED PHOTOPROTEIN:

==> structure_1sl8.pdb <==
HEADER    LUMINESCENT PROTEIN                       05-MAR-04   1SL8
TITLE     CALCIUM-LOADED APO-AEQUORIN FROM AEQUOREA VICTORIA
```

```
==> structure_1sl9.pdb <==
HEADER    LUMINESCENT PROTEIN                      05-MAR-04    1SL9
TITLE     OBELIN FROM OBELIA LONGISSIMA


==> structure_1xmz.pdb <==
HEADER     LUMINESCENT PROTEIN                     04-OCT-04    1XMZ
TITLE      CRYSTAL STRUCTURE OF THE DARK STATE OF KINDLING FLUORESCENT
```

Note that the filename, placed between ==> and <==, is specified before each pair of lines since a wildcard is used to specify more than one file.

The following example will use the file `structure_1gfl.pdb`, but the resulting command can be applied to any of the other PDB files as well. Open the `structure_1gfl.pdb` file from the `examples` folder in a text editor to see the contents. You can interact with a 3-D rendering of the structure online at `tinyurl.com/pcfb-gfp`.

The protein in the example file is a dimer, having A and B subunits of the same molecule. We will create the shell command so that it only gives the results from one of the subunits, but you could just as easily have it return the count of all amino acids from both subunits of the molecule.

Notice that there are many introductory lines and remarks at the top of the file, as well as a few lines at the end, which do not contain the amino acid information we need. In the first pass at isolating the information we do need, we will extract only lines containing the word ATOM using a `grep` command (Figure 16.1). This eliminates many of the irrelevant lines, but leaves in remarks which contain the word ATOM. To remove these particular remarks, pipe the output of the first `grep` to an inverted `grep -v` command, which returns only lines that do not contain the word REMARK:

```
grep ATOM structure_1gfl.pdb | grep -v REMARK
```

Note that you only have to indicate the name of the file in the first command, since the subsequent commands operate on the output of the previous command. In this way you create an ever-more-reduced set of results—in this case, all the lines of the file that contain the word ATOM and not the word REMARK. The maroon area at the top of Figure 16.1 shows an excerpt of this stage.

At each point in this process, to see more clearly the results of the intermediate steps, you can append `| less` or `| head` to the end of the command. This way you won't see all the lines of output, just the first or last few.

After the two `grep` commands, each line of output now contains an atom associated with an amino acid (ALA at the beginning, THR at the end) and the sequential number of that amino acid. You can see that each amino acid is listed across several lines because each of its atoms is listed, but we will want to remove these repeated entries and just leave one line representing each amino acid.

```
grep ATOM
    REMARK   3  NUMBER OF NON-HYDROGEN ATOMS USED IN REFINEMENT.
    REMARK   3    PROTEIN ATOMS              : 3650
    REMARK 470   M RES CSSEQI  ATOMS
    REMARK 500 RMS DISTANCE OF ALL ATOMS FROM THE BEST-FIT PLANE
    REMARK 500 RMSD 0.02 ANGSTROMS, OR AT LEAST ONE ATOM HAS
  grep -v REMARK    ALA A    1    -14.093  60.494  -9.249  1.00 42.10
    ATOM      2  CA  ALA A    1    -14.989  61.651  -8.981  1.00 41.80
    ATOM      3  C   ALA A    1    -14.809  62.769 -10.006  1.00 41.60
    ATOM      9  O   SER A    2    -11.264  62.734 -12.155  1.00 39.50
    ATOM     10  CB  SER A    2    -13.236  65.292 -11.216  1.00 39.90
    ATOM     11  OG  SER A    2    -12.004  65.880 -11.497  1.00 39.90
    ATOM     12  N   LYS A    3    -12.516  63.462 -13.894  1.00 38.90
    ATOM     13  CA  LYS A    3    -11.712  62.828 -14.936  1.00 38.10
    ...                      ...
    ATOM   3644  CD1 ILE B  229     37.302  62.306   9.573  1.00 42.30
    ATOM   3645  N   THR B  230     39.340  65.048   4.879  1.00 48.70
    ATOM   3646  CA  THR B  230     39.969  64.839   3.566  1.00 50.40
    ATOM   3647  C   THR B  230     41.207  63.924   3.637  1.00 51.30
```

cut -c 18-21, 24-26	sort | uniq	cut -f 1 -d " "	uniq -c	sort -nr
ALA 1	ALA 1	ALA	9 ALA	21 GLY
ALA 1	ALA 37	ALA	7 ARG	19 LYS
ALA 1	ALA 87	ALA	13 ASN	18 LEU
ALA 1	ALA 110	ALA	17 ASP	17 VAL
ALA 1	ALA 154	ALA	2 CYS	17 ASP
SER 2	ALA 179	ALA	7 GLN	15 THR
SER 2	ALA 206	ALA	15 GLU	15 GLU
SER 2	ALA 226	ALA	21 GLY	13 PHE
SER 2	ALA 227	ALA	9 HIS	13 ASN
SER 2	ARG 73	ARG	12 ILE	12 ILE
...	...	...	18 LEU	11 SER
ILE 229	VAL 68	VAL	19 LYS	10 TYR
ILE 229	VAL 93	VAL	4 MET	10 PRO
ILE 229	VAL 112	VAL	13 PHE	9 HIS
THR 230	VAL 120	VAL	10 PRO	9 ALA
THR 230	VAL 150	VAL	11 SER	7 GLN
THR 230	VAL 163	VAL	15 THR	7 ARG
THR 230	VAL 176	VAL	1 TRP	4 MET
THR 230	VAL 193	VAL	10 TYR	2 CYS
THR 230	VAL 219	VAL	17 VAL	1 TRP
THR 230	VAL 224	VAL		

FIGURE 16.1 The successive extractions and modifications made by each command in the example pipeline Orange boxes show bash commands and other boxes show the output once those commands have been added to the pipeline.

The first step in doing this will be to use the cut command to extract just the amino acid three-letter code and the numerical position, characters 18 to 21 and 24 to 26. (The intervening A or B indicates which repeated subunit includes that amino acid.) When you create the `cut` command, as we will do next, be sure not to leave any spaces between the numbers, dashes, or commas.

Pipe the output of the two `grep` commands into the `cut` command to get the results shown in the first gray column of Figure 16.1. At this point, each amino acid is represented by a few identical lines, for example `ALA 1` or `THR 230`. Because of the `A` and `B` subunits, there are actually two sets of `ALA 1` lines in the output, one at the beginning and one near the middle of the list. To bring these next to each other, we'll first `sort` the output and then use the `uniq` command to remove all repeated instances of each amino acid. (The number next to the name prevents all occurrences of a particular amino acid from being reduced to a single mention at this point.) The command so far consists of:

```
grep ATOM structure_1gfl.pdb | grep -v REMARK | cut -c 18-21,24-26 |sort|uniq
```

Now there is one line for each unique occurrence of an amino acid in the file. For instance, the nine lines of `ALA` at the top of the list indicate that there are nine total alanines in the sequence. To add up all of these occurrences, you will first cut out the first column, thus removing the numbers from the lines. In this case, instead of using `cut` based on character position, we will extract the first field (the first column) using space as a delimiter, by piping the output through this additional function:

```
| cut -f 1 -d " "
```

At this point we have a list of amino acid names that can be counted using the `uniq -c` command. This function will return two columns of data: first, how often an element is repeated, and second, the name of the element. As a final step, you can sort this list in reverse numerical order to place the most abundant amino acids at the top of the list.

The final command is given below, and the output is shown in the last column of Figure 16.1:

```
grep ATOM structure_1gfl.pdb | grep -v REMARK | cut -c 18-21,24-26 |
  sort | uniq | cut -f 1 -d " " | uniq -c | sort -nr
```

At the end of this chapter you will see how to turn this into a general function, so that you can just type the function name followed by a filename (`countpdb structure_1gfl.pdb`) to get a table of this processed output.

Approximate searches with `agrep`

By now, you are familiar with the `grep` command, which lets you search for lines within a file which either match or don't match a particular string. However, sometimes it is helpful to be able to search for a *nearly* exact match of a string. This can be achieved using a specially modified command called `agrep` (approximate **grep**), which lets you search for matches that include a number of substitutions, matches to entities that span more than one line, and matches to two alternate strings (Table 16.4).

Unlike `grep`, `agrep` is not installed by default. It can be downloaded for all major platforms using the appropriate link at the bottom of `en.wikipedia.org/wiki/Agrep`. (For OS X, choose the Unix command-line version.[2])

One use for `agrep` is to search through many protein or DNA sequences stored in a FASTA file to retrieve a short sequence which is a close match to a query sequence. Recall that FASTA files have sequence names on lines beginning with > followed by lines that contain sequence information.

The command `agrep -d "\>"` indicates that you wish to search in text blocks divided by > instead of by line endings. Instead of lines that match the specified term, the output will now be multiline blocks that contain that term. To avoid interpretation as a redirection symbol, the > character is in quotes as well as escaped with \. When a match is found using this delimiter, `agrep` outputs the sequence name and full sequence, rather than just the single line where the match occurred.

Although you can specify an allowed number of mismatches using a number as a flag, you can also have `agrep` return the best match or matches by adding two additional flags, `-B -y`.

To find the sequence with the best match for the amino acid fragment `CYG` in the file `FPexcerpt.fta`, for example, use the command:

TABLE 16.4 Some options for `agrep`*	
`-d "X"`	Use X as the delimiter between records rather than the end-of-line character
`-B -y`	Return the best match, without specifying an exact number; `-y` tells it to print this best match without asking
`-2`	Return results with up to this many mismatches between the query and the record; the maximum allowed is 8
`-l`	Only list filenames that contain a match
`-i`	Case-insensitive search

*See man `agrep` for a complete list

```
agrep -B -y -d "\>" CYG ~/pcfb/examples/FPexcerpt.fta
```

Additional `grep` tips

Because the `tab` key has several functions at the command line, it is difficult to use it as a search term. In such cases, you can use the `ctrl` V operator. When you press and then release `ctrl` V at the command line, it causes the shell to interpret the keypress that follows literally, rather than carrying out its operation. Even the `delete` and `return` keys will be converted into their text representations when they come after the `ctrl` V sequence. To see that the `return` key is equivalent to `ctrl` M (^M),[3] try typing:

Your keyboard may say `backspace`.

```
echo " ctrl v return "    ← In this case, you do type the quote marks before and after the other keystrokes
```

---

[2] See Chapter 21 for instructions on installing `agrep`.
[3] Recall from Chapter 5 that a caret (^) before a given character indicates that you hold down the `ctrl` key while pressing that character.

Using this method, in order to use tabs in a `grep` search, you can type "⌈ctrl⌉V ⌈tab⌉" and it will insert an invisible tab character into your search string.

Searching for negative numbers with `grep` is trickier than it might seem, because the character that follows a dash is usually considered some kind of modifier argument. For example, if you try to find all the lines containing `-8` with `grep -8` or `grep "-8"` or even `grep \-8`, it won't work. For this search to work, you have to use all your tricks and search with a quoted and escaped version:

```
grep "\-8"
```

Counting words and lines The `bash` command `wc` followed by a file name can be used to print out a count of the lines, words, and characters in a file. When placed by itself after a pipe symbol, `wc` will quantify the components of whatever output is piped to it.

Remember aliases?

At the end of Chapter 6 we gave a brief introduction to shell aliases. These are little shortcuts for commonly used commands. In this section, we present a few especially useful aliases. These examples will help demonstrate the syntax for defining aliases of your own, either at the command line or in your `~/.bash_profile` settings file. Defining aliases at the command line is a good way to create and test them, but unless defined in your settings file, they won't survive once you close your terminal window or log out of a session. If you have an account on a remote machine, you will want to define your favorite aliases in your settings file for that machine as well.

Our first example creates a shortcut for a command which prints a list of the ongoing processes, and then filters this list so that it shows only those processes which include your user name—in this case, `lucy`.[4] (One of those processes will be `grep lucy` itself!) This alias is convenient for finding runaway processes and halting them:

```
alias myjobs="ps -ax | grep lucy"    ← Show the processes matching user lucy
```

This next alias is useful if you regularly log into a remote machine using the `ssh` command (see Chapter 20 for details). You can create a shortcut of just a few characters, so as to save you from typing the full command each time:

```
alias sp='ssh -l lucy practicalcomputing.org'
```

Even the example for `agrep` described in the previous section can be turned into an alias:

---

[4] The `ps` command is discussed at some length in Chapter 20.

```
alias ag='agrep -B -y -d "\>" '
```

Notice that single quotes are used to set the boundaries of the alias definition, with double quotes nested inside to define the delimiter. This alias can then be called using the command:

```
ag CYG ~/pcfb/examples/FPexcerpt.fta
```

where `CYG` is the query sequence you are trying to find, and `FPexcerpt.fta` is the FASTA-formatted data file to search.

Many times you will want to look back through your recent commands to re-execute or copy one of them. With the alias shown here, you can type `hg  pcfb` to find all the occurrences of the string `'pcfb'` in your recent command history:

```
alias hg='history | grep '
```

Finally, to see a list of your currently defined aliases, you can type `alias` by itself. These alias shortcuts, while useful for frequently used operations, are limited in how complicated they can be. Not to worry: you'll learn next about shell functions, which allow you to write miniature programs within the shell to accomplish repeated tasks with ease.

Functions

To create smarter multiline commands than those possible with aliases, you need to use a different style of shortcut, called a **function**. The syntax of shell functions is relatively cryptic, so we will limit this discussion to showing you the most useful basics and providing a few examples.

Functions are defined using the following format, with each such definition being added to your `.bash_profile`:

```
myfunction() {
   first command
   second command
   last command
}
```

The first part of the definition is the function name—that is, what you will type to run it. This is followed by `( )` and an open curly bracket. You can also begin a function definition with this syntax:

```
function myfunction {
```

To replicate the functions of the dir.sh script you wrote in Chapter 6, for example, you could add the same commands to a function definition, here called listall:

```
listall(){
  ls -la
  echo "Above are directory listings for this folder: "
  pwd
  date
}
```

You could of course define this function by typing or pasting these lines at a shell prompt—but as with aliases created in this manner, the effect will only last until that shell session is closed. To have functions available each time you open a new shell window, you must add their definitions to your ~/.bash_profile (or .profile or .bashrc for Linux). In your .bash_profile, the definitions of these functions are written in the same manner as if they would be typed at the command line, so there is no need to include a #! line as you do with shell scripts.

All of the functions described in this section are available in the file ~/pcfb/examples/scripts/shellfunctions.sh. Within shell functions, the echo statement is the equivalent of print in Python. This is used here in the listall function to print out a plain string of quoted text. The shell can also have variables, traditionally named with all capital letters. Remember in Chapter 6 how you set the value of the PATH variable:

```
export PATH="$PATH:$HOME/scripts"
```

In that case, you were setting a new value of PATH while reading the existing values of the variables $PATH and $HOME. In the shell, a variable name with a $ is *read from*, while the same name without a $ is *assigned* a value.

At the command line, try typing:

```
echo $HOME
```

This is the command-line way of saying "Print the value of the variable HOME." System-wide variables such as HOME, PATH, and USER are available within your shell functions. You can also use the special variables $1, $2, and $@ to represent user arguments that follow your command. $1 is a variable containing the first user-specified argument, $2 is the second one (separated from the first argument by a space), and $@ contains all the arguments together.

Create a simple function by typing these next lines in a shell window. Notice that you see a > at the prompt after having typed the first line; this indicates that the shell is waiting for you to finish your function definition with a close bracket:

```
repeater(){
  echo "$1 is what you said first"
  echo "$@ is everything you said"
}
```

With these lines, you have created a miniature shell script that is able to operate based on user input. Test out your function at the command line with some varied input:

```
host:~ lucy$ repeater 1 2 3
1 is what you said first
1 2 3 is everything you said
host:~ lucy$ repeater "1 2" 3
```

Notice that this works exactly like the sys.argv[] variable in Python, as described in Chapter 11.[5]

An immediate use for this would be if you had a frequently used shell command where the filename was followed by a long string of additional values. Typically, only the filename changes each time you use this command, but you can't easily use the ⬆ key to edit the command line, because the change is not at the end.

For example, phyml is a program that infers maximum-likelihood trees from genetic sequence files. It takes the data filename first, followed by a long string of options:

```
phyml sequencesA.fta 1 i 1 100 WAG 0 8 e BIONJ y y
```

When using this command again on different datasets, you would specify the same options, but the sequence filename would change with each use.

For greater convenience and less typing, wrap the command within a function definition and replace the filename with $1:

```
myphyml(){
  echo "myphyml $1 performs 100 bootstraps on AA data"
  phyml $1 1 i 1 100 WAG 0 8 e BIONJ y y
}
```

---

[5] Similar to Python's sys.argv[], the zeroth argument in bash, $0, is the name of the shell script being run. In the case of a function, this is bash itself.

Now you can invoke this function by typing just the following, and the filename (first parameter) will be inserted at the position occupied by $1:

```
myphyml sequencesA.fta
```

If your workflow requires other programs to be run on the output of phyml, you can add lines to your function to invoke those operations and accomplish all of the processing steps with a single command.

What if you want to optionally specify the number of bootstraps to run (currently occupied by the number 100) as a second parameter in the function? You can use the shell's version of an if statement:

```
phymlaa(){
   BOOTS=100
   if [ $2 ]
   then
      BOOTS=$2
   fi
   phyml $1 1 i 1 $BOOTS WAG 0 8 e BIONJ y y
}
```

Notice that the end of the if statement is marked by fi (a backwards if). Indentation does not have special meaning in shell scripts, but do not use tabs in the body of the text if you are going to paste them into the command line—otherwise, the shell will try command completion just as if you were typing them. Spaces are critical within the lines of many shell commands: they must be omitted from assignments like BOOTS=$2, but must be present between the [ ] and $2 or in any logical tests. The open square bracket [is actually a link to the test program, so to get help on logical tests, use the command man test. Comments in the shell language begin with #, just as they do in Python.

In this function, you start by setting a default value for the variable BOOTS. If the user only enters one parameter (the filename) then the default value of 100 is used. If they enter a second parameter $2, this is inserted into the command where the variable $BOOTS sits.

Notice the formulation of the if statement, where square brackets contain the logical expression. In this case, the test is just whether $2 exists, to check for a second argument to the user input.

Even if you don't have the phyml program installed, you can test this function by replacing it with an echo command followed by the remainder of the line in quotes. Try different inputs on the command line to see how this affects the printing of the output lines. Using the echo command is a good way to test your scripts before actually deploying them with the power to overwrite other files.

We will now give some examples of shell function definitions, to give you a sense of what is possible.

Functions with user input

In some cases, you will want to pipe data from user arguments as input to a command in a function. To do this, you can `echo` the user input, add a pipe operator, and follow this with the name of the command. For example, if you have the program `blastall` installed locally on a server, and you wish to quickly do a BLAST search for a sequence that you have copied from a file, you can define this function:

```
myblastx(){
  echo "Blasting protein vs swissprot..."; date;
  echo $1 | blastall -d swissprot -p blastp -m 8 -i stdin
}
```

If you run this at the shell using the command:

```
myblastx "GKCPMSWAVLAPT"
```

then the `echo $1` statement will send the string to the `blastall` program as though it were read from a file.

A dictionary function

To look up words in the system dictionary, you can define a function to `grep` against the list:

```
function dict(){
    grep -Ei ${1} /usr/share/dict/words
}
```

Run this command using word fragments or `grep` wildcards:

```
dict kkee
dict noi | grep  ion
dict ^ct
dict p.z.z
```

Translating characters

To do a batch conversion of end-of-line characters from \r (carriage return) to \n (linefeed) as required by many Unix scripts, you can use the `tr` function, which translates one character to another in the data stream. In some of these examples, the bracket characters {} are placed around variable names, as in ${1}. This helps the script handle cases where the variable includes spaces:

```
unix2mac(){
# this line tests if the number of arguments is zero
if [ $# -lt 1 ]
then
    echo "convert mac to unix end of line"
else
    tr '\r' '\n' < "${1}" > u_"${1}"
    echo "converting ${1} to u_${1}"
fi
}
```

Looping through all arguments passed to a function

Shell functions can also include `for` loops. These operate on a series of space-separated values, the same way that `$1` and `$2` access consecutive values separated by spaces at the prompt. To loop through all the arguments given at the command line, you can use the basic syntax of:

```
for ITEM in $@; do
```

The list of items can also be anything that would be correctly interpreted at the command line. So to loop through all the text files in a directory, you can say:

```
for FILE in *.txt; do
  echo $FILE
done
```

In `bash` commands and functions, a semicolon ends the command in the same way that pressing [return] would. You can use this to join several commands onto a single line. The three-line script above can be rephrased as follows:

```
for FILE in *.txt; do echo $FILE; done
```

In fact, if you use [↑] to step back through your `bash` history, that is how the three-line operation will be presented. Adding other statements after the `echo` statement separated by semicolons can insert those commands into the loop.

The following function loops through a group of filenames indicated at the command line (whether typed individually or signified by `*.txt`) and renames them from `filename.txt` to `u_filename.dat`. Step through the example shown here to see how each file in succession is processed:

```
renamer(){
 # Edit the prefix and extension to change how this works
 EXT="dat"
 PRE="u_"
```

```
# test if there is one or more file name provided
if [ $# -lt 1 ]
then
    echo "Rename a file.txt list as $PREfile.$EXT"
else
    for FILENAME in "$@"
      do
         ROOTNAME="${FILENAME%.*}"

         cp "$FILENAME" "$PRE$ROOTNAME.$EXT"
         echo "Copying $FILENAME to $PRE$ROOTNAME.$EXT"
         done
    fi
}
```

In a shell function, the $@ represents the list of all arguments sent to the script. If you type *.txt at the command line, then $@ will be a list of all the matching filenames. The loop cycles through each parameter contained in the master argument $@. For each filename, the program strips off the extension using the unusual shell expression ${FILENAME%.*} The percent sign starts a search term which gives everything up to the last period and deletes what comes after it. Remember that * is the most inclusive wildcard in the shell, just as .* is in regular expressions. Here, the period is interpreted literally as that character, not as a wildcard. An equivalent operation acting on a full directory name, in which each element is separated by a slash, would be ${RESULT%/*}. This gives everything up to, but not including, the last slash. This is useful for processing directory names to find the enclosing folder.

This function uses the "root name" of the file as the basis for constructing a new filename by joining variables with other bits of text (PREROOTNAME.$EXT includes three variables and a plain text period).

Removing file extensions

Another use of the extension-removing ${VAR%/*} syntax is to find the name of a folder, given the full path to a file. For example, you might want to change into a directory containing a particular program—say, for example, where the grep program is stored. Normally, you would have to type which grep to find the name of the directory, and then cd and retype or paste the part of the path preceding the filename:

```
lucy$ which grep
/usr/bin/grep
lucy$ cd /usr/bin
host:bin lucy$
```

A quick shell function can accomplish these steps in one command: first, finding the location of a command, and then cd'ing to the path with the name of the program removed from the end:

```
whichcd(){
    RESULT=`which ${1}`
    cd ${RESULT%/*}
}
```

This function makes use of yet another shell convention: the backtick symbol ` (usually located near the [esc] key). This shows up in other programming languages as well, and it essentially represents the text that is output when performing a command in a shell window. In this case, if you type the command whichcd zip, it executes a which command on the zip program, and assigns the value /usr/bin/zip to the variable named RESULT. It then strips the program name off the end of the directory name and uses that for the cd command.

MAKING SURE TO ESCAPE Some function definitions will work when pasted at the command line, but they won't when loaded from your .bash_profile. The problem here is typically that characters like & or $ need to be escaped with \ when read from a file. If you are using characters like this in a function without success, try both escaping them and not escaping them, to see which way works.

Experiment with the ${%.*} at the command line by setting up a variable and then using echo to examine the output:

```
lucy$ TEST="/MyDir/MyFile.txt"
lucy$ echo ${TEST%.*}
/MyDir/MyFile
lucy$ echo ${TEST%/*}
/MyDir
```

Finding files

The find command is a way to search your computer for files that match certain criteria. Its syntax, though, is relatively confusing. To create a shortcut for one of its uses—searching all nested folders relative to the current location—you can define this function:

```
findf(){
    find \. -name ${1} -print
}
```

In this shortcut, the command will start at the current location (.) and use the first argument $1 as the name to search for. This provides a more intuitive interface for locating files with the find command.

Revisiting piped commands

Earlier in the chapter you saw how to extract a count of amino acids from a PDB-format protein structure file. To create a function from this command, simply replace the name of the file with $1 and insert it into a function definition. As in Python, you can use a backslash to escape out the end-of-line character and split the single long line into two lines:

```
countpdb(){
  grep ATOM $1 | grep -v REMARK | cut -c 18-21,24-26 | sort | uniq \
  | cut -f 1 -d " " | uniq -c | sort -nr
}
```

A more generic command to enumerate the unique items in a particular column of a tab-delimited table can be defined using:

```
countlist(){
  echo "### Type countlist followed by file name, then column number"
  echo "### Returns a sorted list of the unique items in that column"
  cut -f ${2} ${1}| sort | uniq -c | sort -n -r
}
```

Repeating operations with loops

Imagine you want to test the effects of changing a single parameter on an analysis, but that running through even a single operation takes an hour. Instead of checking on your computer every hour and relaunching the command, or generating a script with all of the commands listed in succession, you can use a shell function to loop through the series of parameters. The basic syntax of a shell for loop is:

```
for k in {1..10}; do
  echo $k
done
```

Note the brackets followed by a semicolon and the word do, to start the loop, then the word done to close the loop.

Now try two methods of creating numeric loops in a shell function. The first increases the value of the parameter x from 20 to 40, and saves the output of the `ABYSS` program in files named `contigs_20.fa`, `contigs_21.fa`, on up to `contigs_40.fa`:

```
for x in {20..40}; do
ABYSS -k$x reads.fa -o contigs_$x.fa
done
```

The second varies x from 30 to 45 by 5:

```
date;
for ((x=30; x<=45; x+=5)); do
    ABYSS -k$x reads.fta -o contigs_reads-k$x.fa;
    date;
done
```

Wrappers

A **wrapper** is a program that controls and expands the functionality of another program. At its core, the wrapper is calling the other program from the command line; however, the wrapper can take care of a variety of other tasks as well, such as reformatting input data, constructing complex strings of arguments (including creating fields that must be calculated), and parsing program output into a more convenient format. Wrappers can make using an existing program a bit more convenient but they can also enable entirely new analyses.

A combination of Python and shell commands can be an excellent way to build a wrapper. Even MATLAB programs can be executed in command-line mode, so they can be included in your automated scripts. The `exifparse.py` script shown here uses the `exiftool` shell program described in Chapter 19 to extract scale bar information that has been embedded in a series of electron microscope images; it also uses Python's `os.popen().read()` function to run a shell command and capture its output:

```
#! /usr/bin/env python
""" Generate a table of pixels per micron for TEM
images, using exiftool. Run from a folder containing
images or subfolders with images...

Requires the program exiftool to be installed and in the path:
http://www.sno.phy.queensu.ca/~phil/exiftool/"""

import os
import sys
```

```
DirList = os.popen('ls -F| grep \/','r').read().split()
DirList.append('./')
#print DirList

for Direct in DirList:
    sys.stderr.write("Directory "+Direct+'----------------\n')
    FileList=os.popen('ls ' + Direct +" |grep tif").read().split()

"""Exif data are in the format:
Image Description : AMT Camera System.11/18/09.14:12.8000.7.0.80.1.Imaging...
-79.811.-552.583..XpixCal=65.569.YpixCal=65.569.Unit=micron.##fv3
"""
    if len(FileList)==0:
        sys.stderr.write("No files found in "+Direct+'\n')
    else:
#       print "Found", FileList    ←Commented out
        for Path in FileList:
            # The statement below runs a bash command and captures the output
            ExifData=os.popen('exiftool '+Direct+Path +' | grep YpixCal').read()
            SecondHalf=ExifData.split('XpixCal=')[1].strip()
            NumberOnly=SecondHalf.split('.YpixCal')[0].strip()
            print Path +"\t"+NumberOnly
```

Thoughts on pipelines

Creating an automated workflow that stitches several programs together, or just does all the dirty work involved in performing the same operation repeatedly, can save a great deal of time. However, there are other benefits to automation that may be even more important in some cases. For one, you know that the task is being performed consistently without deviations in operator input from analysis to analysis. Your automation of a task also serves as a record of what you did. When it comes time to write up your analyses you have a record of how the programs were called, preserved in the very script that called them. This kind of detail is often difficult to document—especially since most biologists are better at keeping notebooks about how they collected their data than they are at recording exactly how the dataset was analyzed.

There is no single solution for automating every kind of task. Some programs are designed so that it is easy for other programs to talk to them directly.[6] Other programs are not designed to interact with other software, but still feature capabilities within the graphical interface that make automation possible. Unfortunately, some programs have user interfaces that make automation difficult or impossible; worst are those with a graphical user interface but no command-line interface, and which don't read external configuration files or have any built-in scripting abilities.

---

[6] When an interface is provided for a program specifically to ease direct communication with other software, this is called an API, or application programming interface.

In general, almost any command-line program can be tricked by another program into thinking that it is interacting with a person. Armed with your knowledge of shell and programming operations, you should be able to accomplish more with less effort.

SUMMARY

You have learned how to:

- Show just head or `tail` of a file
- Use `cut` to extract columns of text
- `sort` lines in a file
- Find and count the unique lines with `uniq`
- Define one-line aliases using `alias aliasname="`*`alias command`*`"`
- Create more involved shell functions using `functionname(){}`
- Create loops in the shell
- Create a wrapper that automates and facilitates the execution of an existing program

Recommended reading

"Bash shell scripting tutorial," http://steve-parker.org/sh/sh.shtml.

Taylor, Dave. *Wicked Cool Shell Scripts: 101 Scripts for Linux, Mac OS X, and Unix Systems.* San Francisco: No Starch Press, 2004.

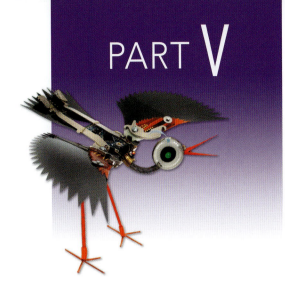

PART **V**

GRAPHICS

## Chapter 17

GRAPHICAL CONCEPTS

Many scientific data are inherently visual, and the ability to clearly communicate using images is central to preparing presentations and publications. Nonetheless, scientists are usually left on their own to learn how to navigate the many trade-offs and pitfalls of the tools, settings, and file types that are available for handling images, as well as how to make basic aesthetic decisions. These decisions can have profound effects on the ability to successfully and effectively communicate, yet scientists often unnecessarily compromise their images because they are unaware of these effects. Before examining the specifics of creating and manipulating images, we will therefore look at some general aspects of images which affect their effectiveness, quality, color, and file size. These considerations will influence how you generate, store, modify, and present your images.

Introduction

Within a relatively short time, image and graphic preparation methods have completely transitioned from Rapidograph pens, rub-on letters, and darkroom chemistry to computer-based illustration and editing. At the same time, journals have shifted the burden of image production to scientists, requiring publication-ready digital files. No longer can you submit a photo that looks right; it has to be a "CMYK image, with 300 DPI at printed dimensions, saved as a TIFF with LZW compression." These variations in the way graphical information can be formatted can be confounding; even journal editors and photo managers sometimes perpetuate misconceptions through the ways they describe image requirements. In this chapter we discuss the technical aspects of images that are relevant to scientific publication, and we introduce tools and techniques to prepare figures for print or display.

General image types

Vector versus pixel

When preparing artwork, the most fundamental consideration is whether it should be a **vector**-based or **pixel**-based image. Vector art is made of independent editable lines, curves, and shapes, all of which are defined by a few key properties. A pixel-based image, also called **bitmap** or **raster art**, is made of a uniform grid of colored dots, the pixels.

In vector artwork, a line can be defined by just two endpoints. If you specify a starting point and an end point, the line will stretch between them. Move one point and all the intervening positions of the line automatically follow. In pixel art, on the other hand, a line consists of many points of a particular color, arranged on the screen next to each other (Figure 17.1). In a way, the pixel line exists only because the colors adjacent to the line are perceived as different; there are no special "line pixels" as distinct from "non-line pixels," except that we see them that way.

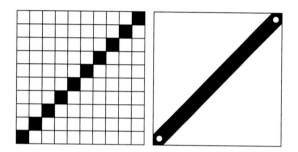

FIGURE 17.1 A 100-pixel (10 × 10) line and a two-point vector line Although it requires 1/50th the number of points, the vector line (right) is usually easier to edit and prints more clearly.

Consider the job of storing the information that describes a straight black line on a white background. For a vector line, you have to save the X and Y location of each end point, the color and width of the line, and the color and dimensions of the background. In this idealized example, there are about nine pieces of information required.[1] This amount of information stays the same regardless of what size the line appears on the screen or paper. It could be scaled up to a billboard and would still only require that handful of descriptors to recreate it. Now imagine the same situation for a pixel-based image. To estimate the amount of information it will take to record the line, you need to determine the dimensions of the image where the line will appear. Is this line on a 10 × 10 grid of pixels, or on a 100 × 100 grid? The information stored in a bitmap is the color of each point in the grid. For an image of just 10 × 10 pixels, this is 100 pieces of information; the color of each point has to be specified whether it is used to draw the line or not. For a 100 × 100 image (still relatively small), this value goes up to 10,000 pieces of information that must be stored. Fortunately, there are ways to compress this information and take advantage of the fact that many of those pixels are the same color.

There are dramatic differences between the ways that you work with vector and pixel art, and in most cases entirely different programs and file formats are used. File formats for storing vector-based images include PDF, EPS, SVG, AI, and

---

[1] For the moment, we are ignoring many subtleties, including that each color usually requires at least three values to describe it.

PostScript. These are discussed further in Chapter 18. Common pixel-based formats include JPEG, PNG, TIFF, BMP, and PSD, which are described in Chapter 19. Particular vector formats such as PDF, EPS, and AI can also store embedded pixel information, so they can accomodate hybrid images. When sending or transferring files, using the proper extension for your file names (`.pdf`, for example) can make it easier for the recipient to correctly identify the format.

A line in a vector-based document can be repeatedly moved, rotated, or scaled up to any size, as well as have its color and width changed, and it will always look perfectly smooth at any magnification. As you zoom in on a pixel-based image, however, the little colored squares that make up the image become more and more conspicuous, leading to jagged edges and other visual imperfections. It is also not always possible to manipulate the width, color, or position of an object once it has been created in a pixel-based world. As we saw above, not only can pixel art require thousands of times more memory, but to moderate this requirement, images are usually compressed in ways that can degrade their quality.

Deciding when to use vector art, pixel art, or both

So why does anyone use pixel art if it has so many potential shortcomings? For one thing, data are frequently collected as a grid of pixels to begin with. A photograph is the obvious example. Pixel-based images also have the advantage that they can contain a large amount of complex and interacting information, which would be hard to represent on a line-by-line, object-by-object basis. However, many scientists default to a pixel-based approach for generating images without having considered whether vector art is more suitable. This is often because they are most familiar with pixel-based art editors such as Photoshop.

A good guideline is that any graphical element you create de novo should begin as vector art and thereafter be preserved as such all the way through publication. Anything that originates as pixel art can be maintained as pixel art, unless it is more appropriate or informative as a vector tracing. Any annotations, arrows, letters, or graphs should be generated as vectors; in other words, if an image doesn't start as pixel art it should probably be created and maintained as vector art. Of course there are exceptions to these guidelines, and it is not hard to find examples of both photorealistic vector art and starkly graphical pixel art.

There are several other advantages to vector art. It is always possible to convert from vector to pixel art, but it is difficult to go the other way. Furthermore, pixel text in figures is not searchable, which means that words in pixel images will not be cataloged in search engines and will be missed by a user who is searching their document automatically. Pixel text can't be easily copied and pasted to another document, either.

Following these guidelines means that, in general, vector art will be used for anything with iconic graphical elements, such as graphs, tables, schematics of experimental setups, diagrams, and flowcharts. Text should be created and kept as vector art. Pixel art is necessary for photographs, gels, most 3-D images, and pictures with subtle tonal gradations or complex visual signals like blurriness or

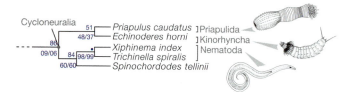
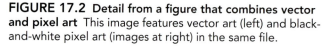

FIGURE 17.2 Detail from a figure that combines vector and pixel art This image features vector art (left) and black-and-white pixel art (images at right) in the same file.

grain. Pixel art might also be called for in plots with thousands of data points, where the vector objects begin to overlap and become cumbersome.

There are many times when it is best to use a combination of both vector and pixel information. Typically, labels for a plate of photographs should be annotated using a layer of vector art. Although the photos themselves cannot be scaled up without compromises, there are still advantages to having the annotations clear and scalable. It is possible to mix pixel and vector art in this way because most vector image file formats (such as PDF) allow pixel art objects to be embedded within them. You can, for instance, import a pixel image such as a photograph into a vector art file, and then label and annotate it with lettering and graphical elements that are themselves vector art. There is no need for all elements of an image to be pixel art just because one element is, so you can have the best of both worlds (Figure 17.2).

Ultimately, any computer-generated image that we see, whether on a display or on a printed page, has been converted to a pixel image to make it visible. Ideally this process, known as rasterization, happens at the last possible moment within the hardware generating the image—just before the ink hits the paper or the dot on the monitor is illuminated. When you are working on an image with vector content, you should defer this conversion to a bitmap as long as possible. This way you will end up with the appropriate resolution supported by the output device, while maintaining the ability to edit the image without loss of quality. Pixel art is overwhelmingly the image currency of the Web, although this may change with increasing adoption of the SVG (vector) image format, which uses XML syntax to describe vector images. In the meantime, it may sometimes be necessary to convert your vector art to an appropriately sized pixel image to post it online. Even so, it is best to delay this translation until all possible editing has been done, so that you maintain a vector version of the final image, and the pixel image is as sharp as possible.

Image resolution and dimensions

Pixel images are made up by a grid of colored pixels—for example, the 10 × 10 grid of dots described earlier. At some point you will have to deal with the question of resolution and size. There are three interconnected values associated with a bitmapped image's size:

- **Pixel dimension**: the number of pixels along the full X and Y axes of the image, for example 800 × 600 pixels

- **Physical size:** the size that the image appears on a printed page, such as 89 mm × 66 mm

- **Resolution**: the size of each pixel, expressed as the number of pixels per unit of physical dimension, usually called dots per inch (DPI) or pixels per inch (PPI)

If you know any two of these descriptors, you can determine the third via the formula: pixel dimension = physical size × resolution. However, of these, the only real dimension describing a bitmapped image is the **pixel dimension**. It is the pixel dimension that determines how much information is contained within an image, which in turn determines file size and the level of detail that can be represented. In effect, the resolution is just an arbitrary multiplier that is stored with the image file, so as to specify the size at which the image should be displayed.

The resolution, and therefore the physical dimension, can be changed at any time without fundamentally impacting the information that represents the image (Figure 17.3). For example, the actual pixels making up an image of a given pixel dimension could be a square micron or a square meter in size; either way, the information content of the image would be the same. Imagine projecting an image of our 10 × 10 line onto a screen and then moving the projector further from the screen. As you moved the projector away, the physical size of the pixels on the screen would increase (in other words, the resolution would decrease), as would the overall physical dimensions of the total image; yet even so, there would still be only 100 pixels being displayed.

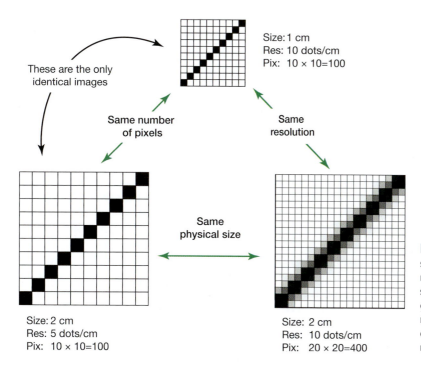

These are the only identical images

Same number of pixels

Same resolution

Same physical size

Size: 1 cm
Res: 10 dots/cm
Pix: 10 × 10=100

Size: 2 cm
Res: 5 dots/cm
Pix: 10 × 10=100

Size: 2 cm
Res: 10 dots/cm
Pix: 20 × 20=400

FIGURE 17.3 The relationships between physical size, resolution, and pixel dimension You can have images of different physical dimension and resolution, yet with the identical information content (pixel number).

Image resizing and the DPI misconception

Given these relationships, it should be clear why it is not informative to ask some-one for an image "at 300 DPI" without specifying the associated physical dimen-sions. At 300 DPI, the image could be 100 pixels wide, 30,000 pixels tall, or any other dimension. There is no way to know, yet journals commonly make such requests.

Before you resize an image, you should consider the largest physical dimen-sion at which you will conceivably use that image. Will it be considered as a cover photo for a magazine, used in a presentation, or only printed in the journal in one 89-mm wide column? Next consider the resolution required at this final dimen-sion (often 300 DPI or 120 dots per cm, but see below). Then multiply the expected physical dimension of the image by the required resolution to get the pixel size of the image. It is best to err on the side of a larger overall pixel dimension if you don't yet know your exact requirements.

The required resolution for an image depends on how it will be shown to the user. Typical resolutions are 72 to 100 DPI for images that will only be displayed on screens (such as web content), 300 DPI for printed color images, and 600 DPI for printed black and white images. The human eye can see finer detail on a printed page than on a screen, and printers typically have much higher resolution than dis-plays. This is why files destined for printing need higher resolution—a 90 DPI im-age that looks crisp on a screen will often seem blocky and fuzzy on the page. The eye can also make out greater detail when the contrast of an image is higher, which is why images need to have higher resolution when they are in black and white.

You can safely change the resolution or physical dimensions of a pixel image without impacting its quality or further limiting future manipulation—this is the equivalent of moving the projector closer and further from the screen. It is easy to undo these changes just by entering a new resolution or physical dimension. If, however, you make any changes that affect the pixel dimensions of the image—in other words, if you decide to resize it—the image will be **resampled** by the edit-ing program (Figure 17.4). There are two possible ways to resample. First, you can downsample your image, by reducing the image's overall pixel dimensions; if you do this, then you will be throwing away information, as existing pixels are com-bined into a smaller number of new pixels. Second, you can upsample the image, or increase the pixel dimensions; here you will be attempting to create more infor-mation through interpolation. There are very few circumstances where it makes sense to increase the pixel dimensions of an image—usually this just unnecessarily increases file size.[2]

If you will be performing any image adjustments to your file, such as adjusting the levels of brightness, do these before you resize the image. This is to avoid in-troducing artifacts, or imperfections that tend to be left over after processing. If you increase the pixel dimensions of an image it can amplify the artifacts that arise

---

[2]One circumstance where upsampling is sometimes necessary is when a photo editor insists on a "higher resolution" of a photo that is already of suitable pixel dimension.

from any earlier adjustments, and if you reduce the pixel dimensions, it will actually reduce the artifacts from adjustments that have been applied prior to resizing.

You should minimize the number of times an image is resampled, and defer resampling as long as possible. Resampling results in a loss of quality, due both to the information that is thrown away and to artifacts of the resampling process itself, and this damage can't be undone simply by returning the image pixel dimensions to their original values. The primary motivation for reducing the pixel dimensions of an image, apart from resizing it for final distribution, has historically been to minimize file size to save disk space and make manipulations less computationally intensive. These concerns are less pressing than they were even a few years ago, given the large performance improvements of personal computers. If you are going to send an image by email or embed it into a presentation, then resampling a copy is good practice. A full screen image for most projectors will be no

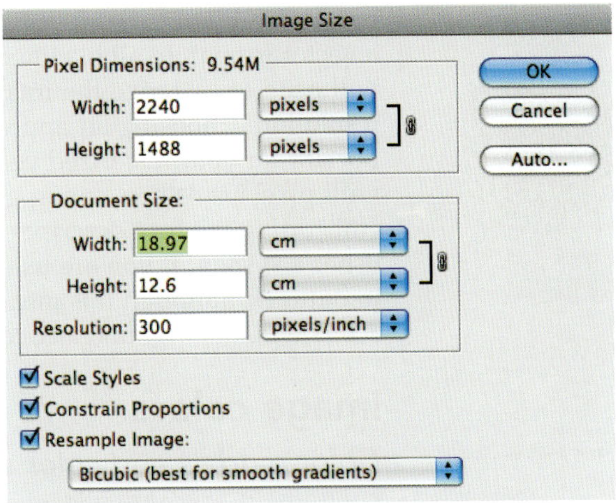

FIGURE 17.4 The Photoshop Image Size dialog box There are options for setting the pixel dimensions, document size, and resolution. Additional options are available to control resampling. If Resample Image is not checked, then your pixel dimension, and thus the true image size, will not change.

more than 1024 × 768 pixels, so you will probably get better results during your presentation if you use a duplicate that has been reduced in size to these dimensions. You will generally want to save a copy of the unmodified image before performing any size modifications or otherwise irreversible changes.

Typically the proportion of length to width is fixed during image resizing to avoid distortion.[3] If you resize the width to 50%, then the proportional value for the length should be determined automatically. In cases of resizing to fit a journal's column, don't worry about the height; just fix the horizontal dimension to the appropriate width. You may get better results resizing by integer multiples (for example, half-size or quarter-size) rather than irregular percentages (for example, 43%); this is because with the former, the program can in effect get rid of every other pixel.

Because vector art is largely resolution-independent, these resizing issues mainly become relevant when a vector image must be rasterized for use on a web page or in a presentation. Some vector resizing operations may not adjust the size of lines or text proportionally, so you can affect the overall look of your image.

A few journals are still reluctant to accept vector formats for submitted graphics, despite their superior printability. In these cases we have had some success

---

[3] There is an impressive method called "seam carving" which can resize images non-proportionally without distorting their content. This is available as **Content Aware Scaling** in the most recent versions of **Photoshop** and as liquid–rescale in the command-line tool **Imagemagick**.

contacting the person responsible for page layout, and seeing if they can handle a vector format. In all likelihood, they will welcome the improvement. If not, you will need to rasterize the image based on the guidelines above for dimension × resolution. Choosing an appropriate image compression format will also be important for your rasterized vector images, as described in Chapter 19 on working with pixel art. When saving graphics files in EPS format, you have the option of embedding the font descriptions into the image, to assure a match with the publisher's system. If you are using non-English typefaces, you can also convert text to outlines; although this means that the text will no longer be editable, the characters will look right on any computer.

Image colors

Color models and color space

In addition to deciding up-front whether an image should be pixel or vector art (or perhaps both), you also need to consider the document's **color model**. The color model establishes the basis of the color system used to describe your image. It essentially determines what the primary colors are and what happens when they are mixed. The two most common color models are **CMYK** (**c**yan, **m**agenta, **y**ellow, bla**c**k) and **RGB** (**r**ed, **g**reen, **b**lue). CMYK is the system used in operations that deal with ink and pigment and is named after the colors of ink that are used in printing. RGB is the standard descriptor for computer displays and digital photographs and is named for the three colors that each element of a display emits.

Many graphics programs allow you to set the color mode of each image file. In Photoshop, for example, this capability is located in the Image ▸ Mode menu, while in Illustrator it can be found in the File ▸ Document Color Mode menu. Other programs, such as Inkscape, work entirely within RGB.

The CMYK color model is based on the absorption of light by pigments. It is therefore a subtractive system, in which higher amounts of each pigment lead to darker colors. A white piece of paper with no pigment has CMYK values of [0,0,0,0], where each value is the percent of cyan, magenta, yellow, and black ink deposited. As you begin to add pigment, color emerges, and the overall brightness is decreased. A CMYK of [100,0,0,0] is cyan, while [100,0,100,0] is cyan plus yellow, resulting in a dark green. Even when you add in 100% of all the colored inks, they can't absorb all the light, so the resulting image is not truly black. This and other darker tints require the addition of black ink to the mix. You can see how there might be several ways to achieve a similar dark color, either by the addition of more of each colored pigment or by the addition of more black. This turns out to be a common problem when converting between color spaces, as we will describe shortly.

The RGB color model is better suited to devices like monitors and projectors, which emit light rather than absorb it. If you look at your monitor with a magnifying glass, or even put a tiny drop of water on it, you will see that what appears to be white is really three brightly lit colored pixels. Higher values indicate more photons of whichever of the three colors are being combined, until the maximum is reached for each of these values; added together, the maxima produce white. For this reason, RGB is called an additive color system. A larger range of colors can be produced by an RGB device, both in the sense of brightness and how saturated the colors look, than can be captured on a printed page (Figure 17.5). RGB values are usually given on a scale of 0–255. In RGB space, [0,0,0] represents black (all colors turned off), while [255,255,255] represents white (all colors turned on at their maximum value). Note that higher values in RGB signify more light, while higher values in CMYK signify more absorption (less light).

ENCODING IMAGE DATA The range of 0–255 for RGB values may seem arbitrary until you think about the binary values used to encode it. One byte of computer memory is 8 binary digits (or bits), and therefore can store values from 0–255 ($2^8 = 256$, with zero being the 256th value). So by dedicating one byte of memory to each of the red, blue, and green pixel values, it makes efficient use of memory in defining the color space. You will see that the number 256 recurs in many places in computer operations, as described in Appendix 6.

With the standard 8 bits per color, a total of 16,777,216 colors (2^{24} or 256^3) can be defined. You may hear about 16-bit or even 32-bit color, as supported by some cameras and image-editing programs. Here, two bytes (65,536 values) are used to define each RGB value, leading to a very large tonal range. With a high tonal range, the subtlety of gradations that can be captured for each channel is greatly increased; as a result, features like details in shadows can be seen. However, image compression methods such as JPEG can alter this range and noticeably affect the quality of an image.

Color space is the actual range of real colors, independent of how they are described. Neither RGB nor CMYK can describe every color in the full color space, which is why the color model you select can have a big influence on the appearance of your final image. RGB describes a larger portion of color space than does CMYK, and the colors that can be described with CMYK are a subset of those that can be described with RGB.

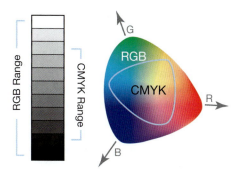

FIGURE 17.5 Schematic representation of color range In RGB, increasing values correspond to brighter pixels and thus lighter colors, while in CMYK they represent more ink, thus darker colors. An RGB model can represent colors that are both brighter and darker than possible in CMYK. Another way to look at it is that the 2-D color space represented in CMYK is a subset of that found in RGB.

Converting between color models

When converting a file from RGB to CMYK, colors in RGB that are out of the range of CMYK need to be compressed, or translated, to colors that are within CMYK. There is no one right way to do this, and the way this compression is done can have very different results. Publishers request CMYK files because they want to make sure that color conversion decisions meet author's intentions and needs. Color conversion can render a figure unintelligible if elements that had different colors in RGB are translated to the same color in CMYK (Figure 17.6). Photographs may no longer bear a reasonable likeness to the original subject (especially if they have saturated colors), or images may just appear ugly with a slightly different color scheme. Making the conversion before the files leave the author's hands avoids these problems and reduces the number of changes that would need to be made when proofing pages.

Bright colors and blacks in particular are difficult to translate from RGB to CMYK. Figure 17.6 demonstrates a particular case where the translation changes the meaning of the figure. We can't accurately show you an example of the RGB image on the printed page; however, Figure 17.6 is also available as an RGB image in `examples/MapColors.pdf`, so that you can see the actual colors on your computer display. Compare the three pairs of color values in the white boxes. Each is distinct in the RGB version (the red is somewhat subtle), but they are nearly indistinguishable in the CMYK transformations. When someone downloads a PDF of your paper and prints it out, this is how the image will look to them. Knowing this in advance, you can control the transformation of colors, and design your images (and your color schemes) to look good and convey information both on a display and in print.

When converting from RGB to CMYK, a variety of options are available that provide control over the translation process. The default settings in a photo editing program will often have adjustable limits on the amount of black ink the program assigns to a color, and this provides one of the most important controls for adapting the translation process to an image. Some programs like Inkscape don't support CMYK colors, though, and MATLAB can only export CMYK in certain formats (TIFF and EPS).

When working with photos it is best to stay in RGB-space for as long as possible. Many of the adjustments you might want to make are exclusively or more effectively implemented in the larger RGB color space. If your image is destined for display on the Web or in a projected presentation, you can leave it as RGB. For an image to be printed, however, it must ultimately be converted to amounts of ink on white paper, meaning the RGB values will probably need conversion to a CMYK color space.[4] If you are handling printing yourself, contact the prepress or art de-

---

[4]Some newer printers will accept RGB files directly, but you will still be relinquishing control over the conversion process. If you are printing a design with only a few colors, you can use process colors, such as the Pantone matching system. These special ink mixtures are reliably referenced to a book of printed swatches so that colors match exactly with your expectations.

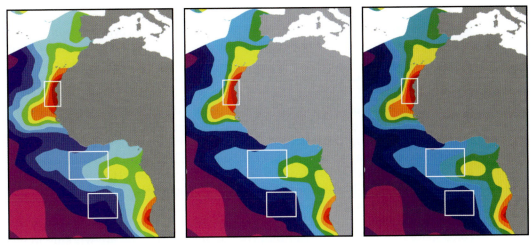

RGB CMYK
Coated paper

CMYK
Uncoated paper

FIGURE 17.6 The same image represented in different color spaces To see the true RGB color, open the example file `MapColors.pdf`. RGB is how the image appears on the screen, CMYK with coated paper is how it would print in a journal, and CMYK with uncoated paper is how it would appear on a laser printer. Given this selection of colors, information about the oceanographic features in the white boxes is lost due to color mapping issues.

partment to determine their requirements for output. They can inform you of the proper formats, color profiles, and color conversion settings for their facilities.

Color gamut and color profiles

The **color gamut** of a display is the number and range of different colors it can produce. A display with a high gamut can produce many colors, with fine gradations between them, while a display with a low gamut produces fewer colors. It can be difficult or impossible to distinguish small variations in color and intensity on a low gamut display, especially in the case of red or dark blue on a black background, and cyan or yellow on white backgrounds. One may spend hours working on a presentation on a standard computer display, and then stand up in a room full of people and realize that critical colors are indistinguishable on the projector. This is because many projectors have a shallow color gamut, while even inexpensive computer displays can show millions of colors.

Dealing with variable color gamuts raises many of the same issues as dealing with different color models—in both cases, the colors of an image are translated and compressed to a smaller part of color space. In fact, different color models are used precisely because they anticipate the reduced color gamut of particular output devices, such as printers in the case of CMYK.

Even though an RGB value of [255,0,0] represents "pure red," it will not look exactly the same on different displays. Similarly, 100% cyan ink will appear different depending on the paper and printing process used (see Figure 17.6). It would be quite cumbersome, however, if a different color model had to be used for every single model of computer display, projector, printer, and mobile phone screen.

Information about the color space of an output device can be stored in a standard format known as a **color profile**. Using color profiles, it is possible to specify how the same image will look on different types of output devices. If you know the color profile for a particular printer, for instance, it is possible to anticipate how an image will look when printed to it. In Photoshop, you can get a preview of what the image will look like in various target color spaces using the Proof Setup and Proof Colors items of the View menu.

In addition to color models and color profiles, it is also possible to specify the color space of a file. For present purposes, to insure that you are working in the largest possible color space, you should adjust your color settings in your graphics programs so that you are using the selection in Color Settings... called **Adobe RGB 1998** by default. Web applications cannot represent this larger color space, so web images should be viewed in the commonly used sRGB color space. As for converting a file's color profile, we won't address the mechanics of that here, but you can find out more by searching the Web and reading about "rendering intent." In Photoshop, conversion settings are found in the Edit ▸ Convert to Profile... menu option.

Color choices

At least 7% of males have some degree of color blindness, which usually affects the ability to tell red and green apart. For this reason, you should avoid using red and green to distinguish features of your figures. If you would like to use colors similar to red and green, the simplest fix is to use magenta, which can be differentiated from green, in place of red. You can check how your image will appear to the color blind at vischeck.com if you are uncertain about your color selection. The goal is not to create a figure that will look the same to everyone, but to make sure that the information that is conveyed through color can be perceived as distinct. Versions of Photoshop including CS4 and higher also have a filter built in for a color-blind simulator.

At the same time, you may want to consider how your image might appear in grayscale. Some journals offer the cost-saving option of printing the figures in grayscale in the journal itself, but retaining color for no additional cost in digital reprints. Another reason to consider the implications of grayscale is that people may also have to photocopy your paper from a journal. You should check the overall tonality of your color choices to see how they translate into lighter and darker grayscale values. Although there are many options for converting color to black and white, you can get a quick preview of this conversion using the same Convert to Profile... option used in Photoshop to convert from RGB to CMYK.

In addition to the RGB versus CMYK issues illustrated in Figure 17.6, you should also consider color gradients when choosing a color scheme. Color gradients are often used to convey information in a figure, and to be clear what values in your figure are higher than others, you should choose a color scheme with an intuitive progression from cool to warm colors (or dark to light). Some programs have default color palettes in which similar colors are used to represent the lowest and highest values, or in which the similar colors reappear through the palette. These cause the data to be difficult or impossible to interpret at a glance, or lead to confusing banding patterns in contour plots. See the file examples/ODVpalette.pdf for one particularly problematic color scheme. There are many online and desktop color palette generators that can help you create color palettes, including kuler.adobe.com, mypantone.com, colorbrewer2.org, and degraeve. com/color-palette.

HEX VALUES FOR COLORS You can refer to RGB colors by their numeric values (for example, [9,141,255]) but in editing programs and on the Web, you will often see color encoded using the corresponding **hexadecimal** values (in this case, #098DFF). This seeming gibberish is actually a straightforward way of representing the same information, but using three base-16 or hexadecimal digits (09, 8D, FF), rather than the normal base 10. (See Appendix 6 for an explanation of hexadecimal notation.) The maximum value of a two-digit hex number is FF, which represents 255. It is not a coincidence that this is also the maximum value for each RGB color. It is relatively easy and efficient to represent three numbers ranging 0–255 using only six characters, in a way that corresponds to how the computer stores the values in memory. You can see the hex values for colors in the "web palettes" of many color pickers, including the system color picker available in TextWrangler.

Summarizing the decision-making process

While there are many options and potential file formats for any particular graphics job, most of the decision-making process can be distilled into just a few points. They are recapped here, and portrayed in Figure 17.7 as well.

Use RGB vector art wherever possible, making conversions to CMYK and pixel art as necessary for final presentation. It is often useful to generate diagrams and graphs for publications using a CMYK color palette from the start, so that you have fewer surprises when your fluorescent green and hot pink color scheme becomes disappointingly non-distinct upon printing.

FIGURE 17.7 Graphical options for a variety of scientific uses

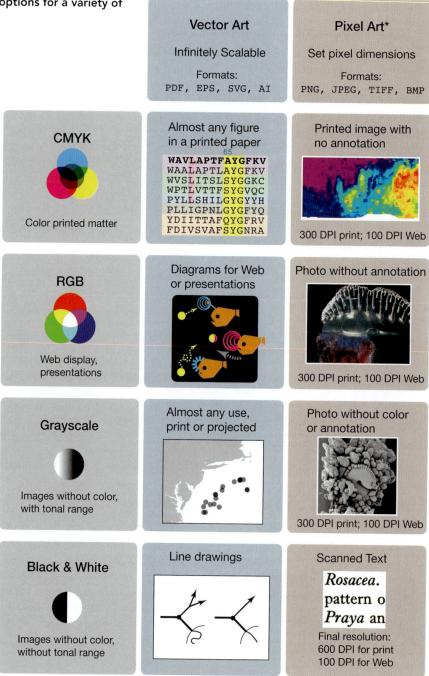

*Note that pixel art can be included as an object within vector art compositions, but the rules of resolution still apply

Layers

Layers in image editing programs can be thought of in a very physical sense, as if they were transparent pages in a sketchbook. Each graphical element is on one of these pages. The layers can be reshuffled to bring some groups of elements in front of others.

Layers are most useful, though, as containers for organizing the objects that make up an image. Often, text annotations will be on their own layer, graph data points on another, and graph axes or scale bars on another. You can place scratch work or other items you don't want to include in the final image in their own layer, and then make this layer invisible. This allows you to preserve reference images that help you remember how the image was generated, or to save rejected bits of a figure that you may want to use later, without managing different files. Layers can also be locked, so that you can edit some elements of an image without modifying other elements you wish to remain unchanged. Locking some layers also makes it easy to quickly select a subset of graphical elements from a complex image.

Even though layers greatly improve the efficiency of drawing, they are often underused by people new to image editing. When layers are used to their best advantage, a typical scientific figure may end up with a dozen or more. In fact, it is difficult to have too many layers. Layers are supported by all but the most basic graphics editors; however, they are not preserved when an image is saved to JPEG or PNG file formats. This means that maintaining layers while working on pixel images constrains your selection of file formats to either PSD (the default format in Photoshop format), TIFF (also available in Photoshop), or XCF (a format found in GIMP, an editing program discussed later Chapter 19).

General considerations for presenting data

There is, of course, much more to preparing an image than decisions about color space, vector versus pixel art, and file format. None of these technical issues will make a difference one way or the other if the image isn't designed to communicate effectively in the first place. There are a variety of aesthetic issues that arise repeatedly in scientific illustrations. When poorly addressed, these problems can make it difficult and time consuming for your audience to grasp what you are trying to say, or your audience may miss your point entirely. These issues are critical to keep in mind when you are acquiring image data, drawing new artwork from scratch, and laying out figures and slides.

Eliminate visual clutter

One of the most common problems with scientific illustrations is clutter—distracting graphical elements that don't add information. Extraneous graphics are sometimes added deliberately with the intent of making the image more appealing, while at other times they may be the passive result of not optimizing graphs or other artwork generated by an analysis program. Bar charts in 3-D might be common in annual reports, but are a liability for scientific papers (Figure 17.8).

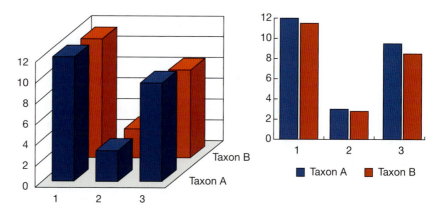

FIGURE 17.8 Bar graphs of the same data in 3-D and 2-D Although most scientists stick to the clearer and more informative 2-D versions, it is not uncommon to encounter 3-D plots produced by spreadsheets in the literature and especially in slide presentations. The plot at the left obscures the fact that the data from Taxon A, in blue, are slightly higher than the corresponding values from Taxon B.

Clutter is often added in an attempt to compensate for existing clutter. If you find yourself thinking that you need to add an element to an image to improve readability, think instead of ways to achieve the same goal by removing existing elements or reorganizing what is already present.

The default settings of spreadsheet programs can generate graphs with jarring combinations of visual elements: darkly pigmented background colors, extraneous borders, thin hairlines which don't print well, and a default panoply of colors for the lines and symbols of the graph which don't convey additional information.

Use transparency for overlapping data

It is often a struggle to represent large amounts of data on a single plot. The number of lines or points on the graph can obscure other values that lie behind them, leaving an amorphous blob of data that can be difficult to interpret (Figure 17.9A,C,E). In some cases, when the density of points is an important aspect of the data, it will be worthwhile to make a contour plot for the data—essentially a top-down view of a 2-D histogram—in which the color of the plot represents the density of points in the plot. An easier way to present many overlapping values is to make the values partially transparent (Figure 17.9B,D,F). This will allow the lines and points in the background of the figure to show through, so that where there are more data points, the colors will be darker. This imparts a histogram-like perspective onto a traditional 2-D plot, and it can turn a messy figure into a compelling one.

Make effective use of space

Screen presentations Whether you are laying out a slide that many viewers will see from twenty meters away, or designing a figure that is part of a paper with a firm

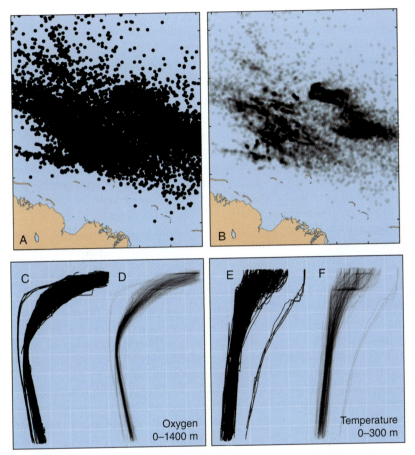

FIGURE 17.9 The added information available when using transparent points and lines in a plot The same data series are shown in each pair of plots (A and B; C and D; E and F), but structure which had been obscured in the opaque versions can be seen in the transparent versions. For example, in B, showing many thousands of geolocated whale calls, the area in the upper right has a higher density of call locations—a distinction that is not visible in A. In F, showing temperature profiles obtained during submarine dives, the depth where the vehicle sits and the thermometer equilibrates is visible as a horizontal band of dense coverage near the top of the profiles.

page limit, making effective use of space is critical. A common mistake is to bunch up elements in a figure, leaving areas of empty space that, if used, would have made the image much clearer without increasing its overall footprint. In presentations, scientists sometimes make graphical elements too small, then apologize by saying, "I know you can't read this, but..." When you are putting together your presentation, a good rule of thumb is to back up from the computer—at least three meters away—and see if you can quickly read and recognize text and labels on your slides. The projection screen for many audience members will be about the size of your fist held at arm's length, so view your screen from the same perspective.

Borders cut into the available image area, and should therefore be avoided. Three-dimensional shapes are sometimes used to present two-dimensional data, but this takes up added space without adding information. Edges and drop shadows can not only add visual noise, but may distort the conclusions of the data.

As discussed earlier, pictures used in slides of a presentation do not need to be much larger than 1024 × 768 pixels (thus, less than 1 megapixel in file size), although even camera phones can take images larger than 5 megapixels. The presentation will take more memory on disk and may stutter during projection if you don't use smaller copies of your photos. Some presentation programs can even support vector images, allowing you to project text and graphics crisply at any resolution when you embed PDFs directly into your slides.

Print figures Before starting work on what will be a printed figure, it is good to know the physical dimensions of the figure as it will appear in print. Nearly all journals provide figure size guidelines in their online instructions. When you first create a figure, pick the color space, set up a bounding box of the desired size, and create your figure to fit within that box. You can put the bounding box on its own locked layer (see Chapter 18) so that it doesn't interfere with your workflow. It's no use to work on details of your masterpiece filling a full sheet of paper, only to have it shrunken so those features become nearly invisible in the journal.

When operating within this constrained space, for most purposes the line size should be no smaller than 0.5 points. Hairlines generated by some programs are 0.1 pt, which is beyond the limit of the 0.35 pt thickness that most printers can render accurately. A line with a thickness of 1 point is 1/72nd of an inch wide, so 0.25 pt is 1/288th of an inch—roughly equivalent to two tiny pixels on a 600 DPI printer— and near the limits of visibility. Thin lines like these might look okay on your screen, but that is because the program is using a full screen pixel to represent something that is actually much smaller. View at about 300% to 400% magnification to get a more accurate idea of how things will look when printed. At this magnification, one pixel on the screen will correspond to one pixel printed at 300 DPI.

Given current practices for accessing scientific publications, the resolution at which the journal will print your article is almost irrelevant. Most users will either see the image on their screen, or print a PDF copy to their not-as-fancy printer, so aim for maximum visibility under those conditions. This is also a reason to try to get the figures embedded into your article as vector art rather than a bitmapped conversion: then they will look good at any magnification.

Consistency

You will often have little control over the original layout of artwork, which makes it difficult to have consistent colors and aesthetics across the figures in a paper or presentation. Artwork may be from a collaborator or generated by analysis tools that have limited display options. It may be from a program that has lots of options, but all of them are bad. By default, the lines may be too thick or too thin, the text poorly proportioned or in a different typeface, the colors inconsistent, and some graphical elements superfluous. To give your figures a consistent appearance and fix sub-par starting material, you will need to edit almost every image.

If you can get the artwork in vector format, you are in good shape. You can run it through a vector art editor (discussed in the next chapter) and adjust the appearance of each element as much as you like, so as to improve clarity and tweak the style (provided that this doesn't alter the scientific interpretation of the image, of course). If the program you are using doesn't provide an option for exporting vector art directly to a file, you can almost always generate a vector art file by printing the artwork to a PDF. If you can't obtain the original image in a vector format, it is often worthwhile to take the time to trace it, or add vector elements to the basic data plot. (Tracing will be explained in the next chapter.)

Consistency also refers to the relationship between graphical elements and the data they represent. It is confusing to the reader when in one figure, red is treatment and blue is control, but in a later figure, green is treatment and red is control. Because image editors give you complete control over the appearance of your plots, you are not constrained by the default settings used by any given plotting program. You can add information that may not be easy to display in the original program—for example, white circles to represent daylight data, black for night, and grey for crepuscular data. You can take different types of plots generated by several programs, all of which may have very different appearances, and edit them to create an integrated and consistent feel. This will help your reader focus on differences in data and results, rather than on puzzling stylistic elements.

Maintaining data integrity

Any tools that make it possible to prepare and edit scientific images also provide the opportunity to fundamentally alter the interpretation of the data. Editing graphs to eliminate extraneous information, such as irrelevant data series, or to alter the aesthetics of the graph, such as changing the color of data points, isn't usually a problem, but deleting or moving points within a series certainly would be. Likewise, cropping pictures usually isn't a problem, but erasing or adding photographic elements within them clearly would be.

Intentionally misleading your audience is, of course, outright fraud, but even without malicious intent, certain image modifications can unintentionally alter the interpretation of the data. Many journals now provide lists of best practices or else strict rules for how images (and pixel art in particular) may be modified. Be sure to pay close attention to these guidelines and rules. When it comes to pixel art, transformations that uniformly impact the entire image, such as changes in brightness or contrast, are usually not problematic. However, modifications of parts of an image, such as the use of the clone tool in Photoshop to clean up some elements of a picture, are often expressly prohibited—even for seemingly innocent tasks like removing specks and dust particles.

As opposed to nonuniform changes, fewer issues arise with uniform transformations; however, even these can sometimes be problematic. The most common issues arise when the saturation of an image such as a fluorescent micrograph is increased to a misleading point: either the noise is brought up to look like part of the signal, or low-level fluorescence is suppressed, making the signal appear artificially prominent. Keep track of the modifications you make, and if you think there is any chance that a modification might alter a reader's interpretation of an image, be sure to explicitly state the modifications in the legend.

Why you should avoid PowerPoint

You may have noticed that PowerPoint has not been mentioned anywhere, either as an editor or an image format, even though it is frequently used by scientists to edit and share images. This is not an accidental omission on our part. The primary reason you should avoid using PowerPoint for tasks other than presentations is that color modes, image compression and resolution, and other key factors are not immediately obvious when images are embedded in PowerPoint presentations, and file types and quality can be automatically converted in nonintuitive ways. If you send a PPT or Word file to a colleague, you won't know how it will appear on their screen or printer, and they will be largely unable to open the images it contains or integrate them into any other useful program. It is best to use editors and file formats that don't obscure these critical image attributes or add layers of unnecessary complexity.

If you need to create a multi-page document containing several figures, do not embed them into a Word or PowerPoint document because the figures will become uneditable. Instead, you can open one of the figures in Preview on OS X, or Adobe Acrobat on other systems. In Preview, if you show the sidebar, which contains thumbnails of the document's pages, you can drag other images into that container, rearrange their order, and save them together as a single PDF file.

SUMMARY

You have learned:

- The difference between vector and pixel art

- Why to use pixel art for photographic images, and prefer vector art for most everything else

- The relationships in pixel art between pixel dimension, physical dimension, and resolution

- Why to use RGB color for Web content, photos, and presentations

- What layers are, and what they can be used for

- General considerations for graphically presenting scientific data

Moving forward

- When you see distinctive graphical elements in journal articles, Web sites, or news stories, try to figure out what it is that visually works and doesn't work with them. Use these inspirations for your own visual communication tasks.

- Open some of your images in an image editor and see how changing the color space affects them.

- In your computer's display settings, for whatever operating system you are using, take note of the color profile you are using; then try changing it to see what effect the change has on how images look.

- View the gallery of data visualization approaches at `http://vis.stanford.edu/protovis/ex/`.

- Be inspired by the stunning graphs and uses of transparency at `tinyurl.com/pcfb-flights` and at `tinyurl.com/pcfb-eigen`.

Recommended reading

References are given in order from the most highly recommended, general texts at the top to the more specialized books at the bottom.

Tufte, Edward Rolf. *The Visual Display of Quantitative Information*. Cheshire, CT: Graphics Press, 2004.
This classic is required—not recommended—reading for any scientist.

Wainer, Howard. *Graphic Discovery: A Trout in the Milk and Other Visual Adventures*. Princeton, NJ: Princeton University Press, 2005.
A historical perspective on the progression of approaches to data graphics.

Nelson, Roger B. *Proofs without Words: Exercises in Visual Thinking*. Washington, DC: The Mathematical Association of America, 1997.
Elegantly mute graphical proofs of mathematical principles. Inspires novel ways to present relationships between data.

Cleveland, William S. *Visualizing Data*. Summit, NJ: Hobart Press, 1993.

Cleveland, William S. *The Elements of Graphing Data*. Summit, NJ: Hobart Press, 1994.
These two books present data in several different ways to highlight how the clarity of the message is dependent on its presentation. Available at `hobart.com`.

Steele, J., and N. Iliinsky, eds. *Beautiful Visualization*. Sebastopol, CA: O'Reilly Media, 2010.
A collection of chapters written by scientists explaining the methods and motivations for some of their complex graphics generated to display data. In some cases, the examples include the programs used.

Chapter **18**

WORKING WITH VECTOR ART

There are many software applications and file formats for generating and editing vector art. Here we introduce some of these applications, discuss key principles in creating and organizing vector elements of images, and explain the basic illustration tools that are used to create and edit vectors. This will allow you to modify and annotate vector art generated by other programs, begin drawing new images from scratch, and create tracings of photographs.

Vector art mechanics

Like some other topics covered in this book, there are dozens of guides that go into exhaustive detail about using various vector art editors. Because many of these attempt to explain every feature in minute detail, it can be difficult to sift out the parts that are most relevant to scientific illustration. Science curricula, however, rarely include instruction in using graphics programs, and usually skip the topic entirely. To help close this gap we focus here on the aspects of editing vector art that are most directly relevant to the scientific illustration tasks encountered by biologists.

File formats

The primary vector image formats are PDF (Portable Document Format), PS (Post-Script), EPS (Encapsulated PostScript), and SVG (Scalable Vector Graphics). AI (used by the program **Adobe Illustrator**) is also a common format, although proprietary. PDF is the most broadly implemented and frequently used of the vector art file formats, and **Adobe Illustrator** can also use PDF as one of its native formats. SVG is a completely open standard that is gaining popularity in web design and is supported by a wide variety of programs. In general, PDF and EPS files will address most of your needs. They are both supported across many different software applications, and accepted by most journals.

Generating vector art

After all the discussion in the previous chapter of why you should use vector art wherever possible, you might wonder how to generate it in the first place. There are three common ways to create vector art: (1) export an image, such as a graph or phylogeny, from another program, (2) draw an image from scratch, and (3) trace an existing piece of pixel art.

Exporting images from another program

Vector images can often be exported directly from your analysis programs. Many programs, including MATLAB, can export graphics straight to PDF format, and some can also save to other formats such as SVG and EPS. Even if there is no option to export images directly as vector art, PDF files can be generated from within almost any Print dialog box by choosing the option Save as PDF. As a last resort, exporting EPS or PostScript files, which are the formats understood by printers, is almost always an option even if PDF is not available on your system. Be aware, though, that just because a file is PDF format does not necessarily mean that the image has the best vector format. Remember that PDF files can store pixel images too, and some export methods, as described below, capture the vector properties of a graph better than others.

Choose Print to File and select EPS as the format.

In MATLAB, do not use the File ▸ Save As... dialog box to export a figure. Instead, with the desired figure in the foreground, use the command `print('-dpdf', 'myfigure.pdf')`, where `myfigure.pdf` is the name you would like to give your new image file. If you intend to edit the figure in Adobe Illustrator, use the same command, but with `-dill`, in place of `-dpdf`. This gives better results than PDF upon import.[1] Use the PNG format (`'-dpng'`) for exporting pixel images. In R, do not use the Save file option that is available in the figure window for your plots. Instead, before making your plot, use the command `pdf("myfigure.pdf")` to open a printer-like output "device." Once you are done with your plot commands and adjustments, close the file with `dev.off()`.

We run almost every figure through a drawing program prior to publication. It is often simpler and more effective to modify image color, line weights, and lettering with a drawing program than within the software that generated the image. Making these changes in an editor can improve clarity and stylistic continuity across figures. Sometimes vector art that is exported from analysis programs will have extraneous invisible elements that complicate editing,[2] and sometimes entities that appear as a single object are really complex compositions of many different elements. These situations create a bit of extra work, but usually they are relatively straightforward to handle.

---

[1] Some programs and formats will split lines and objects into many separate segments, or even divide text up into individual characters. In MATLAB, certain types of objects (arrows, ovals) are always exported as pixel objects even if you use a vector file format.

[2] In Illustrator, you can choose View ▸ Outline (⌘Y) to see the outlines of invisible and obscured elements for easy deletion.

Drawing new images

Vector images are commonly created by drawing them from scratch. This may not sound very easy, but because you can adjust the location and properties of each line and object after creating it, this approach is almost always simpler, and the results look better, than performing the same task in a photo editor. You could be tempted to fire up PowerPoint to draw vector art, and this might be okay for adding some As, Bs, and scale bars to a figure. However, if you limit yourself to this program, you will ultimately have little control over the final look of your image.

There are several subgenres of programs for drawing. If you are trying to create a diagram with 3-D aspects, you can generate images with correct perspective in applications like Blender, Cheetah3D, or the free Google SketchUp, and then export your image as vector format or trace it, as described below (Figure 18.1). For flowcharts, organizational diagrams, and images with many interconnected and repeated elements, programs like OmniGraffle can produce uniform vector diagrams with relative ease.

Tracing photographs

The raw data representing the results of a study may be a pixel image of an organism, a micrograph, or a photo of the experimental set-up. These images typically contain details that are irrelevant to what the picture needs to convey. In these cases, a simplified line drawing can be a more effective communication tool. This is why drawings, not photos, are favored by descriptive morphologists and systematists. The same principle applies to many other situations, such as depicting the setup of an apparatus. To go from a photo to a drawing, tracing the picture using a vector editor is an excellent approach.

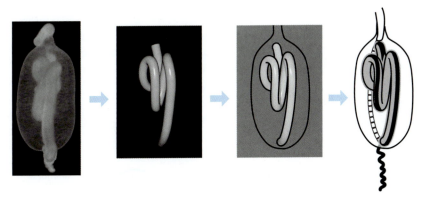

FIGURE 18.1 Stages of creating a 2-D schematic from a photograph of a complex 3-D object With a photograph as the guide, a general 3-D model is created of the tentillum from the siphonophore *Resomia dunni* and imported into a vector drawing program, where it is made partly transparent and then traced. Using an illustration allows you to highlight important elements that may not show up equally clearly in a photograph.

To start the process, scan or photograph the image, import it into your illustration program, and move it to its own locked layer. Then create drawing layers above the locked image and trace the image using the pen tool, creating Bézier curves as described in the next section. You can hide or delete the layer containing the original image when you are done. It can also be helpful to make the background image invisible or semi-transparent, to quickly check the overall look of your tracing. A drawing tablet is often more convenient to use for tracing and drawing than a mouse or trackpad, especially for bigger projects. You would not want to use a pixel art program such as Photoshop for tracing, as this would result in a simplified pixel image of the original pixel image, rather than a clean vector art representation.

Some programs have an auto-trace feature that can help automate vector tracing, though the results almost always require subsequent adjustment. The Live Trace feature of Adobe Illustrator lets you dynamically control the threshold for identifying which image boundaries should become lines. To create single outlines instead of filled objects, choose Technical Drawing from the Live Trace pop-up menu. Once you have an object you like, you can select Object ▸ Live Trace ▸ Expand to turn the trace into an editable outline.

Anatomy of vector art

Bézier curves

At its core, vector art is composed of anchor points and lines that connect anchor points. These lines are known as **Bézier curves** (pronounced bezz-ee-ay). Vector drawing programs have the standard suite of tools for creating boxes and straight lines, but to get full control over your illustrations, you will need to learn how to manipulate these curves. While they seem confusing at first (and they are hard to describe in words), once you understand them you will find it much easier to draw what you see in your mind.

A Bézier curve is a line that intersects a series of **anchor points**. Anchor points are sort of like pins stuck through a very flexible rod, which represents the line. The line must pass through each anchor point, and its shape is controlled by the

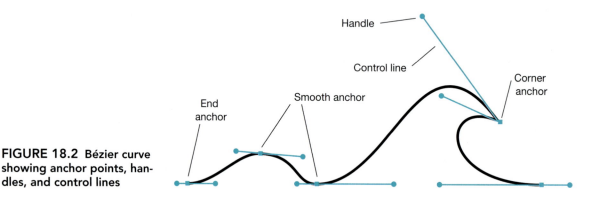

FIGURE 18.2 Bézier curve showing anchor points, handles, and control lines

number and position of the anchor points, as well as by control handles, which can be independently extended, retracted, and rotated about the anchor points (Figure 18.2). These handles control how the curve leaves the anchor point in each direction. Curves always intersect the anchor point tangent to (that is, parallel to) the control line. The length of the control line determines how "attracted" the curve is to that handle, as well as how far the curve extends along the handle in that direction. The handles for a given anchor point can be linked to each other by a straight line which passes through the anchor point (making a smooth intersection), or each control line can intersect at an angle, forming a sharp kink at the anchor point (making it a corner point).

Any two-dimensional shape can be composed of a series of Bézier curves. They therefore provide a general tool for creating and modifying vector elements. A Bézier curve can contain any number of anchor points along its path. It can also be either open, in which case it has two endpoints, or closed, in which case the starting and ending anchor point is the same. The property of being open or closed affects how the curve is filled with color.

Stroke and fill

Each line in a vector art image has several attributes associated with its **stroke** (the line itself) and its **fill** (the space bounded by the line). Both stroke and fill have color attributes, as well as transparency settings. The stroke has settings for width, which is given in points (a point is a standard printing unit that is usually defined as 1/72nd of an inch); whether it is solid or dashed; and the shape of the end cap of the line, which can be rounded or squared off. In addition, strokes can have paintbrush-like attributes, or can exhibit an ornamental pattern made of repeating elements. Fills are usually solid colors, but they can also possess transparency, patterning, or gradients.

Working with vector art editors

The different programs for drawing and editing vector art have similar workspace layouts and tool selections, but they vary in feature richness, stability, and ease of use. In the world of vector art there is a powerful (and expensive) program that can't be ignored: Adobe Illustrator. If you can afford to purchase Photoshop for your science, you should go the extra step and get Illustrator as well. (Many universities have site licenses and academic discounts.)

As an alternative, the open-source vector art editor Inkscape is gaining traction among graphics artists and web designers. There are a variety of features Inkscape does not yet support, such as CMYK color mode; the OS X version of the program has a very non-OS X user interface; and there aren't as many shortcuts as in Illustrator for speeding up common tasks. Even so, Inkscape is well suited to many of the tasks that it is designed to accomplish, and more than adequate for common graphical tasks. It is a good place to get started if you don't already have access to Illustrator, especially since it works on all platforms and can be downloaded for free from `http://inkscape.org`. Another free option is sK1; it includes a few technical features that Inkscape lacks, and is available from `sk1project.org`.

In this chapter, we provide an introduction to both Illustrator and Inkscape. They share many features, although these shared features are often implemented slightly differently. There are many features in both programs that we don't cover here, including some tools that exist in both programs but are only described in detail for one of the programs. This chapter is intended only to give you an initial taste of vector editing and get you headed in the right direction. You are encouraged to look at additional tutorials and resources once you get up and running.

Selecting and manipulating entire objects

The solid arrow ▲ (called the Selection Tool in Illustrator and the Select and transform objects tool in Inkscape) can select, deselect, move, rotate and resize entire objects or groups of objects.[3] This arrow is used to adjust how objects are arranged relative to each other, modify their proportions (making them taller or wider), or select one or more objects so that their properties can be changed. Open the file examples/LEDspectra.pdf to try some of the behaviors described next. Don't save your changes (or if you do, use a different file name), so that the files will be useful for future examples.

There are a couple of different ways to use the solid arrow. One is to simply click the object you would like to select. Depending on the structure of the image, the object could be a single line, a simple shape composed of multiple line segments, or a large complex set of objects that are grouped together. If you didn't create the image by hand, it is hard to expect what will be selected with this tool. Vector art exported by some programs is lumped together into a single large object, in which cases the solid arrow isn't useful for editing parts of the image. In that case you could ungroup the objects or use the second arrow described in the next section to select subsets of the group.

You will often want to select several objects, but if you click on one object and then another, the object that was selected first will be deselected. To get around this, hold the ⎡shift⎤ key as you click; this will add subsequent objects to those already selected. Holding the ⎡shift⎤ key while you click an already-selected object will deselect it. You can also select multiple objects by drawing a box around them with the solid arrow; this can be much easier than selecting many smaller objects. You can use these different behaviors of the solid arrow in combination to save time. You may, for instance, want to select most of the items in the region of an image, but not all of them. To do this, draw a box around the entire region with the solid arrow, then hold ⎡shift⎤ and deselect the items you don't want. This is often quicker than selecting the items you do want one-by-one.

Once one or more objects have been selected, the solid arrow can be used to drag them to change their location. It can also be used to resize the entire set of ob-

---

[3] If you select an object with this tool but don't get a rectangular frame around it, check that Show Bounding Box is selected in the View menu in Illustrator.

jects by grabbing and moving the corners and sides of the selected area. To maintain proportions, hold [shift] while resizing.

Selecting and manipulating parts of an object

The second type of arrow, the hollow arrow ⬉ in Illustrator (the Direct Selection
tool) and the elongate triangle in Inkscape (the Edit paths by nodes tool), selects
and modifies parts of an object. Most newcomers to vector illustration get confused as to why there are two different types of selection tools and stick with the
solid arrow, not realizing that the **second arrow** (as we refer to this tool here) is
more convenient for most jobs. In fact, the second arrow is probably the most-used
tool for editing vector art.

The second arrow selects individual anchor points and line segments within an
object. It shares some features with the solid arrow, but allows better control of
individual components of a line or shape. The second arrow has slightly different
behaviors for different object types in Inkscape. In Inkscape, anchor points are
shown only for some types of shapes. Any shape that doesn't have anchor points
in Inkscape can be converted to one that does with the convert to path shortcut,
[ctrl][shift] C.

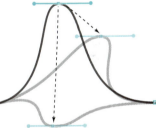

Try out the second arrow using the LEDspectra.pdf file. The
curve near the title of the graph has only three anchor points. Select
the uppermost point by clicking on it or dragging a box around it
with the second arrow. By moving just that point, you can produce
a range of curve shapes. Also notice how easy it is to keep a uniform smooth shape of the curve when there are few anchor points,
compared with reshaping the other colored curves below.

The second arrow can select or deselect single or multiple anchor points in much the same way as you can select and deselect
entire objects with the solid arrow through the use of the [shift] key and by lasso-
ing objects with a box. If all the anchor points in an object are selected using the
second arrow, you can move the entire object. If only some of the anchor points in
an object are selected, though, moving them will change the shape of the object.
This allows you to refine and adjust your artwork. In a typical illustration, you
will rough-in the general shapes with the pen tool, creating all the lines and anchor
points you will need, and then use the second arrow to carefully lay them out,
reshape their handles, and align them.

Creating Bézier curves with the pen tool

Both Inkscape and Illustrator have a pen tool ♀ for drawing Bézier curves. Unlike
the pencil tool, which leaves a trail behind while you hold down the mouse button
and drag the cursor, the pen tool defines individual anchor points of the Bézier
curves. This means that it takes several clicks of the mouse to draw a single line
with the pen tool. Drawing in this way is different than using any physical tool,
but it will quickly become second nature. You skim the surface of the new shape,
with a glancing click at each inflection point—the places where the curve bends,

kinks, or changes direction. This dragging motion sets the length and angle of the handles for each anchor point, which changes the degree of curvature of the line. While generating a line, clicking without dragging creates a corner point in which two line segments bend sharply where they intersect at the anchor point.

To draw an open shape, press return after you have created all the anchor points needed to define your line. To create a closed shape, draw your anchor points and place the last point on top of the first anchor point. The two ends of the line will be fused.

When creating Bézier curves, don't worry if things aren't exactly correct on the first pass—just keep moving along. The beauty of vector graphics is that you can always nudge a point or convert between anchor types until things are just how you want them.

If the object you are making is radially or bilaterally symmetrical, or contains repeating elements, then don't try to draw the whole thing. Draw a fraction of it and then use other transformations to rotate, reflect, duplicate, and join those elements.

Modifying Bézier curves

There is more to Bézier curves than creating anchor points with the pen tool. You will also need to edit the anchor points and control handles of your existing objects.

In Illustrator, the following apply:

- Click on a selected curve with the pen tool to add an anchor point.

- Click on an existing anchor point of a selected path with the pen tool to delete the anchor point.

Ⓦ Ⓛ ▶

Windows: use the alt key.

- Hold the option key while the pen tool is selected to get the Convert Anchor Point tool, or **control handle modifier** ⌐. This tool is also available in the toolbar if you click and hold on the pen tool icon. Click a smooth anchor point with this tool and release without dragging to turn it into a corner point. Using this tool, you can also create a corner anchor point by grabbing the control handle of a smooth anchor point, which allows you to move it independently of the other control handle on the anchor point. If you click and drag out from a corner anchor point, it will make the curve smooth, adding control handles at the same time. Play with this tool using the curves of the example, or with shapes you have drawn. See Figure 18.3 for examples of different curves.

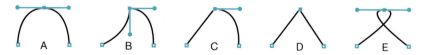

FIGURE 18.3 Bézier curves The three anchor points of these curves are all the same, but their control handles give them very different shapes: (A) smooth anchor; (B) corner point; (C) one handle retracted; (D) both handles retracted; (E) smooth anchor point, but with the control handles rotated 180° relative to (A).

The tools for modifying Bézier curves in Inkscape are associated with the second arrow, rather than with the pen tool as in Illustrator. When you are using Inkscape's second arrow, a toolbar with buttons to add, remove, and otherwise modify anchor points will appear at the top of the screen.

The Join function

Objects may become fragmented into individual line segments, especially when figures are exported from graphing programs or spreadsheets. You may also want to create an object by combining shapes into one. The Join function in Illustrator is one way to achieve this. Select two endpoints on independent curves and then Object ▸ Path ▸ Join (⌘ J). If the points are superimposed, they will be welded together into a single anchor point. If the points are far apart, then a straight line will be added between them. The file examples/LEDspectra.pdf includes curves that have been joined in this way, and you can try merging line segments into a single object using that image.

The other way to achieve similar effects is through the Pathfinder tool. The first button in this palette will merge overlapping filled shapes into a single shape that

takes the composite outline. Click that button and then click Expand to finish the conversion. This tool is extremely useful and likely overlooked during normal image construction. If your objects are not filled shapes, but are single lines (which don't work well for Pathfinder), you can convert the lines to filled shapes using the Object ▸ Path ▸ Outline Stroke menu. Once the strokes are converted into filled shapes, they can be easily merged.

Stroke and fill

Stroke and fill colors (the outline and inside of an object) are set with the **color palette** (Figure 18.4). It is located in different places in Illustrator and Inkscape and is used a bit differently.

In Illustrator, you can hide or show the color palette with the menu option Window ▸ Color, or simply click on the two diagonally arranged boxes near the bottom of the tool palette. In this palette you can use any of several color pickers:

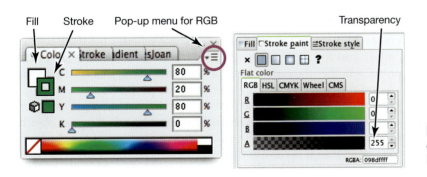

FIGURE 18.4 Color palettes for **Illustrator** (left) and **Inkscape** (right)

RGB, CMYK, grayscale, or others (see Figure 18.4). You will probably want to use whatever color picker matches the color space you are using for your source document. Switch between color pickers by using the pop-up menu at the upper right of the palette.

To set the fill, click on the solid fill box at the upper left of the palette to bring it to the foreground, and then select a color. To change the stroke, bring the box with the little square in it to the foreground. The color of text in most programs is set using the fill option, and there is usually no outline stroke to text. In Illustrator, a separate **stroke palette** (Window ▸ Stroke) controls the width of the line (expressed in points), and this is also where you can create dashed lines and arrowheads. (In versions of Illustrator prior to CS5, arrowheads are generated separately through the Effect ▸ Stylize menu.)

In Inkscape, the color palette can be shown by double-clicking either the fill or stroke color swatch shown at the bottom left of the window. This palette has separate tabs for fill and stroke colors, as well as a third tab for stroke properties, such as line width and control of dashed lines.

If you need more than two or three distinct colors, CMYK and RGB color pickers get cumbersome. They aren't particularly well suited at finding colors that are easy to tell apart but look good together. There is another palette called HSB (**hue**, **saturation**, and **brightness**, or sometimes value) which allows you to directly select the hue of the color from within the RGB space, rather than create it with combinations of other colors. Hue is usually presented in degrees, either as a wheel or a rainbow-colored slider from 0° to 360°. If you want two colors that are as different as possible, you could select any two colors that are 180° degrees apart (such as red and cyan). If you need six colors that are as different as possible, you can pick any six colors that are 60° from each other. You can rotate these colors, while maintaining their distance from each other, to change the composition of the palette for aesthetic purposes or to avoid selecting red and green colors that would be difficult for the color blind to differentiate.

Transparency is also handled differently between Illustrator and Inkscape. In Illustrator it is assigned layer-by-layer or by object using the Transparency palette, while in Inkscape it is also part of the fill and stroke color controls.

Layers

Layers are a critical tool for organizing objects in your images and simplifying your image editing. A layer palette (Figure 18.5) is available in both Illustrator (Window ▸ Layers) and Inkscape (Layer ▸ Layers...). These palettes list the layers in the document, allow you to lock them (with the padlock button to the left of the layer name), control whether they are visible or not (with the eye icon, also to the left of the name), and provide access to a variety of other properties.

A key task for working with layers is to move objects from one layer to another. You can move selected objects between layers in Illustrator by selecting the objects you want to move, then dragging the little colored square to the right of the layer name in the Layers palette so it is next to the name of another layer (the red circle

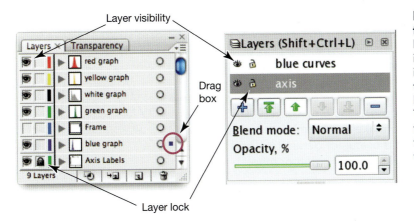

FIGURE 18.5 Layer palettes for **Illustrator (left)** and **Inkscape (right)** Layer visibility is toggled by clicking on the eye icons. Layers can be locked so that their objects will not be selected or moved during editing. To move selected objects between layers in **Illustrator**, drag the colored box at the right of the palette to the layer desired.

in Figure 18.5). In Inkscape, you can move objects between layers using the Move Selection to Layer... options in the Layers menu.

Related to layers is the ability to move objects in front of (above) or behind (below) other objects. This is done with the options in the Object ▸ Arrange options in Illustrator and the Raise and Lower options in the Object menu of Inkscape. Moving things in this way only rearranges them within the layers, so you may have to move the layers themselves to get the desired arrangement.

Illustrator *tips*

Illustrator has a variety of features and shortcuts that may seem a bit esoteric at first, but that can greatly speed up your work.

Illustrator keyboard shortcuts Illustrator keyboard shortcuts go far beyond the usual cut, copy, paste, and undo tasks: the keyboard lets you move between the various Illustrator tools on the fly, change your view of the document, and modify what the mouse clicks do, all without navigating the menus. Working quickly in Illustrator is a two-handed job, one hand on the mouse and the other positioned on the lower part of the keyboard ready to hit the space, Command (⌘), option, and shift keys. Most other keyboard shortcuts are shown in the menus, and they also pop up when you hover over a tool, so you can quickly learn the shortcuts for the ones you use the most.

Use ctrl in place of ⌘.

- To temporarily switch from the current tool to the most recently used arrow, hold the Command key (⌘). This is particularly helpful when drawing Bézier curves. You can use the pen tool to rough out the line and press ⌘ at any time, to use the hollow arrow to fine tune the anchors and their handles. When you release the ⌘ you will get your pen tool back, all without needing to move your mouse to the side of the screen and click an icon. If you get the solid arrow instead of the hollow arrow when you press ⌘, select the second arrow in the tool menu and then reselect the pen tool.

- To duplicate an entire object, select it with the solid arrow key and drag it while holding the $\boxed{option}$ key. This saves several keyboard strokes relative to copying and pasting. The $\boxed{option}$ key also modifies the behavior of many tools, including rotate and reflect, so that they copy the image as they transform. To repeat the duplication with the same spacing, choose Transform Again ($\mathcal{H}$ D). This is a quick way to make a uniformly spaced row of objects and is useful for generating grid lines, tick marks, and labels on a graph or map.

- When copying and pasting objects, you may find it frustrating that your copy is displaced an unpredictable amount from its original position. To paste a copy in the exact position that you copied it, use Paste in Front ($\mathcal{H}$F) instead of the normal paste command. This is useful for creating overlays or multipanel plots.

- To constrain the current tool to operate only in fixed directions (up, down, left, right, and at 45° angles), press and hold $\boxed{shift}$ after pressing the mouse button but before releasing it. This is helpful for laying objects out in precise geometric patterns. Holding $\boxed{shift}$ also maintains proportions when resizing. If you press $\boxed{shift}$ before pressing the mouse button, the effect will be different, as described in the earlier section about selecting and deselecting multiple objects.

- To drag your entire workspace around in the window, press the $\boxed{space}$ and move the page with the hand tool that appears. This can be easier than using the scroll bars to reposition your view.

- There are a variety of keyboard shortcuts for zooming in and out. $\mathcal{H}$= (holding $\mathcal{H}$ and pressing the Equals key) zooms in, while $\mathcal{H}$– zooms out. $\mathcal{H}$ $\boxed{space}$ gives the magnifying glass. Clicking on the image with this tool will re-center and zoom in. Drawing a box with it will zoom in on the boxed region. Add the $\boxed{option}$ key to this combination ($\mathcal{H}$ $\boxed{option}$ $\boxed{space}$) to get the zoom-out tool. Clicking with this tool re-centers and zooms out.[4]

Illustrator view options   There are a couple of view options that are particularly useful. View Outlines ($\mathcal{H}$ Y) reduces your artwork to a collection of simple wire frames for each object. This can help you understand what objects an image is made of and reveal invisible objects that might be interfering with your selection tools. The Hide Edges ($\mathcal{H}$ H) view mode doesn't change the properties of the objects themselves, but makes the selection highlighting invisible to keep it from obscuring your artwork. This is particularly helpful for adjusting the stroke of a

---

[4]One warning: By default in Mac OS X, the Spotlight search tool conflicts with the magnifying glass shortcut. You can change Spotlight's shortcut to $\boxed{ctrl}$ $\boxed{space}$ or a similar equivalent in the System Preferences ▸ Spotlight.

selected object, since the selection lines could cover up the very lines you need to see while you make these changes. Remember to unhide the selections (⌘ H again) when you are done adjusting things. It's easy to get confused when you try to select something later and nothing appears to be happening.

Selecting objects with the same properties in Illustrator One of the most useful and overlooked features in Illustrator is the Select ▸ Same series of menu options. Scientific illustrations are often composed of different types of elements, and there may be many of each of these. Common tasks include things like selecting and deleting all tick marks in a graph, or selecting all the tiny line elements that make up the curve for one of many data series in a plot. Manually selecting each of the tens or thousands of elements would be tedious. Instead you can just select one, then use Select ▸ Same ▸ Fill & Stroke to select the entire bunch. You can then change their properties, such as stroke weight or color, en masse. To remove them you could just hit ⌐delete⌐, or, better yet, create a new layer, drag them all to this layer, and then make the entire layer invisible. You could use the same general strategy to separate the elements of a graph, such as data series, axes, and axis labels, into separate layers. This makes future manipulations simpler since you can lock all layers except the one containing the elements you want to edit.

Creating a keyboard shortcut in Illustrator When you find a command like Select Same Fill and Stroke useful, you can create a keyboard shortcut for the operation, either through the main System Preferences panel, or within Illustrator itself. To do the latter, choose Keyboard Shortcuts... from the very bottom of the Edit menu, and find the corresponding menu command (Figure 18.6). This preset will be saved on your system and will be available to you within future documents.

Inkscape *tips*

Inkscape also has a variety of time-saving shortcuts and useful unique tools. Being a younger, community-driven program, though, it doesn't yet have as many as those available in Illustrator. A helpful introduction to keyboard

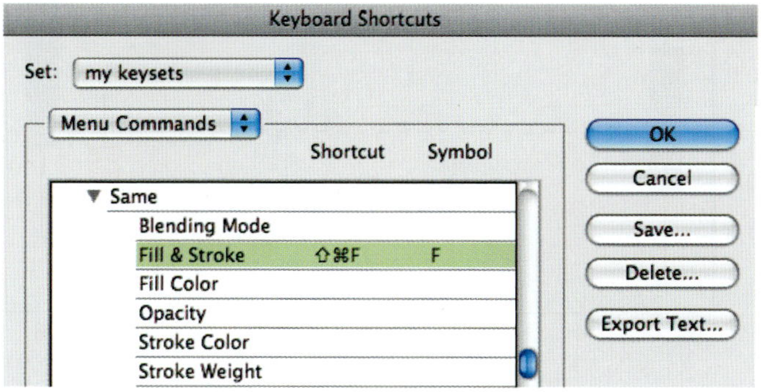

FIGURE 18.6 Creating keyboard shortcuts for **Illustrator** commands

shortcuts and frequently used methods is available at `www.inkscape.org/doc/advanced/tutorial-advanced.html`.

Many of the same general shortcuts are available in Inkscape as in Illustrator, but they are often controlled with slightly different keyboard commands. For instance, you can constrain the geometry of a tool to help draw objects at precise angles by holding the ctrl key once you start using the tool. This has the same effect as holding the shift key after you start using a tool in Illustrator.

A typical workflow

You can try your new skills out with the example graph (examples/ScatterandBoxPlot.pdf) shown in Figure 18.7.

This graph combines plots from several analyses—something which is difficult to do with most graphing programs. It is relatively easy to make the plots separately and then combine them in a drawing program as long as you keep the data together with their associated axes at all times. To create such an image from your own graphs, export them as PDFs, import them, and place each plot on its own layer. Choose one to be the reference, then resize the axes of the secondary graphs so they match exactly with it. Then you can move the unused axes labels to their own layer and make it invisible.

With a graph like the example, you might want to see how it looks with slightly larger or smaller data points. In Illustrator, the PDF file will preserve the layer information, since PDF is a native format. Lock the other layers and select all the dots. To make the results of your experimentation more clearly visible, you can hide edges (either through the menu with View ▸ Hide Edges, or with the keyboard shortcut ⌘ H). In Inkscape, to get to this same point you can select the groups of

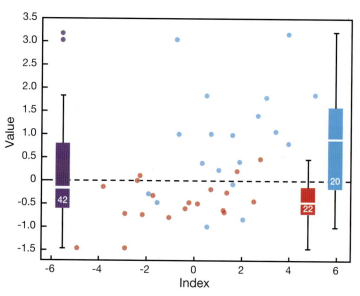

FIGURE 18.7 A graph image created by merging plots generated by different programs

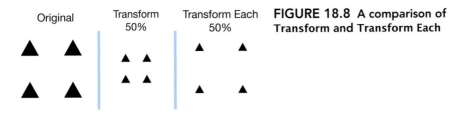

Original   Transform   Transform Each   **FIGURE 18.8 A comparison of**
                    50%   50%   **Transform and Transform Each**

points and move them together to a new layer. Then ungroup them all so that they appear as individual points, and lock the background layer (the layer that doesn't have the data points any longer).

In Illustrator, now choose Object ▸ Transform ▸ Transform Each. In Inkscape, choose Object ▸ Transform and check the Apply to Each Object Separately box. Instead of transforming the whole view as one, this option changes each element individually. This command only works as expected if the objects selected are not part of a larger group. Select with the solid arrow and choose Object ▸ Ungroup before trying to transform items individually. (You can maintain the ability to select a group of objects together if you move them all to their own layer.) You can see the difference between scaling using Transform and Transform Each in Figure 18.8.

Within the Transform Each dialog, try scaling the dots up to 150% or 200% of their original size, both horizontally and vertically. Undo that change (⌘ Z), and then, try 75%. There are also some overlapping points in the plot. To make these more distinct from each other, you can change the fill color so it is partially transparent. Test this when the dots are larger and overlapping to see the effect better.

At times you might need to rescale the graph non-proportionally to fit into a page layout, or to merge with another image. This operation maintains the quantitative nature of the data, but distorts the symbols in the process, making your circles into ovals. To recover proportionality, you can use Transform Each with the inverse of the scaling ratio—for example, if you shrank the x-axis to 75% of its original size but left the y-axis at 100%, then transform each symbol horizontally by 133.3% (=1/.75) to regain the original proportions.

When working with quantitative data in an illustration program you have to be very aware that the transformations you make are uniform and consistent. You can rescale the x- and y-axes independently, but any associated or overlaid graphs have to be resized to exactly match.

Creating regularly arranged objects

A common task in scientific illustration is to generate a set of consistent, regularly spaced, regularly aligned objects. These could be tick marks on a graph, a panel of photographs, or an illustration of a replicated experimental design. Creating regularly spaced objects is very easy using the Align and Distribute commands. A palette with tools for automatically aligning and distributing objects is available in both Illustrator (Window ▸ Align; Figure 18.9) and Inkscape

(Object ▸ Align and Distribute...). Here we only describe how to use the Illustrator tools in detail, but the Inkscape tools are similar.

In this example, you will make a vertical grid marking every 50 nm along the x-axis for the graph of LED spectra. On a new layer in Illustrator in the LEDspectra.pdf document, draw a vertical straight line with the pen or line tool. Give this line the stroke color and size that you want for the grid (you can try gray, dashed, etc., from the stroke and color palettes). Then click and hold the line with either arrow, and press and hold ⎡option⎤ (clone) and ⎡shift⎤ (constrain) simultaneously. If you hold the ⎡shift⎤ key before you click on the line, it may cause it to be added to any existing selection. Drag the line horizontally to approximate the spot where the next grid line should fall. Repeat this step to create as many lines as you need, without worrying about their exact positions. (You could also use the Transform Again shortcut, ⌘D, to generate additional copies after you make the first copy.) Now move the rightmost and leftmost gridlines precisely where they should go (lined up with the 400 and 800 nm tick marks), select all the lines, and click the Distribute Objects Horizontally button on the Align palette (see Figure 18.9). The lines will jump into evenly spaced positions between the two end lines. There are also buttons to distribute objects vertically, and you can control whether the spacing is relative to the center of objects or one of their edges.

In addition to uniformly distributing objects, as you might suspect, the Align tools can also automatically align objects so that they are flush. If entire objects are selected (either using the solid arrow, or with all anchor points selected using the second arrow), then the Align Top button will move the objects so their top edges are flush. If the second (hollow) arrow is used to select only some anchor points within each object, then the alignment tool will just move those anchors into line.

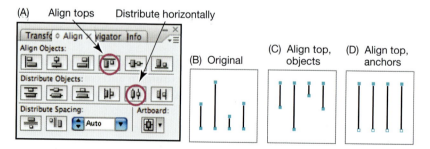

FIGURE 18.9 Alignment tools and examples The Illustrator Align palette (A) and examples of the effects of aligning objects vertically. If the original lines (B) are selected with the solid arrow and aligned vertically, each object is moved so that its upper end is flush with the top of the object that had the highest point (C). If the second arrow is used to select just the upper anchor point of each line and the same alignment is used, the length of each line changes when the anchor points are made flush (D).

This can be used to resize lines and shapes so that they have the same length or width.

Best practices for composing vector objects

A few common problems often crop up with the drawing and presenting of vector art. These can be distracting and reduce the clarity, editability, and professionalism of the final piece. Developing good practices from the start will save time in the long run.

You should use the minimum number of points necessary to define a curve. If you have too many extra points, the curve will tend to look irregular and will be more difficult to modify since excess adjustments will be required. The number of anchor points required to define a shape is often much less than one might imagine at first. For example, a circle can be defined with just two anchor points, although four are created by most circle tools to provide additional points of control.

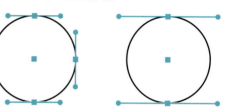

If you are drawing several straight lines and intend them to be parallel or at fixed angles, use guide tools (such as **Align** or shift to constrain) as described above to ensure that they are correctly arranged. You can also use a built-in underlying grid as a guide for arranging your objects by turning on **Snap to Grid** in the **View** menu.

Lines that terminate on other lines should intersect exactly, with the endpoint of one line lying along the other line so that there is no space and no overextension beyond the point of intersection. Corners require special attention and overlapping endpoints should be typically joined into a single path.

There are of course many valid stylistic reasons for breaking these rules, but that is different from not taking these issues into consideration in the first place. One irony of computer illustration tools is that while they make precise drawing accessible to untrained illustrators, the very regularity of the tools and precision of the final product are unforgiving of even very small errors. This means that would-be illustrators must pay careful attention to detail, particularly with regular geometric illustrations.

SUMMARY

You have learned that:

- There are many software environments and file formats for handling vector art

- Vector art can be generated by exporting from analysis programs, by making new drawings, and by tracing other images

- Bézier curves are a simple and powerful way to build up complex shapes

- Minimizing the number of anchor points used to define shapes makes them more regular and simpler to work with

- Layers are a critical tool for organizing image elements, as well as for locking them and changing their visibility

- A small number of tools and shortcuts within Illustrator and Inkscape will serve many of your drawing needs

Moving forward

- Open a PDF file in a vector editing program, and see what elements are grouped together, how to select similar images, and what it would take to modify the color scheme.

- Experiment with transparency in some of the graphical elements to see how this might be used to convey more information.

- Explore some of the Adobe Illustrator tutorials available for free online, for example at `http://lynda.com`. (Free tutorials are underlined in their directories. Others require a subscription.)

Chapter 19

WORKING WITH PIXEL IMAGES

Pixel images are a frequent product of scientific research, whether they are generated by normal photography or as the output of a more sophisticated imaging process such as radiography. Here we explore the tools and concepts needed to edit pixel images and to integrate them with vector art in figures, as well as how to evaluate the most appropriate trade-offs when selecting file types, resolution, and compression.

Image compression

General principles

In addition to the resolution and dimension issues discussed in Chapter 17, when handling pixel images one must also consider compression strategies. Much of the information within pixel images is redundant in a technical sense: large blocks of pixels may all be the exact same color, there may be repeating elements that only need be stored once, or only a small subset of possible colors may be used. Through compression, programs and file formats reduce the amount of memory needed to store an image.

There are three image-storage strategies: no compression, **lossless compression**, and **lossy compression**. Each is associated with different file formats. At times, especially for small images, there are few or no performance costs to using uncompressed files. Lossless compression reduces the memory needed to store a file without actually losing any information, hence its name. Like an accordion, the image is compressed into a smaller space and returns to its same condition when expanded. A program achieves this by identifying and consolidating redundant information in the file, be it colors or patterns. This provides reduced file size at no cost to the appearance of the image, but it might not reduce the file size very much, particularly if the image is complex.

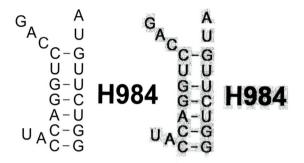

FIGURE 19.1 Comparison of PDF (left) and JPEG (right) versions of a figure Notice the cloud of gray "dust" that surrounds the objects in the right figure, typical of JPEG compression. (The original artifacts in the compressed image above have been darkened for visibility.) This is more a function of compression methods than of insufficient image resolution.

If small file size is a more pressing concern than perfect image reproduction, you can use lossy compression. Lossy compression permanently throws away some information when the file is compressed. Similar colors may all be considered as a single color, decreasing the color depth of the image, or regions of the image may be approximated, changing the texture and lowering overall picture detail. In addition to reducing the amount of information in an image, lossy compression can introduce artifacts when the image is re-expanded (Figure 19.1). In many cases the differences introduced by lossy compression are imperceptible, but sometimes they can seriously impact the readability of an image. The JPEG format offers different compression levels, where you can choose to have smaller low-quality files or less compressed higher-fidelity images.

Implications for image workflows

Lossy image formats become problematic if an image is repeatedly opened, saved, and closed. With each iteration of this cycle, more and more image detail can be thrown away, and artifacts can accumulate. For this reason, lossy compression should not be used for images that are being repeatedly changed.

For practical reasons, most cameras and instruments that generate pixel images save them using lossy compression. This is often appropriate since far more images are generated than will ever be used, and the deficiencies of lossy formats are not particularly acute if only one round of compression is applied. Once a subset of image files is selected for further processing, they should first be resaved in a lossless image format. This will ensure that the images can be opened, edited, and resaved as many times as needed without degrading their quality.

If specimens are of particularly high value or every bit of resolution is important for the task at hand, then you should consider saving images in a lossless format immediately upon acquisition. The default file format used by most cameras and instruments can be easily changed to a lossless format.

Pixel image file formats

There is a wide and often confusing variety of pixel image file formats. They differ in several key technical respects. Some don't provide any compression; others support lossy or lossless compression. Some support transparency, while others do not. Some can store only a single image; others can support a series of images and video. Some allow more than one layer within one file but others allow only a single layer. Pixel image formats also differ in licensing. Some are open formats

TABLE 19.1 Selected image formats and their feature sets

Format	Compression	Layers	CMYK	Transparency
		| Supported features		
TIFF	None, lossy, or lossless	Inconsistent	Yes	Yes
JPEG	Lossy	No	Inconsistent	No
PNG-24	Lossless	No	No	Yes
PNG-8	Lossy (via color)	No	No	Yes
GIF	Lossless (color options)	No	No	Limited
PSD	Lossless	Yes	Yes	Yes
XCF	Lossless	Yes	No	Yes
RAW	None, lossless	No	No	No

that are free for use by any program and any user, and were designed specifically to improve the interoperability of software and portability of data. Others are closed formats, owned by companies which in some cases charge royalties to developers who use these formats in their software.

Table 19.1 lists common pixel art file formats. The file formats you will encounter most often are PNG, JPEG, TIFF, PSD, and RAW. Most journals will request pixel art in TIFF files, which can be saved with no compression, different types of lossless compression, or JPEG compression within the TIFF file. For RGB and grayscale images without layers, PNG is an excellent lossless file format, and JPEG is a very widely supported lossy format.[1]

CMYK is rarely used as the native color model for a pixel image, but it is sometimes necessary to convert to a CMYK space prior to publication. PNG does not support CMYK, and JPEG supports it inconsistently, so that CMYK files generated by one program may show up in another program as color negatives.

PSD is the native file format of Photoshop, and XCF is the native file format of GIMP. Both are largely specific to these programs, and you will usually export images to another file format when publishing them, sharing them with others, or transferring them between programs. These files can store many types of data, including text, some vector information, sophisticated layer properties, and other information that helps provide fine-scale editing control.

If you are using high-end digital cameras, you are also likely to encounter images in the RAW file format. This isn't a single file format, since each camera manufacturer has their own RAW format—for example, Nikon's RAW format is

---

[1] PNG-8 is a lossy subset of PNG that uses 256 or fewer selected colors within the whole image. It is still widely used for web icons and non-photographic images.

NEF and Canon's is called CR2. RAW refers to pixel files that contain minimally processed raw data from imaging sensors. Some information is discarded when converting the sensor data to a JPEG or other standard image format, so saving RAW files provides greater control over downstream image adjustments such as white balance and exposure modifications. It is almost as though you can retake the photo after the fact, using different camera settings. RAW images typically allocate more information to storing each color value (the bit depth), allowing them to support a higher dynamic range. See Appendix 6 for more on colors and memory.

BMP or GIF file formats should be used only as a last resort when software has limited options for import or export. Both are relatively old and don't support many features or perform especially well across programs.

Transparency

Some programs and image file formats support the **alpha channel**, which encodes the transparency of each pixel. By default, all pixels are fully visible. It is often desirable, though, to make the portions of an image transparent when combining images, or just to simplify the composition of a figure. In vector art you can easily do this by copying the objects by themselves. In a pixel-based image, you have to edit the alpha channel mask that indicates which pixels should be visible. This is typically an additional 8-bit channel (256 levels, just like R, G, and B) stored along with the three color channels. Partial transparency is useful to convey time (for example, superimposing subsequent exposures) or spatial relationships between images (for example, a fluorescence overlay on a white-light image).

Pixel art editors

There are many pixel art editors, and, like vector art editors, each has its own set of trade-offs. Photoshop is the industry standard, especially when it comes to editing photographs. It is made by Adobe, the same company that produces Illustrator, and many of the shortcuts and navigation tools are the same in both programs. This is helpful when you frequently switch between them.

GIMP is the leading open-source pixel art editor, and is freely available for major operating systems at www.gimp.org. It is a bit more mature than its vector art analog, Inkscape, and can even do some vector art illustration. Like Inkscape, GIMP does not have quite as many features as its commercial counterparts, but is sufficient and easy to use for many common tasks.

Working with pixel images

Masks and nondestructive editing

Many programs that handle both vector and pixel art, including Illustrator, Keynote, and PowerPoint, support image masks for cropping and resizing images. **Image masks** are essentially little windows that can be modified to show only a particular region of a picture. The effect appears the same as cropping the image, except that you can edit the mask later to show parts of the image that were previously excluded. In Photoshop, you can select a portion of an image, and subsequent adjustments can be applied only to the pixels within this masked region. Adjustment layers are also very useful in Photoshop. These apply an adjustment to all underlying layers or to a masked region, but with the advantage that the modification exists on its own layer and can be turned off or readjusted later in the process.

Some image editing and photo management programs, such as Aperture and Lightroom, implement more sophisticated forms of nondestructive image editing. In these programs, image manipulations—such as cropping, resizing, levels adjustments, color changes, and touchups—are stored as a series of commands associated with an image, rather than actual modifications to the image itself. These commands are reapplied each time the image is displayed or exported. This avoids the problems associated with resaving files in lossy formats, since the original file remains unchanged throughout the editing process. It also makes it easy to revert to any previous version of the image.

Levels adjustment

Many factors can cause images to come out with an exposure that is not quite optimal. In your image editing program, the best way to correct these is not by using the Brightness/Contrast menu item that most people gravitate to, but by using Image ▸ Adjustments ▸ Levels... instead. (You might also take the approach of using an adjustment layer, in this case choosing Layer ▸ New Adjustment Layer ▸ Levels....) This dialog box presents a histogram of the brightness of the pixels in your image (Figure 19.2).

Along the left side of the curve are the black pixels, in the middle the graph shows the number of grey pixels, and along the right side are the white pixels. In Figure 19.2 you can see that the image is probably a bit underexposed: most of the values are distributed between black and gray, with very few pixels brighter than grey and none approaching white. The white point slider, circled here in red, sets the level of brightness above which pixels are considered fully white. To adjust this picture for a better balance and to bring up the overall brightness while remapping the pixels across a broader intensity range, you would slide the white point slider to the left until it is near the right edge of the

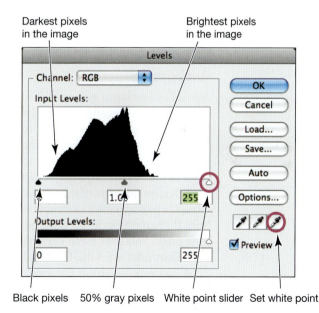

FIGURE 19.2 The Levels dialog box in Photoshop This box shows a histogram of how many pixels have each level of brightness. Other programs will have very similar views.

pixel histogram. This means that those pixels at the upper edge of the distribution would now be white instead of gray.

If there is a region of the image that you know is white, you can use the Set White Point eyedropper tool, also circled in Figure 19.2, and click inside the white part of the picture. This will set the white point to match that pixel. The white point tool also adjusts the RGB channels of the image independently, so it can change the white balance of your photo (sometimes for the better, and sometimes in inappropriate ways).

As long as these transformations are carried out as described above, and applied uniformly to the entire image, the transformation is a linear remapping of values. Therefore, this is very similar to changing the exposure on a camera and should be fully within the restrictions on image adjustment mandated by journals.

Grayscale images

In preparing scientific figures, it is sometimes clearer and almost always cheaper to produce figures in black and white—in this case meaning grayscale—rather than in color. Using grayscale images can also help your annotations to stand out against a monochromatic background. When converting to black and white in Photoshop, avoid using the Image ▸ Mode ▸ Grayscale menu item. Like an autoexpose function, this uses a weighted average of the red, green, and blue channels of your image, but it is not necessarily tailored to scientific images with non-standard color composition. Instead, try Image ▸ Adjustments ▸ Black and White and choose from the preset adjustments to see what produces the best results. This effect is very similar to the way that microscopists use (used to use?) color filters to improve the contrast when taking black and white images.

Antialiasing

A pixel image has a given resolution, and converting to a different resolution typically involves resampling the image data. Because vector art effectively has infinite resolution, converting it to pixel art is similar to downsampling a higher resolution pixel art image. Artifacts arise in this translation process when there are hard edges on a solid background, such as the edges of text characters or lines. The edge of the vector art won't exactly line up with boundaries of the pixels. Objects with perfectly horizontal and vertical edges may end up slightly smaller or larger, which is particularly noticeable for shapes that are thin to begin with. Curves and lines that aren't perfectly horizontal or vertical end up with a staircase effect along their edge, and may have variable thicknesses. If an object is thinner than a pixel it may disappear entirely. These graphical glitches are known as aliasing.

A common solution for these artifacts is **antialiasing**, the use of intermediate color values between objects with hard edges (for a simplified example of this, see the lower right panel of Figure 17.3). If the boundary of the vector lies exactly at the boundary of the pixels, then nothing need be done. If the boundary of the vector passes through a pixel, then that pixel would have an intermediate color that depends on how much of the pixel is occupied by the vector and other image elements. Up close, where independent pixels can be discerned, antialiased images appear fuzzy. From a typical viewing distance, though, they appear much smoother and don't have the harsh artifacts that arise when converting vector art without antialiasing.

Most vector art applications that have the option of exporting pixel art images provide the option to perform antialiasing during the conversion. Recent versions of Illustrator give the option for a "pixel preview" and take into account the underlying pixel grid when translating text and vector art into pixels.

Layers

Layers are extremely important in pixel editing. Without layers, if you create a new object to annotate or cover the image, it will overwrite the original pixels, making it impossible to move or modify your annotation. Even effects like levels and contrast can be applied to an image through layer properties rather than changes to the image itself, making it easy to remove or adjust this effect later on.

GIMP and Photoshop both have a similar layer palette, with a blank-page icon to create a new layer and an eye icon to toggle the visibility of a layer. A dropdown menu specifies how the layers are combined; normally they are collapsed from top to bottom to create a single composite image, but you can set some layers to affect the way others are displayed or combine them in different ways. In Photoshop, when you first create or open a new document, the background layer will be locked with regard to transparency. Deleting portions of an image will revert them to the background color rather than leaving a transparent gap. To unlock, double-click on the layer named Background and select unlock.

When creating and moving selections in Photoshop, some potentially confusing behaviors may be observed. First, dragging the selection with the hollow arrow merely relocates the selection border, and not the pixels beneath it. To move the pixels themselves, you have to either hold the Command key (⌘) or switch to the black arrow tool. Second, an operation only applies to the layer highlighted in the Layers palette (and within that layer, only to selected pixels). So if you are editing without noticeable effect, make sure the correct layer is highlighted.

Feathering is a term used to describe gradually transitioning from selected pixels to non-selected pixels using a boundary of partially selected (semi-transparent) pixels. This can be an important way to modify your selection and smoothly compile images from several sources into a single image. All considerations about data integrity of course apply to use of these tools in merging images.

Colors in GIMP

GIMP has an innovative color picker, with several color space tools all in one window. When one slider is changed, the sliders for other color spaces are dynamically modified. This is nice because you can change the way you browse for colors without switching between color space windows, as is necessary in most other graphics programs. It is also very informative to see directly how changes in one color space map to changes in other color spaces.

One problem with all of this, though, is that GIMP doesn't support the CMYK color space natively. Color conversions which you can try in Photoshop will not be easy to achieve in GIMP without additional plug-ins.

Photoshop *shortcuts*

Like Illustrator, Photoshop has a rich set of well-thought-out keyboard shortcuts that can greatly accelerate your workflow. Many of these shortcuts are the same in Photoshop as in Illustrator, but there are additional ones as well:

- The Rectangular Marquee Tool (the dashed rectangle on the tool palette) selects rectangular regions of your image that you can then copy, cut, delete, or transform in other ways. Like the other selection tools, you can hold shift while you drag to select multiple regions simultaneously. If regions are overlapping they are joined together, which can allow you to select complex non-rectangular areas.

Use alt in place of option .

- Holding option while using a selection tool will *subtract* the newly selected region from the previous region, which can create holes in the middle of an existing region or remove chunks from its perimeter. You can also invert the selection, to then delete everything from a layer except the area of interest.

- As with Illustrator, holding down space turns the cursor into a hand that moves the canvas around, and ⌘ space will bring up the magnifying glass. When using the Rectangular Marquee Tool to drag a rectangular selection, holding space after you start dragging will allow you to drag so as to move the point of origin for the selection box. Letting up space will resume dragging out the edges of the selection box.

- Photoshop initially appears to have only one level of Undo. If you repeatedly press ⌘ Z, it will just keep undoing and then redoing the last command. To go back further in time, choose Window ▸ History, and you will see a chronologically arranged palette displaying the commands you have issued.[2] Click on a command in the list to jump back directly to that point in your editing process.

- You can also use the keyboard shortcut ⌘ option Z to step back through your history.

Command-line tools for image processing

It might seem counterintuitive that command-line tools, which entirely lack graphical user interfaces, are sometimes the most convenient way to work with image files. They are, however, particularly useful for automating repeated image manipulation tasks and for quickly extracting information about images. They also make it simple

---

[2] Multiple Undo commands first became available in Photoshop version 5. Before that time, image editing could be a stressful activity.

to script routine operations, such as performing a series of standard transformations to photographs of electrophoresis gels.

The `sips` program

This command-line image editor is included with OS X. It can retrieve a variety of image properties, rotate images, and perform cropping and resizing operations. In the scheme of image editors, it has limited functionality, but can work well for some common tasks.

The following commands create a new folder named `converted`, then resample each JPEG in the current folder to a width of 470 pixels, and place the new file into the `converted` directory:

```
lucy$ mkdir converted
lucy$ sips *.jpg --resampleWidth 470 --out converted
```

Here, `*.jpg` refers to the input file list to use, and `converted` specifies the directory in which to save the images that are created. Use caution, because `sips` will overwrite images of the same name in the destination directory. To see a full list of `sips` options, type `man sips` at the command line.

ImageMagick: `convert` and `mogrify`

ImageMagick does not come with OS X, but it can be installed on OS X, Linux, and Windows. The installation procedure is described in detail in Chapter 21. You can also install it with the `port` command if MacPorts[3] is installed on your computer:

```
sudo port install imagemagick
```

In Linux, you can use the command:

```
sudo apt-get install imagemagick
```

The installation is not a single program, but a collection of command-line and graphical utilities, and it has many more features than `sips`. It can adjust the levels of colors, brightness, or contrast, and perform other transformations like inverting an image. If you have a photographic data set which needs to be batch processed for analysis, this tool could save a great deal of time.

You will usually run ImageMagick by typing the `convert` command. For example, to convert a gel image from its native TIFF format to a half-size, inverted (black on white) and compressed PNG format, you could type:

```
convert gelscan.tif -resize 50% -negate gelscan.png
```

---

[3]You should install MacPorts from `macports.org` to make it easy to install and update command-line programs and all their dependencies. See Chapter 21 for more information.

For doing batch conversions of several files, you can generate a shell script containing several `convert` commands, as shown in Chapter 6, or you can use the `mogrify` function, which is basically `convert` applied to a group of files. The same operation above can be applied to convert all `.tif` files in a folder using this command:

```
mogrify -format png -resize 50% -negate *.tif
```

Be careful when using `mogrify` and always work on a copy of the original files, since it will overwrite without checking. A more complete description of the `convert` command and other ImageMagick functions can be found by typing `man convert` or `man mogrify`, or by referring to these sites:

www.imagemagick.org/script/convert.php
www.imagemagick.org/Usage/
www.fmwconcepts.com/imagemagick/

ExifTool

Many media files, such as photographs and audio tracks, have embedded metadata that describe how the file was created, as well as recording such other properties as time and location of creation, exposure settings, and camera type. This embedded metadata is often stored in a standard format called Exif. The Exif data can also hold tags (snippets of text that help categorize your images) and these can be accessed from database programs. TIFFs and JPEGs support Exif metadata.

These Exif data are accessible from within many image editors, but you will sometimes want to take a quick peek at many files without opening them all up. The free command-line program `exiftool` reads and writes Exif and other metadata formats. It is available for download at http://www.sno.phy.queensu.ca/~phil/exiftool/.

ExifTool is also a great way to recover the original creation information for photos, and could form part of an automated cataloging pipeline which sorts images without user intervention. To extract the location from several images, you could create an `exiftool` pipeline as described in Chapter 16. Some electron microscope programs will embed data about the magnification used when an image was captured. If you forget to add scale bars to your image, explore the Exif data to see if you can create a tool to print out a table of magnification values, as shown in Chapter 16 and in the corresponding example `scripts/exifparse.py`.

Image creation and analysis tools

ImageJ

ImageJ is an image analysis program developed by the United States National Institutes of Health, and is freely available for OS X, Windows, and Linux at `rsbweb.nih.gov/ij`. It can import a wide variety of image types, and provides

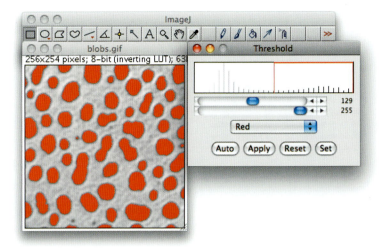

FIGURE 19.3 Setting the threshold for black and white conversion in ImageJ

a rich set of features for extracting quantitative data from them. This includes measuring the length, perimeter, and area of objects. It can also perform more specific tasks, such as gel electrophoresis analysis and 3-D reconstruction of optical slices. A wide variety of customizable plug-ins are also available, and tutorials are linked online. ImageJ is scriptable, so that macros can be recorded and then edited to replay repetitive analyses.

One very useful feature is the ability to automatically count particles of a certain size. This tool can be applied to ultrastructural studies, cell or organism counting, quadrat surveys, and morphological studies. To try it out, download ImageJ and choose File ▸ Open Samples... ▸ Blobs. This is one of the many demo images that are linked from within the program.

To count the blobs, you have to convert the image into a black and white image. In this case, that means just black or white pixels, with no levels of gray between. The most flexible way to do this is with the Threshold command. Choose Image ▸ Adjust ▸ Threshold... from the menu, or use the keyboard shortcut ⌘ shift T to bring up the threshold dialog box (Figure 19.3).

Within this dialog, you can use the two sliders to set the brightness range of the image that will be converted to black (shown by the red box in Figure 19.3) or set to white. Play with the sliders to see how the red area expands and contracts, and when you have the blobs distinct from the background, click Apply and then close the Threshold box.

Your picture should have a white background with dark black blobs on it. Depending on the threshold you chose, some of the blobs might be tiny specks. Not to worry: the next step will let you filter out these non-target particles.

Now to count and measure all these particles, get out a pad of paper and a ruler. Or alternatively, choose Analyze Particles... from the Analyze menu. If you try to run this analysis without having converted your figure to black and white, you will be reminded that this only works with thresholded images.

FIGURE 19.4 The results of the Analyze Particles operation in ImageJ

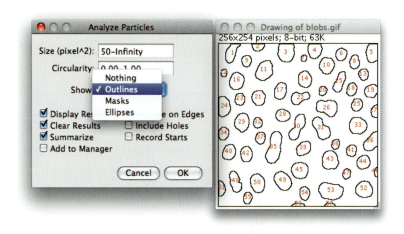

To filter out the small black specks, set the minimum Size to 50 (corresponding to the total area in pixels of the blobs to count), and to visualize your results afterward, select Outlines from the Show pop-up menu (Figure 19.4). You can also choose to ignore blobs that touch the edge, or to count the total area if there are holes in your blobs. Click OK, and you should get a couple of tables summarizing the number of blobs in your image, as well as a new picture showing each blob labeled with a number. From this and from the table summarizing particle area, you can decide whether you want to count certain blobs (for example, blob 30) as representing two actual particles.

MATLAB

In a data storage context, pixel images typically consist of three 2-D matrices, with each matrix corresponding to one of the RGB values of the image. The x-y size of the matrix matches the pixel dimensions of the image. In a standard 8-bit image,[4] each value in an array is a number from 0 to 255 representing the brightness of that color pixel. Because of their fundamentally numerical nature, images can be sliced, sectioned, analyzed, and processed relatively easily using the array tools in MATLAB. Use the `imread()` function to load an image into an x by y by 3 array. To visualize a matrix (whatever its origin) as a pixel image, use the `image()` function on the array. These can be saved with the `print('-dpng','-r300','myimage.png')` command. The extra `-r300` parameter in this example indicates the resolution to use at the given screen size.

R

The analysis system R can also import JPEG files with the `ReadImage` package, and TIFF files with the `rtiff` library, installed separately. It can interface with ImageMagick (see above) using the `EBimage` package. It can also export image files from its command line or as part of a script.

---

[4]See Appendix 6 for the explanation of 8-bit images.

Animations

It is easy to create animations from individual images by compiling them as consecutive frames. This is ideally done with the output from a MATLAB or R program which automatically saves a uniquely named PNG each time through a loop. There are many programs which can open a batch of images in a folder and treat the files as animation. QuickTime (`www.apple.com/quicktime`) and the free command-line tool `ffmpeg` are two examples.

Photography

The visual documentation of organisms, experiments, and data through photography is an important aspect of biological research. Although this is not the venue for a detailed photography lesson, there are a few tips that can greatly improve the scientific pictures that you take.

Aperture and exposure time

Aperture and f-stop The aperture is the effective size of the hole through which light enters the camera, telescope, or microscope. The aperture of most cameras can be opened up to let more light in, or made smaller to let in less light. (Some cameras, such as those typically found in mobile phones, have a fixed aperture.) It may seem strange that you would ever want to exclude light from your camera in this way, but there are other critical trade-offs that are related to aperture size. The smaller the aperture is, the better **depth of field** your image will have. Depth of field describes the distance—foreground to background—over which objects will be in focus. When using a microscope, you can get a sharper view with more depth of field if you close down the aperture in the light path. This is usually controlled with a dial or sliding knob on the microscope body between the eyepieces and the objective lens. The aperture's relationship to depth of field even applies when using an electron microscope.

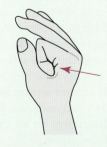

GETTING A FEEL FOR HOW AN APERTURE WORKS Make a circle with your thumb and first finger and bring your hand up close to your eye, so that you are looking through the hole you have created. Collapse your index finger down until you can see just the tiniest possible pinhole of light coming through. This is what we will call your "manocle." Now bring your other hand up, so that it enters half the field of view quite close to your face, and look partly at your hand and partly at the scene in the background. Your hand and distant objects should both be in fairly good focus. Move your "aperture" quickly out of the way to see your hand go out of focus. You can also look through your manocle at the screen of the computer from about an inch away, seeing every pixel, then keep looking while you move your hand away to see things drop immediately out of focus.

In photography, the effective aperture size is quantified in **f-stops**, with values for a typical lens ranging from 2.8 to 32. This value is *inversely* proportional to the open area of the aperture, so the higher the number, the less light and the better the depth of field. *F*-stops are marked in unusual increments: 1.4, 2, 2.8, 4, 5.6, 8, 11, 16, 22, 32, 45. These intervals are selected so that each increment lets in one-half the light of the one below it; thus, $f/16$ lets in about half the light of $f/11$. For most scientific photography, you want the aperture to be on the small side (higher number) so that most of the subject will be in focus.

Exposure The **exposure**, the amount of time that the shutter stays open, is usually displayed by the camera as the denominator of a fraction of a second; for example, 250 means 1/250th of a second. Just like the trade-offs encountered when changing the aperture size, there can be direct effects on both the amount of light and the sharpness of the image when modifying the exposure. The reason is entirely unrelated to depth of field; rather, longer exposures let in more light but make the shot more susceptible to motion blur caused by vibration, an unsteady camera, or a moving subject.

In most situations, you should avoid using exposures slower than 1/60th of a second when the camera is handheld or the subject is moving. Longer exposures may work if the subject is absolutely still and the picture is being taken with a camera on a tripod or through a microscope. For fluorescence images, you may even need exposures of several seconds. When light is a limiting factor, as in these cases, it is usually worth the loss in depth of field to open the aperture and keep the exposure time within reasonable limits. If you have plenty of light and can operate faster (most cameras can go to 1/1000th of a second without a problem) then do it. If you are shooting macro subjects (that is, images which are very close-up) and need better depth of field, then you can use a higher *f*-stop (up to $f/32$), but optical effects (diffraction bands and edge distortion) will start to intrude.[5] If you know what concerns will be most important for your photograph (for example, avoiding motion blur or maintaining depth of field), you will be able to adjust your settings to account for the conditions.

ISO for sensitivity If you find that the aperture is all the way open and your exposure is so long that everything is coming out blurred (the lab of a rolling ship comes to mind), you have another option at your disposal to get more out of the available light: you can adjust the sensitivity of your camera using the **ISO setting** (also called ASA or film speed). ISO formerly referred to how sensitive the film for your camera was; now it also refers to the sensitivity of a digital camera sensor. ISO typically ranges from 50 to 400, but usable speeds of 3,200 and beyond are available on good consumer cameras. The ISO scales directly to the sensitivity,[6] so an ISO of 200 is twice as sensitive as an ISO of 100, and requires half the exposure for an equivalent image.

---

[5] Using your manocle, you can see the effects of distortion and diffraction as well. Look at your screen or a book from up close, and move your view around. You will see a fish-eye effect around the edges.

[6] Finally, not an inverse relationship!

As you have probably guessed, there are trade-offs for increasing the ISO, and—in an interesting parallel—they play out nearly the same for digital and film cameras. As you increase the gain and sensitivity, the sensor also becomes more susceptible to noise, visible as grainy textures and bright speckles. These may not be immediately apparent on the camera's small viewfinder, but they may be very noticeable when your photo is blown up. When light is not limiting, you should keep the ISO low (50 to 100), but as your subject matter becomes darker, faster moving, or more dimly fluorescent, you may need to bump up the sensitivity.

Illumination The best way to avoid the limitations that arise from all of these trade-offs in image quality is to provide more light, although this may not always be possible. If you are taking photographs through a microscope, check the light path to make sure there are no unneeded filters and that the illuminator apertures are open. For stereomicroscopy or macro photography, an off-camera flash on an extension cable or remotely triggered flash makes it possible to provide light directly to the specimen and greatly improve the lighting of the scene. With this additional illumination, you can achieve much faster shutter speeds, especially useful with live specimens.

Overexposure and underexposure In addition to checking your images for focus, you should make sure that the exposure is not clipping, or failing to register fully, the light or dark values of your picture. You want to capture the full range of values in the scene, and not have areas that are overexposed ("blown out") or underexposed. Most cameras can display a histogram of the relative number of pixels at each level, from black to white, present in your image; this is similar to the Photoshop levels dialog box in Figure 19.2. Figure 19.5 shows one such plot for an overexposed image. Unless this is a picture of an object sitting on a pure white background, you can tell from this graph that the highlights are extending off the right side of the plot. You can see that there is large number of pixels with a value of pure and near white, and that these are not distributed uniformly. In other words, the brightest portion of the image is well off-scale, and other regions of the image that would normally be some shade of gray are also being shown as pure white.

There is often a temptation or a tendency to adjust exposure times so that microscope images, especially of fluorescence, look bright and saturated. If you are fully saturating pixels on your detector, though, you are losing a great deal of information that will not be recoverable. It is often better for an image to be a bit *underexposed*, rather than overexposed, because information can be recovered up to the point that the detector becomes saturated and the values get

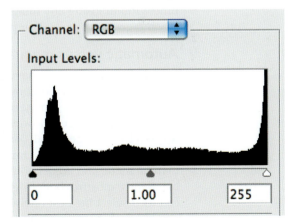

FIGURE 19.5 Levels display from an image with overexposed highlights

clipped at the upper threshold. As long as your intensities aren't going beyond the edge of the levels diagram, you will be able to adjust the exposure after the fact, through linear remapping as described earlier in this chapter.

Color balance

When taking photographs, especially through a microscope, the most common error is to have the camera's **white balance** set incorrectly. Sources of illumination have different intrinsic temperatures—the orange glow of a candle is cooler than the blue flame of a gas torch or stove. In the context of photography, incandescent lights have a yellow tint compared with the bluish light of a strobe or sunlight, so cameras have built-in settings to correct for these differences.

White balance is usually found under the WB or AWB menu on a camera. Set it manually according to the demands of your lighting setup. Use the little light bulb icon if you are using the dimmable light source on a microscope or other incandescent lights, and use the lightning bolt if you are using a strobe. Outdoor or natural light should of course use the setting for sun or clouds, since daylight has a strong blue component to the camera sensor. For other illumination sources, such as LEDs, you may have to experiment a bit. The fluorescent bulb setting was designed for use with fluorescent room lighting, but it actually works well for capturing fluorescent microscope images, since it falls somewhere between the yellow of incandescent bulbs and the blue of a strobe.

Once you are aware of these white balance issues, you will notice right away from the blue or yellow tint of your photos and the photos of others if the white balance was set incorrectly. This sometimes happens when moving back and forth between taking bright-field and fluorescent scenes on a microscope. If your images have already been captured with an incorrect color balance, you can still make adjustments to the hue with careful processing in an image editing program. If your images were taken in RAW format, then you can set the white balance at a later time without losing any image quality. RAW, despite its larger size, also has other advantages, such as a larger potential color space.

Automatic versus manual operation

By default, most cameras have the capability to automatically navigate the settings and trade-offs discussed above, including white balance, aperture, exposure, and ISO. In fact, at times it may be difficult to find out how to change settings manually. Scientific images, however, often present unique challenges that aren't addressed by one-size-fits-all consumer options. Automatic settings, especially white balance and exposure, will rarely give accurate results. You will need to switch to a manual mode and adjust your settings. It is good practice to bracket your options by taking exposures both brighter and darker than you think you might need. You can then select the best one when you are viewing the images on a large monitor and not rushing to finish an experiment.

SUMMARY

You have learned:

- The general strategies for compressing pixel image data, and their implications for working with images

- The trade-offs between the common pixel art file formats, including JPEG with lossy compression and PNG with lossless compression

- The basics of two popular pixel image editors, Photoshop and GIMP

- Tools to edit images at the command line, including ImageMagick and `sips`

- How to extract quantitative data using image analysis tools like ImageJ

- Some principles for taking photographs

Moving forward

- Identify recurrent uses of photography in your research (gel documentation, microscopy, etc.), and see if there are ways to improve or automate repetitive portions of the workflow.

- Open one of your images in Photoshop or GIMP and try adjusting the levels controls to get comfortable with how they affect your picture.

- Visit the gallery at `processing.org` to see examples of interactive plots and try out Processing, a programming environment for graphics. Processing has a low barrier to entry, and can interact with hardware like Arduino as well (see Chapter 22). Other related images are shown in the gallery section of `nodebox.net`.

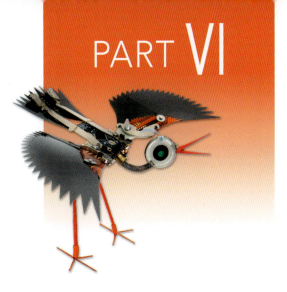

PART **VI**

ADVANCED TOPICS

Chapter 20

WORKING ON REMOTE COMPUTERS

The final chapters of this book delve into more specialized subjects, some of which won't apply to everyone, and others of which are unlikely to be used every day. First among these topics is working on remote computers. Remote tools are necessary if you need to conduct analyses that require more processors, more memory, or more software than you have on your personal computer. The remote computer can be a large cluster shared by many investigators, or a single desktop computer in your lab. We also introduce tools that control how programs run, including their priority and whether they should stop when you log out.

Connecting to a remote computer

Many biologists venture to the command line only when forced to, usually because they have to run analyses on a remote computer that does not provide a graphical user interface. It can be daunting to learn how to use the command line at the same time you are figuring out how to connect to other computers, transfer files to remote accounts, launch programs on large shared clusters, and manage program execution. This is one reason that it is so valuable to become familiar with the command line on your own computer—it provides a foundation upon which you can incrementally add the skills for remote work. In this chapter, which builds directly on the command-line lessons of Chapters 4–6, and 16, you will learn how to get command-line access to remote computers, transfer files, and control analyses. Some of these skills will also be relevant to how you run larger analyses on your own computer.

Clients and servers

You will often hear the words **client** and **server** when discussing network connections. In most situations the client is the computer where the user is sitting and where the connection is initiated from, while the server is the remote computer that accepts the connection and provides a service. There are many kinds of servers. Physically, a server can be anything from a regular personal computer—even a laptop that just happens to be configured as a server—to a massive data center occupying several city blocks. A server can serve many different things to its clients, including disk space, web pages, databases, and computational power. Each such service has its own particular connection needs and therefore requires different software and protocols (i.e., agreed-upon standards for how to interact with other computers).Though client and server speak to each other in a common language, they too require different software and different configuration settings. Finally, for each network protocol, a computer may be set up as a client, a server, or both.

Almost all scientific servers run a version of Linux or Unix optimized for large-scale, computationally intensive tasks. These operating systems and the programs bundled with them can differ in subtle ways from the command-line interfaces of Mac OS X and Ubuntu Linux which we have focused on in this book. The operating systems on remote servers have a longer history and are more specialized than those found on personal computers, and are therefore more varied. As a result, there are some limitations as to the generality of what we can present here. Even things like using the ⬆ key to cycle through your command history, or using the backspace key to delete characters, may not work as expected on a remote system. If you find that you have questions about a particular server, or that the material in this chapter doesn't seem to apply, look for help on the server's support Web site (if there is one), or contact the administrator for the machine you are working on.

Typical scenarios for remote access

There are a few typical scenarios that require the remote access tools described in this chapter. An overview of a couple of these at the outset will help you understand which parts of the chapter are most relevant to you, and how each component fits into the overall objective.

Running analyses on a large cluster First scenario: You need to perform an analysis that would take weeks or months on your laptop. After investigating the computational resources at your university, you find that you can use a Linux cluster available on campus. The cluster consists of a group of computers linked together via a local area network, effectively comprising a single computer with several hundred processors. You contact the system administrator, who confirms that the software you need is already installed on the cluster and is configured to use multiple processors at once.

The administrator creates an account for you, and provides all the information you need to log in remotely. You verify that you can log in. On your laptop you prepare the material you will be working with, making sure that you have all the files you need, that their data are correctly formatted, and that folders are in place to hold both data and analyses. You compress all this into a file archive and upload the archive to the cluster via a secure file transfer program. You then log into the cluster at the command line and expand the file archive. Following the instructions on the cluster Web site, you start the analysis with a special shell script that provides access to multiple processors. You log out, and several days later get an e-mail notifying you that your job has completed. You log into the cluster, compress the result files into an archive, and then transfer that archive back to your laptop to investigate the results.

Checking in on a lab computer from home Second scenario: Imagine that your lab has a large multiprocessor desktop computer off in a corner. It is connected to your favorite instrument, which is automated and takes a few days to collect data on your samples. The lab manager has configured this computer as a server, so that lab members can gain command-line access to it and transfer files remotely. You live a half hour from the lab. One Sunday morning, you wonder how the current run on the instrument is going. You log in to the lab computer and check the data files. Everything is on track, and you can go to the North Shore instead of making an unneeded trip to the lab just to check the status.

Finding computers: IP addresses, hostnames, and DNSs

In order for computers to connect to each other on the Internet, they must be able to find each other. Understanding a bit about how this is done will make the process more transparent and help you troubleshoot some of the most common connection problems.

All computers on the Internet—whether a web server or a laptop computer connected by WiFi at a library—have a network number, known as an Internet Protocol address. Traditional **IP addresses**, such as `173.194.33.104`, consist of four 8-bit numbers—that is, four numbers ranging from 0 to 255.[1] You can think of the progression of these four numbers as successively narrowing the address down to a particular location. The first number in the series is the broadest, and the fourth and final number is the most specific, identifying an individual network connection (e.g., a computer, instrument, smartphone, etc.). Your school, company, and even your local coffee shop might therefore have addresses where the first few numbers are the same, and only the final one or two numbers differ between computers.

IP addresses are cumbersome for humans to remember and type, so most computers are also assigned a text-based hostname. The hostname is familiar as the base of a Uniform Resource Locator, or URL, in Web addresses, such as `sinauer.com` or `practicalcomputing.org`. In order to connect to a computer with a given hostname, however, the hostname first has to be translated to an IP address. When you type a URL into your web browser, your computer finds the IP address for the specified hostname and then makes a request for the indicated files from the computer identified by the IP address.

Translation between names and addresses is the job of Domain Name System servers, which collectively function as the Internet's equivalent of a phone

---

[1] This traditional scheme for encoding IP addresses is known as IPv4. Under IPv4, each IP address is a 32-bit number (four 8-bit numbers together). However, only about four billion devices can be addressed under this scheme, and these addresses have nearly all been used. To get around this problem, a new protocol called IPv6 has been developed which uses up to 128-bits per address; this will provide more than 3×10^{38} addresses altogether.

book. There are many DNS servers, each with their own copy of the master DNS database. This database is updated many times a day as new hostnames and IP addresses come online. Often your Internet Service Provider will specify a DNS server to use, or the local network may be configured so that computers can automatically connect to the local DNS server without custom configuration. There are also several publicly available DNS servers, such as Google Public DNS (IP address 8.8.8.8), which will work from almost anywhere.

As the Internet has grown, more flexibility has been added to the relationship between hostnames and IP addresses. Multiple hostnames can point to the same IP address; among other things, this allows more than one Web site to be hosted on the same computer. Because there is not always a one-to-one correspondence between IP addresses and hostnames, when you look up the IP address from a hostname, and then do the reverse search, you may get a different IP address than you started with. The situation is further confused by the fact that many hostnames are aliases (shortcuts) to other hostnames.

Translating between IP addresses and hostnames The command-line program host allows you to find the IP address for a hostname. (The program nslookup does the same thing.) In some cases, it can also find a hostname if you give it an IP address. This command is dependent on having a network connection, because it sends your query out to a DNS server and returns the result:

```
myhost:~ lucy$ host www.jellywatch.org
www.jellywatch.org has address 134.89.10.74
myhost:~ lucy$ host 134.89.10.74
74.10.89.134.in-addr.arpa domain name pointer pismo.shore.mbari.org.
```

Sometimes, the name or IP number you are trying to find with the host command won't have any available information and the lookup will fail. Even when host fails, however, you can get some general information with the whois command. The whois command requires that you provide the name of an online whois database with the –h parameter, which is then followed by the address you want more information for:

```
whois -h whois.arin.net 75.119.192.137
```

This will return the registration information for the entire block of addresses from 75.119.192.0 to 72.119.223.255, which includes the IP address you specified.

The special hostname localhost The hostname localhost always points to the computer that you are logged onto at the moment, and it always has the IP address 127.0.0.1. Try checking this using the command host localhost. If you have web sharing enabled on your computer, http://localhost can even be used as an address in your web browser. (You will get a "could not connect to localhost" error if you don't have web sharing turned on for your computer. For OS X, this can be turned on in the Sharing pane in System Preferences.)

Security

When working on computers remotely, security is of paramount concern. You should be thinking of security every step of the way. It should go without saying that you never give your password to anybody, not even system administrators. If a system administrator needs access to your account, they can reset the password themselves, and you can change it again later. This is standard etiquette, and any deviation from it may be grounds for concern. Use different passwords for different computers so that they won't all fall victim if the password for just one is compromised.

If you are configuring a server, keep in mind that you are opening it up to possible attacks. Don't enable any network protocols or services that aren't essential, as each protocol is a possible vulnerability. Make sure that all software is kept up to date, since many updates patch security vulnerabilities that will shortly become common knowledge to hackers. One common strategy to prevent unwanted access is to restrict the IP addresses that can access particular services, for instance by limiting access to only computers on the same campus. The method for doing this depends on the operating system and network protocols you are using.

Security should also be a concern when you are logging into a remote server from your own computer. Keep your own computer software up to date; this will help ward off keyloggers and other malicious programs that could steal login information and then compromise the other computers you connect to. Wherever possible, use encrypted connections that prevent others from listening in on your connection to the remote computer. Avoid connecting to servers over open network connections, such as unsecured WiFi at a coffee shop, without turning on tools (such as VPN, described later) that encrypt your connection.

Secure command-line connections with `ssh`

Much of this book has focused on learning to interact with the computer in front of you, using the command line. You are now familiar with the process of opening a terminal window, which launches a shell program and gives you a prompt, and then using it to navigate and control the computer attached to the screen and keyboard. As you will see, it is just as simple to use the command line to interact with a remote computer over a network connection. You will still sit at your own computer with your own screen, keyboard, and terminal window, but now the window will be relaying your commands to the remote computer and returning the results. It is as if you were physically sitting at the remote computer (which, in reality, may not even have a keyboard or screen).

Remote command-line logins are made with a network protocol called `ssh` (secure **sh**ell). The `ssh` program creates an encrypted connection between your client computer and the remote server, so that when you type at the terminal window in your computer this window is a connection to the remote computer. The data transmitted via `ssh` are encoded so that it would be difficult for anyone listening in along the way to figure out what is being said.

The `ssh` command

Since we don't have a server that we can open up to every reader of the book, it won't be possible to follow along with remote access commands in this chapter unless you have your own server access. If you are affiliated with a university or company, you probably already have an account on a server, though you might not know about it. The end of the chapter has information on setting up a server, which you can use to configure a server of your own to test with some of the examples here.

You need at least three pieces of information to log in to a remote computer with `ssh`. First, you need the remote computer's address. This can be the IP address or a human-readable machine name, such as `myhost.myuniversity.edu`. Second, you need a username, sometimes called an account name. Third, you need a password. The system administrator will create your account and provide these pieces of information to you. Don't be surprised if they won't email them to you—the content of an email is visible to many computers en route and is a dangerous way to send security-critical information such as login credentials. Some administrators will email your username and the address of the computer, and then call you with the password.

The first thing you should do when you log in is change your password with the `passwd` command. You can follow the simple prompts to accomplish this. The command to launch a connection to a remote server with `ssh` has the following general structure:

```
ssh username@address
```

where `username` and `address` are specific to the server you are connecting to. After entering this command and pressing [return], you will be prompted for a password. Enter the password provided by the administrator, and press [return]. If your login credentials are correct, you will then be greeted by a command-line prompt served up by the remote computer.

The first time you connect, the server will give you a warning that it has received a new key, and ask if you want to add it to your keychain. This key is a bit of information that allows the two computers to set up an encrypted communication channel. Accept the new key. If you are ever prompted to accept a new key for the site again, contact the administrator for the remote system. It could simply be a change to the remote computer that necessitates a new key, but it could also be an attempt by a nefarious party to intercept your communication.

Troubleshooting `ssh`

When logging into a remote system, you might get this error:

```
ssh_exchange_identification: Connection closed by remote host
```

This is usually an indication that the host on the other end did not approve of the address of your computer. Depending on what the system allows, there are two possible solutions. One is to set up a Virtual Private Network (VPN) connection to make your computer's address seem as though it is located internally alongside the system you are trying to connect with; see the section about VPN later in this chapter. If this capability isn't available, the other solution is to contact the system administrator of the remote machine and get him or her to add your address (or the range of addresses from which you might connect) to the list of allowed clients.

Once you have logged into a server from one computer, if you later log in from a different computer you may get an error of varying severity. For example:

```
IT IS POSSIBLE THAT SOMEONE IS DOING SOMETHING NASTY!
Someone could be eavesdropping on you right now !
It is also possible that the RSA host key has just been changed.
The fingerprint for the RSA key sent by the remote host is
...
Host key verification failed.
```

In such cases, the secure key passed between the systems has changed. Edit the `~/.ssh/known_hosts` file in your home directory to remove the line for the host you are trying to connect with. When you reconnect again the next time, it will generate a new key pair for you.

Working on the remote machine

The `ssh` program provides seamless command-line access to a remote computer, in the sense that it is no different from being physically located in the same room with the server. You may find, however, that your command-line experience on the remote computer is quite different from working on your own computer. This is because the remote computer may be running different software and have unfamiliar configurations. One of the most common surprises is that the shell program itself may be different. Even if you are using the `bash` shell on your local machine when you launch `ssh`, you will interact with the remote computer using its own default shell, which may or may not be `bash`. This shell starts once the `ssh` connection is active. If it is a different shell than `bash`, you may notice some subtle differences in the way it responds and how the prompt is presented. It is therefore a good idea to check which type of shell it is, by typing `echo $SHELL`. If the remote shell isn't `bash`, but instead one of the common alternatives such as `tcsh`, `sh`, or `zsh`, consult the online documentation for that type of shell to familiarize yourself with the most critical differences between it and `bash`.

Your account on the remote machine will have a home directory, independent of the home directory on your own computer. When you first log in you will have default shell configurations (these will be located in `.bash_profile` if the remote shell is `bash`); your `PATH` will only include directories set at the system level; and your home directory will be largely empty.

Once you have command-line access to the remote system, you can interact with the remote file system with the same command-line tools you use on your local computer. For instance, you can create folders on the remote computer with `mkdir`, change directories with `cd`, and list folder contents with `ls`. One of the first things you'll want to do upon logging into a remote system is explore the files and folders in your home directory (there is often a README file that explains system policy and use) and set up new folders to store data and analysis results.

While `ssh` provides a way to log in and operate a remote computer, it doesn't provide direct access to files on the remote machine. You will have to use other tools, as described in the next section, to transfer files back and forth. You can accomplish other tasks entirely on the remote machine. For example, to make small changes to a remote file, it is often more convenient to edit it with a command-line editor such as `nano`, rather than transfer the file, edit it on your own computer, and then transfer the edited file back to the remote machine. (See Chapter 5 or Appendix 3 for a `nano` refresher.) Being comfortable with a command-line editor when logged in via `ssh` will be an important skill for your remote computing.

Transferring files between computers

Once you are able to use the command line on a remote machine, you will probably want to get some work done. First, however, you have to be able to transfer data and analysis files between your computer and the remote computer. There are several options. For the most part, these options are interchangeable, but you may find that some are more convenient than others for your particular situation.

File archiving and compression

You will often want to transfer a collection of files rather than one single file. Not all file transfer tools can be used to transfer directories of files. You could transfer the files one by one, but this quickly gets tedious as the number of files grows. If the tools are available on your server, it is often easiest to use the `zip` command to merge the files into a single archive and compress them before transferring them. Try the `man  zip` command to see some of the compression options. If `zip` is not an option, then a solution that is assured to be present on all Unix servers is creating a **tarball**. You can combine multiple files into a single file archive with the command `tar`:

```
myhost:~ lucy$ tar -cf ~/Desktop/scripts_25Jun2011.tar ~/scripts
```

The example above creates a file archive called `scripts_25Jun2011.tar` in the `Desktop` folder. The `-cf` argument specifies that you want to create an archive with the name `~/Desktop/scripts_25Jun2011.tar`. (By convention, archive files are given the extension `.tar`.) The files and directories you would like to archive are listed last; there can be one or more entries, separated by spaces. In this example, the directory being archived is your `scripts` folder.

File archives created with `tar` contain all the data in the original files; the files are just joined together into one big file. Making an archive with the command above leaves the original files in place, exactly as they were. It is good practice to include the date in the name of the archive. This makes it easy to understand at a glance what the archive contains, and allows you to store archives of the same files in the same folder without any name conflicts.

A `tar` archive is roughly the same size as the total size of the files included in it. To save Internet bandwidth and hard drive space, archives are often compressed with the command-line program `gzip`. Applying `gzip` compression to ordinary files produces files with a `.gz` or `.Z` file extension; applying it to a `tar` archive typically produces a file with a combined file extension of `.tar.gz` or `.tgz`.

The following command will compress the archive you created above:

```
myhost:~ lucy$ gzip ~/Desktop/scripts_25Jun2011.tar
```

This creates a new file on the Desktop, `scripts_25Jun2011.tar.gz`, and deletes the original uncompressed file. The compressed file will typically be much smaller than the original file.

The `gzip` command has a counterpart, `gunzip`, for uncompressing whatever has been compressed. Likewise, the `tar` command has arguments which allow an archive file to be expanded back into its constituent files and directories. So it is quite common to use these two commands in sequence to get the original files back out of a compressed archive:

```
myhost:~ lucy$ cd ~/Desktop
myhost:~ lucy$ gunzip scripts_25Jun2011.tar.gz
myhost:~ lucy$ tar -xvf scripts_25Jun2011.tar
```

The `tar` argument `-xvf` specifies that you want to expand the indicated file archive, which in this case is `scripts_25Jun2011.tar`. Once expanded, the directory and file names will be the same as before they were archived. (That is why we created the archive on your Desktop: we didn't want to clobber your original scripts directory). When restoring files, the archive is not deleted after it has been expanded.

Together, `tar`, `gzip`, and `gunzip` constitute a powerful trilogy of programs for storing and compressing all types of files, one you will likely find useful in many situations. You may frequently use these programs before transferring data to and from servers. Many data and software files you download from the Internet will be gzipped tar files, and you will use `gunzip` and `tar` to expand them. Compressed archives are also a convenient format for sharing collections of files with colleagues.

File transfer with `sftp`

The command-line program `sftp` (**s**ecure **f**ile **t**ransfer **p**rogram) can be used to transfer files over an `ssh` connection, via Secure File Transfer Protocol, or SFTP. Because it uses an `ssh` connection, the transfer is encrypted, you don't need to enable other types of connections to the remote machine, and it can be used anywhere you have `ssh` access. Opening an `sftp` connection is similar to running `ssh`, but once connected, a link is maintained between the local and remote file systems, allowing you to use the commands get and put to move files back and forth. To transfer files, move to the directory on your local machine where you want to send or receive files. Then connect to the remote server with the command:

```
myhost:~ lucy$ sftp lucy@practicalcomputing.org
Connecting to practicalcomputing.org...
lucy@practicalcomputing.org's password:  ←Type password
sftp> ls
```

Once connected, you will have a new command-line prompt, provided by the `sftp` program. The most common commands for `sftp` are listed in Table 20.1. The get and put commands can take wildcards such as `*.txt` as part of the file-name descriptors. If you get an error when trying to transfer several files, try the command `mput` or `mget` instead, which is set up for transferring multiple files.

Copying files with `scp`

It can be confusing to use a command line to simultaneously navigate through remote and local file systems, so you may prefer using the secure remote **c**o**p**y command `scp`. This functions just like the `cp` command, except that instead of just providing paths, you provide a username and a server name as well. The basic format for downloading a file is:

```
scp user@hostname:directory/remotefile localfile
```

TABLE 20.1 Some common commands when using `sftp`

Command	Usage
cd	Change directories on the remote machine
get A B	Download file A from remote machine and save as B locally
put B A	Upload file B from the local machine to the remote file A
lcd	Change directories on the local machine
!ls	List the files on the local machine
!command	Perform the specified shell *command* on the local machine
exit	Close the `sftp` connection

For uploading, you just specify the hostname as part of the second argument:

```
scp localfile.txt user@hostname:remotefile.txt
```

Here you can move to the appropriate local directory before running the command, so that you don't have to type the full path to the localfile. You will however have to specify the path to the file on the remote system, giving the directory name relative to your home directory on that system.

Other file transfer programs using SFTP

An even simpler SFTP option is to use a GUI program to handle file transfers and renaming, allowing you to drag files as you might in the Finder. You can try the shareware programs Cyberduck or Fetch, or the commercial program Transmit. All of these will present a window that will let you drag files to and from your remote machine. In Linux, you can get equivalent capability using FileZilla, which is available for Windows and OS X as well.

Other file sharing protocols

Many operating systems have built-in file sharing protocols which let you connect to a remote computer as though it were a disk drive. In OS X, you can use the Finder menu bar (or ⌘K) to select Go ‣ Connect to Server and enter the address of the remote computer. Windows computers can be accessed using the prefix `cifs://`, and other Macs can be accessed using `afp://` followed by the `username@servername` format used with `scp`. The Connect to Server dialog box can also be used for VNC connections, as described in the next section.

Full GUI control of a remote computer with VNC

Although we are focusing on command-line interaction with remote computers, it is possible to access the complete graphical user interface of a remote computer using Virtual Network Computing, or VNC. While `ssh` enables you to work as if you were looking directly at a terminal window on the remote computer, VNC enables you to work as if you were not only looking at the entire screen of that remote computer, but physically sitting in front of it and able to use its keyboard and mouse. This extra capability can be useful for modifying system settings, or for running programs unavailable at the command line. It can also be used to access your own desktop system while traveling.

VNC has a few limitations. It requires much more network bandwidth than `ssh`, and if there is a slow connection between the computers it may be so sluggish as to be entirely unusable. Some implementations of VNC do not encrypt data before it travels over the network, meaning that someone could eavesdrop on your activities and even control your computer. (One of the authors lost several weeks' worth of work when a computer was hacked in this way.) Be sure you have encryption enabled. Also, some remote servers, particularly large clusters, don't have a GUI installed in the first place—and VNC can't provide access to a remote GUI that isn't there.

There are two components to a VNC set-up: VNC server software running on the remote computer, and VNC client software on your own computer that lets you interact with the remote server. Such software is available for all platforms, even including smartphones. In OS X, VNC capability is built into the file sharing system. You can set up your computer as a VNC server that supports incoming VNC connections using the Sharing panel in System Preferences, enabling either Screen Sharing to allow people to connect to your computer, or Remote Management to enable screen sharing in addition to other remote login options. You must set up a password for access at the same time, to keep people from controlling your computer without your permission. For Windows, RealVNC offers both a server and a client. For Linux, VNC server capability is also available, but to configure it securely follow the instructions at `help.ubuntu.com/community/VNC`.

Starting up a VNC server opens port 5900 (a portion of your network connection to the outside world) for traffic. Once you have turned on the VNC server on a computer, you will be able to connect to it using its IP number or address. Before you try connecting from an outside client, you can test whether the server is set up correctly by visiting this address in a web browser:

> `http://www.realvnc.com/cgi-bin/nettest.cgi`

It will tell you what your IP address is, and will try to send traffic to you through port 5900.

To view the screen of a remote computer running a VNC server, you access the connection as a client. VNC client software is available in OS X without downloading additional software, although you can also download and install a third-party program such as Chicken of the VNC. To connect using OS X's built-in client, use the Finder to select Go ‣ Connect to Server (or ⌘ K); in the dialog box which appears, type the prefix `vnc://` followed by the address of the remote machine. Although you can mouse around and otherwise operate a remote computer with VNC, you can't typically use it to transfer files.

Troubleshooting remote connections

Getting local with a Virtual Private Network (VPN)

Some services (e.g., access to an institutional library's Web site) and some network connections (e.g., `ssh`) block computers attempting to connect from outside the local network. Because every local network uses a particular range of IP addresses, your IP address off-campus will be different than the one you have on-campus. If you are off-campus, servers can see this and deny you access, if configured to do so.

A way around this is to use a virtual private network, or VPN.[2] Upon verification of your credentials, a VPN connection will pass your network traffic through

---

[2] VPN and VNC are often confused with each other in the heat of battle, but they are very different network tools.

a computer on the local network, effectively giving you an on-campus IP number. This will allow to access resources otherwise forbidden to computers off campus. VPN has the added benefit of being encrypted. If you are in a coffee shop using an unsecure WiFi connection, you can fire up the VPN to encrypt all Internet traffic between you and the VPN server on campus. This will prevent anyone from "sniffing" your passwords and other data over WiFi.

On OS X, set up a new VPN connection in the Network panel in System Preferences. Click the + below the list of settings, and choose VPN from the pop-up list. For VPN Type, you can try selecting PPTP and typing `vpn.` before your institution's domain name (e.g., `vpn.yourschool.edu`), but if that doesn't work, you will have to search your computer support pages or contact a system administrator to get the exact settings to use. Some VPNs require that you install special software.

Mapping network connections with `traceroute`

A useful command for diagnosing your inability to connect to a remote site—whether you are attempting an `ssh` connection or accessing the site via the Web—is to trace the network steps between your system and that address. This can be accomplished with the `traceroute` command. Simply type `traceroute remoteaddress` in a local terminal window, and you will see the hops between you and the server at your remote study site on a tropical island a hemisphere away. Test this by tracing the route between yourself and some of your favorite Web sites.

Configuring the `backspace` key

Your `backspace` key may not behave as expected when you are logged into some servers. Why? Because as with end-of-line characters, two different characters have been historically used to indicate a deletion or backspace. If you see ^H (`ctrl` H) whenever you try to delete something at the command line, try changing the backspace character using the `stty` command.

Begin by typing:

```
stty erase '
```

After the single quote, press the `backspace` key on your keyboard. Instead of deleting the quote, it should cause the terminal to display the ^H or ^? (a single keypress will cause these two characters to appear) that you were getting previously. Close the string with another single quote mark and press `return`. Your ability to backspace should be restored.

To make this change permanent on the remote machine, use `nano` to edit your `.bash_profile` (or whatever the appropriate configuration file happens to be, if `bash` is not the default shell on the remote machine). Add the line:

```
stty erase '^H'
```

To type the [*ctrl*]H within this file as you edit it, you will have to use the trick of typing [*ctrl*]V first before the [*backspace*] key. Remember from Chapter 16 that this forces a literal interpretation of the next key typed, whether a [*return*], [*backspace*], or [*esc*].

Two other features are sometimes missing on remote machines. The first of these is tab auto-complete at the command line, which is sometimes disabled on servers. Like the [*backspace*], enabling it requires shell-specific modifications. The second is the use of the [↑] key to step back through your command history; this is controlled by the readline library, and it may be missing entirely or simply not yet configured. Search for the version of the system software you are accessing, or contact your system administrator to get these features configured to your liking.

Controlling how programs run

In most cases when using a remote server, you will transfer files to the remote machine and then launch computationally intensive analysis programs. How you launch and manage programs on the remote machine will depend on the type of computer you are accessing, who manages it, how many people are using it, what the system policy is, and how it is configured. Remotely managing analyses on a dedicated lab desktop computer is different than launching programs on a large cluster with many users, where special job management software ensures fair access to the processors and keeps different analyses from interfering with each other.

In this section, we describe tools for managing how programs run on regular computers—in other words, computers that aren't large shared clusters. Tools more relevant to large-scale analyses are described in the next section. Some of the tools addressed here will be helpful even for controlling how programs run on your own computer, so they really aren't specific to remote access. We then follow up with a brief overview of running analyses on clusters. It is difficult in this context to provide anything more than general advice, since there are so many different types of clusters.

To help demonstrate how to play with operations on a regular computer, we will use the `sleep` command as a stand-in for an analysis program. This program does nothing for the specified number of seconds, then stops. Try it out by typing:

```
host:~ lucy$ sleep 15
ls
```

Notice that when the `sleep` command is running, you don't have a shell prompt, and you lose the ability to run other commands until the operation is complete. (For example, type `ls` while you are waiting.) Any commands you type while `sleep` is running will be run after `sleep` quits.

Terminating a process

A running program is referred to as a **process**. The ability to terminate a process is an important skill for remote operations. This is because you may not be able to physically access that machine if something goes wrong, yet will still need to be able to stop a crashed program or a faulty analysis that is taking up valuable computing resources.

We have mentioned ctrl C a few times in passing. This is the interrupt command. It tries to terminate a program that is active at the command line. You type it by holding the ctrl key while pressing the C key. Because ^ represents ctrl in shell operations, this command is frequently abbreviated ^C.

Try typing the sleep command below and then pressing return :

```
host:~ lucy$ sleep 1000
```

This sleep command, if left undisturbed, would stop after 1000 seconds. To interrupt the running process before then, type ctrl C and you will get your shell prompt back. Some programs—for example, nano, and of course anything on Windows—interpret ctrl C differently. It isn't a universal command, but it is very widespread.

Starting jobs in the background with &

Usually when you run a program, you lose the command line for as long as the program will take to execute. However, if you type an ampersand (&) by itself at the end of the command line before pressing return , the program will run in the background, and you will get your shell prompt back immediately. Although these backgrounded commands are running unattended, they would cease operation if you were to close the window.

Try a sleep command with & at the end of the line:

```
host:~ lucy$ sleep 15 &
[1] 4990
host:~ lucy$
```

When you background a task like this, the prompt is immediately available, but the program is still running in the background (in this case, with job number [1] and the unique process ID number 4990) until the 15 seconds have passed. This is very different from terminating a process, in which case the process isn't running anymore at all. You can't use & to send a process to the background after it has begun, only at the time you start it.

Checking job status with ps and top

To check the status of all programs that are running, including those in the background, you can use the commands ps and top. The ps program gives a snapshot of current processes. Try entering your `sleep` command as above, and then at the command line, run `ps`:

```
host:~ lucy$ sleep 15 &
[2] 4992
host:~ lucy$ ps
  4800 ttys001    0:00.02 -bash
  4992 ttys001    0:00.00 sleep 15
```

First you are informed of the process identifier, or PID, for the `sleep` process that was put into the background. Every running process has a unique PID that can be used for job control operations. The `ps` command then tells you the PID for the processes that are running in the current shell. One of these is `sleep`, and the other is the `bash` shell itself. This list is not a full list of everything your computer is doing. To see that list, use the command:

```
ps -A
```

The output of this command will likely be cut off at the right edge, since many of the program names are quite long. This command shows all the processes that are running, including all the housekeeping utilities that do things like listen for when you plug in a new mouse and check periodically for new network connections. Even the Finder and your graphical programs are in this list; this may seem strange at first, but they are processes running on your computer as well.

To find a certain process in the long list, you can pipe the output of `ps  -A` to a `grep` command to look for the name in the output stream:

```
ps -A | grep Finder
```

This should show you just one line of output with information for that program.[3] This command is most often used to find runaway processes that you might want to stop, using the `kill` command described later in this section.

The `top` program provides a real-time (continuously updating) view of the same list of processes as `ps`. To see a basic list of processes, sorted by PID, just type `top` by itself. You can quit the `top` display by typing q, as for the program `less`. To see a list sorted reverse-alphabetically by command names, use the `-o` modifier and specify that you want to sort by command name:

```
top -o command
```

Other options for `top` are shown in Table 20.2.

Other systems may use lower-case −a.

---

[3] In Linux or Cygwin, you won't have a Finder program, but the command should work the same with other program names.

Open a second terminal window, and place it to the right of the window running the above command. Now you can use the left window to watch your activities in the right window. Rerun the `sleep 15` command in the right terminal window. You should see the `sleep` program listed alphabetically in the top window for 15 seconds before it eventually finishes and disappears.

In addition to the basic list of processes, `top` provides information on how much memory and Central Processing Unit (CPU) power each process is taking. This is a convenient way to keep an eye on computationally intensive programs and quickly get a sense of system status. If you use `ssh` to log into a shared lab computer to start an analysis, the first command you should run is `top` to make sure that someone else isn't already using all the processors for other analyses. If your own computer is getting hot for no apparent reason, you can run `top -u` and see what the cause might be.[4] We will show you in a bit how to terminate these processes.

TABLE 20.2 Options for the `top` command

`top -o`	Sort by the parameter that follows; default is `pid`, but with `-o`, command will sort by program or process name
`top -u`	Sort by CPU usage; useful for detecting runaway processes
`?`	While `top` is running, show a list of display options
`q`	While `top` is running, quit

Suspending jobs and sending them to the background

If you run a command without the final `&` and only later realize that you want to send it to the background, you can first suspend the operation by typing [ctrl] Z. This is different from [ctrl] C in that the program is merely frozen, not terminated. It is also distinct from putting a job into the background in that no operations are proceeding. Try this out with the `sleep` command, set for a relatively short duration, followed by [ctrl] Z:

```
host:~ lucy$ sleep 15
^Z
[1]+  Stopped                 sleep 15
host:~ lucy$ ps
  PID TTY           TIME CMD
 5760 ttys001    0:00.02 -bash
 5954 ttys001    0:00.00 sleep 15    ← This job is stopped, not just backgrounded
```

---

[4] Ahem, Flash content?

The output looks very similar to what you got when running `ps` after you launched `sleep 15` `&`. In both cases the process is listed along with its PID, and you are able to interact with the shell even though sleep was started and has not finished. However, the big difference is that if you keep checking `ps` for more than 15 seconds, the process will still be listed; it is suspended and not running to completion in the background.

To list all the background processes, use the `jobs` command. First add another `sleep` operation so that it is running in the background:

```
host:~ lucy$ sleep 20 &
[2] 5989
host:~ lucy$ jobs
[1]+  Stopped                 sleep 15
[2]-  Running                 sleep 20 &
```

Notice that our new operation gets job number `[2]`, and that it is listed as `Running` while the first `sleep` job that was suspended with `ctrl` Z is listed as `Stopped`. To resume job `[1]` (the stopped `sleep` command in the background), use the `bg` command followed by the job number:

```
haeckelia:~ haddock$ bg 1
[1]+ sleep 15 &
[2]    Done                    sleep 20
```

Our updated job list now shows `sleep 15` now running in the background, while job `[2]` has already finished.

There is also a `fg` command that will take a suspended job and move it into the foreground. You can use this combination of `ctrl` Z, `jobs`, `bg`, and `fg` to move processes between active and inactive states and between the background and foreground.

Stopping processes with `kill`

Once you send a job into the background with `&` or `bg`, it will be impervious to interruption by `ctrl` C. This is because keyboard input can only be passed to programs that are in the foreground. You will also not be able to interrupt it by normal means. The command to terminate any process is suitably named `kill`. This operates using the PID as displayed by the `ps` and `top` commands (aha!).

To test your powers, use a background `sleep` process, but this time set it to go for 60 seconds. Then use `ps` or `top` to find out the PID it has been assigned. Finally, `kill` the process, and then check again with `ps` to see that it is indeed gone:

```
host:~ lucy$ sleep 60 &
[1] 5563
host:~ lucy$ ps
  PID TTY            TIME CMD
 5089 ttys000     0:00.06 -bash
 5563 ttys000     0:00.00 sleep 60     ← The first number is the PID
host:~ lucy$ kill 5563
host:~ lucy$ ps
  PID TTY            TIME CMD
 5089 ttys000     0:00.06 -bash
[1]+  Terminated              sleep 60
host:~ lucy$
```

On some occasions, a recalcitrant program may not be terminated by the `kill` command. When this happens, the way to force it to quit is to use the top secret command `kill -9`. If the process still can't be terminated because you don't have permission to interfere with it, you can also override this with `sudo`. The combined effect is lethal:

```
sudo kill -9 5563
```

That will stop just about any process, so use restraint (especially when the processes belong to other people on a shared computer).

There is another `kill` command, `killall`, that lets you interrupt programs by name, rather than by PID. This includes forcing termination of GUI programs that are running in a Finder window. This command too must be used with caution because it will kill every process whose name matches. For example, if you run several `sleep 30 &` commands in a row, a single `killall sleep` command will terminate all of them. If you are running OS X, see what happens when you type this:

```
killall Dock
```

Under certain circumstances, the `kill` command can save you from losing data or unsaved work, even when working in a GUI and not via the terminal. When a computer freezes up due to a certain program, such as a web browser, you are often still able to log into that computer remotely using `ssh`, as long as it was set up to do so in the first place. Once connected, you can then use `top -u` to find the PID or name of the offending process (look for the CPU hog or the name of the program you know is frozen). The `kill` command issued from this remote machine should be able to free up the system and let you save work in other programs that were running at the time.

Keeping jobs alive with nohup

When you terminate a shell by closing a terminal window or logging out of an ssh session, all the programs running in the shell stop too. This happens even if the processes are in the background. It means that when you connect to a remote server to start a program, the program will stop when you log out. If the program is likely to take hours, days, or even weeks to complete, needing to keep your personal computer continuously connected to the server for that entire time would be a huge inconvenience.

 To get around processes terminating when the shell closes, you can insert the nohup command (**no h**ang-**up**) before the name of your program when you launch it. This indicates that the program should keep running even when the shell used to start it is long gone.

Using nohup is only one of the steps you need to assure that your program keeps running. If your program provides output or error messages along the way, when they are generated but their destination (your terminal window) is no longer present, that can cause the job to cease. We have not often asked you to use a command without explanation, but describing the inner workings of the full command we are about to present is beyond the scope of this book. If you find it useful, and it works, then you can just take it on faith that there are reasons for the strange syntax that follows.

Imagine that your program would normally be run with the following command, redirecting the output to the file log.txt:

```
phrap infile.sff -new_ace > log.txt
```

To rephrase that so it will run unattended, insert your command between the bolded code like this:

```
nohup phrap infile.sff -new_ace > log.txt 2> /dev/null < /dev/null &
```

The initial nohup command lets the program run without interaction after you log out. The rest of the text following 2> sends any error messages to an imaginary place called /dev/null—essentially redirecting them to nowhere. Lastly, the ampersand at the end causes the job to run in the background.

It is often the case that a process takes much longer than anticipated. If you started it on a remote computer and expected it to take only a few minutes, you probably wouldn't have launched it with nohup. If it were still running several hours later when you needed to pack up your computer to go home, you might not want to terminate the process and lose all the progress it has made.

Fortunately, there is a way to modify a process after it has started so that it won't terminate when the shell closes. First, suspend the process with [ctrl] Z, and send it to the background using bg 1 (or whatever actual job number is reported).

Although the process is in the background, it still belongs to the shell and will close when the shell does. To free it from the shell so that it doesn't close when the shell does, run the `disown -h` command. You should probably test this command before leaving an important job for a long time: if the program is producing output and wasn't invoked in the output-suppressing manner of `nohup` above, then it may still terminate when you log out.

Changing program priority with `renice`

Unless an analysis program is limited by other factors, such as hard disk speed or RAM, it will use all the processor power it can until it has completed. Many programs can only use one processor at a time, regardless of how many processors the computer has, meaning that they will max out a single processor. Other programs are designed and configured to use multiple processors at a time. In either case, if there are multiple programs running at the same time, they can start to compete for resources. When the computer is saturated with running programs, each new program that is started will slow down the rest.

However, usually some processes are more important than others. This being the case, you can adjust the priority of each process with `renice` so that it gets a larger or smaller fraction of processor power than other programs. The `renice` command takes two arguments: the priority you want to assign to a process, and the process's PID, which you can get with `ps`. The priority can be an integer running from `-20`, the highest priority, to `19`, the lowest priority. Processes are assigned a priority of 0 by default.

A typical `renice` event looks like the following:

```
lucy$ ps
 PID TTY           TIME CMD
29932 ttys000   0:00.01 -bash
29938 ttys000   0:00.06 raxmlHPC -b 13 -# 100 -s alg -m GTRCAT -n TEST
lucy$ renice 19 29938
```

In this case, the `raxmlHPC` process is reduced to the lowest priority.

If you are running long analyses on all the processors of a computer that is intermittently used for other purposes, such as taking pictures at a microscope, you should `renice` the analyses to the lowest priority, `19`. The analyses will then take all the computer power they can when the processor isn't being used for other tasks, but they will cede that power to any other programs that request it (unless other system resources are limiting).

High-performance computing

In some cases, the remote server you are connecting to will be a large cluster specifically designed for high-performance computing needs. Running software on these machines can be different than launching analyses on your own computer or on a small server. This is because the analysis software must by configured and called in such a way that it can use multiple processors, with processes strictly controlled by job management tools that allocate the available resources to many different users.

Here we simply describe some of the issues at stake and mention some of the common tools used to address them. Consult the system administrator of the cluster you are using for details specific to your system.

Parallel programs

Some analysis tasks can only be run on one processor at a time. This is because each calculation requires the result of the calculation that came before it, so spreading the task across processors wouldn't speed anything up. However, many analyses can be divided up into pieces to be analyzed in parallel on multiple processors. Each piece might be an independent replicate in a statistical analysis, or a single part of a very large summation. Tackling a problem with multiple processors requires more than the problem being of the sort that can take advantage of multiple processors: the software itself must be written to use multiple processors.

In the simplest case, an analysis involves going line by line through an input file and performing some calculation that depends only on the contents of that line. This analysis can be made parallel by the poor man's strategy of just breaking the input file up into chunks, analyzing them separately, and then combining the results. You can even do this by hand. If you have four microprocessor cores on your computer, for example, you could break the input into four equal parts, launch four analysis processes (one on each part of the input data, using an ampersand at the end of each command to get the prompt back), and then combine all the result files.

In most cases, parallel analyses are more complicated, and require extensive communication between the processors to coordinate calculations and exchange data. Rather than write software from scratch to handle all these complicated tasks, parallel data analysis programs usually draw on third-party software. The reason this is relevant here is that many software tools that would be used for large analyses on remote computer clusters fall into this category, and it affects the way the analyses are started. Rather than call the analysis program directly, it may be necessary to call the parallel software and tell it the name of the program you want to run, including all the relevant parameters for your analysis. It must be emphasized that these tools cannot magically turn any analysis into a parallel analysis; the analysis software must have been designed from the ground up to be compatible with and use these tools.

One of the most widespread software tools for facilitating parallel analyses is the Open MPI library (`www.open-mpi.org`). To launch a parallel analysis with Open MPI, you call the `mpirun` or `mpiexec` command, depending on how your system is configured. It is possible to install and use Open MPI on your own computer if it has multiple processors and you have programs that can use it. Other such technologies include Pthreads and the multithreading libraries in Boost C++.

Job management tools on clusters

If everyone on a large shared cluster were free to log in and start software however they liked, pandemonium would ensue and nothing would get done. Job management tools provide a way to structure how analyses are run and resources are allocated. If you are running your analysis with a job manager, you will not need the general commands for controlling programs described above (e.g., `renice`, `nohup`, and `bg`). The job manager takes care of all these issues, and using these other commands could actually interfere with your analyses as well as those of others.

When you log into a remote cluster with `ssh` you will have access to many, or even all, of the standard command-line tools you are already familiar with. While it is possible to start programs as you would on your own computer by typing their name and specifying the required arguments, it is standard practice that they will be terminated if they take too long or use too many resources. It is fine to run `gunzip` for a couple of minutes to uncompress your data files, but it isn't okay to call `phrap` at the command line for a five-day genome assembly. Instead, you need to prepare a configuration file for each analysis and submit it to the job manager.

This configuration file is often a shell script that includes supplemental commands to be read by the job manager. There are variables that determine how many processor cores the analysis needs and how these should be distributed across computers. The file sets up log files, and arranges other housekeeping. You then add one or more lines with the commands for the analyses you want to do.

Once the configuration file is ready, you submit it to the job manager. This is akin to getting in line. If there are enough free resources to start your job, it will begin immediately. Otherwise it will wait until other analyses finish.

Two of the most common job managers on academic clusters are Portable Batch System, also known as PBS, and Oracle Grid Engine, previously known as Sun Grid Engine and still commonly called SGE. Each has its own format for configuration files, as well as different commands for managing jobs. If you are using a cluster, ask the system administrator for an example job submission script that you can modify for your own analyses, and for the appropriate commands for submitting and monitoring jobs.

Setting up a server

Up until this point, this chapter has focused on how to connect to and use a remote server. In some cases, though, you will want to build your own server. You may want to configure your personal computer as a server to test and learn particular software packages, or you may want to configure a lab desktop as a server to run analyses or provide remote access to instrument data.

Building a server isn't necessarily as daunting as it might sound. Remember that a server is just a computer that accepts requests for a particular service and then delivers that service. The main tasks in setting up a server are as follows:

- Install and configure server software for the service you want to provide, i.e., ssh access or a web server. Some of the most common server software, such as that for serving web pages and ssh access, is installed by default on many operating systems.

- Configure the firewall on the server to allow access to the service you will provide, and to prevent access for services you are not using. A firewall provides an added level of security against attacks aimed at unused services, but can interfere with proper function of active services if not configured.

- Make sure your local network (i.e., the network in your building or on your campus) allows connections to computers from the outside world. Most local networks have firewalls that protect the entire network, just as each computer's firewall protects that computer. This provides an added level of security, but if all incoming requests are forbidden, you will need to work with your network administrator to allow remote access for particular services on your computer. In extreme cases, institutional policy may forbid remote access to all machines under all circumstances.

- Get the IP address or hostname of your server so that other computers can find it.

Building a server can consist of just configuring and turning on an already installed piece of software. In other cases, things may be a bit more complicated. You may need to install software or work with your network administrator to configure access to the server. Below we walk through some of these tasks in greater detail, focusing on ssh access. The material is relevant to configuring other types of servers, including web servers.

Configuring the ssh server

The ssh server is not installed by default on Ubuntu, but openssh-server can be added with the Synaptic Package Manager.

Before you enable remote ssh login, make sure that you have chosen a login password for your computer, and that this password is not something that can be easily guessed. This same password will be used to authenticate ssh access. A weak password is a serious security vulnerability.

OS X comes with the ssh server software already installed. To activate it, open System Preferences and select the Sharing pane. Click the Remote Login checkbox. This will start the ssh server software and enable remote ssh connections.

The OS X firewall is configured via the Security pane in System Preferences. If your firewall is running, any service enabled in the Sharing pane will automatically be added to the list of permitted services to which the firewall allows access.

Finding your addresses

In order to connect to a computer, you will need to know its address. During your shell session, the hostname of your computer is stored as the system variable HOSTNAME. To retrieve this name, simply type echo $HOSTNAME. To find your IP address, use this variable as input to the host command:

```
myhost:~ lucy$ host $HOSTNAME
myhost.practicalcomputing.org has address 100.200.10.20
```

There are other ways to find your hostname. You can use the shell command ifconfig:

```
myhost:~ lucy$ ifconfig
...
en0: flags=8863<UP,BROADCAST,SMART,RUNNING,SIMPLEX,MULTICAST> mtu 1500
     inet6 fe80::11a:bbff:33bb:aa11%en1 prefixlen 64 scopeid 0x5
     inet 100.200.10.20 netmask 0xffffff00 broadcast 100.200.10.255
     ether 00:11:22:bb:cc:ee
...
```

Because you probably have several kinds of connection options on your computer (Ethernet cable and wireless card, for example) this command typically returns a lot more information than just your IP address, meaning you have to dig through the results. The cabled network address should be found in the lines below the en0: entry, next to where it says inet. A WiFi address will be listed in the lines following the en1: entry. (Unfortunately, the addresses do not appear on the same lines as the en0 entry, so they are not easily obtained by piping through a grep command.)

In our fictitious example, the IP number for the system is 100.200.10.20. This number may change each time you connect to the network, because your local system may dynamically assign you a new number. (You will hear this called DHCP, which stands for Dynamic Host Configuration Protocol.) This number may change every day or so, as well as every time your computer is unconnected and then reattached to the network. This is less than ideal since the number may change while you are away, meaning you will have no idea what the IP address is when you try to log in remotely. There are a couple of ways around this. First, you can use your system name as returned by $HOSTNAME. This is more stable than a dynamic IP address. Second, you can make a request to your network administrator for a fixed IP address. A fixed IP address is dedicated to a particular computer and does not change, unlike an address assigned by DHCP.

There is another potential address complication for remote connections. Some routers, including most WiFi base stations, are configured so that a single IP address is assigned to the router, with the router then setting up a subnetwork that shares this connection among all the computers. The outside world sees only one machine with one address (the router) while the router assigns each computer on the subnetwork its own IP address that only works within that subnetwork. This means that the IP address of the computer on the subnetwork is not the same as the IP address the computer has on the Internet. You can recognize subnet address-es because they usually take the form `10.1.X.X`, `172.X.X.X`, or `192.168.X.X`. In order to remotely connect to a computer on a subnetwork, you will need to specially configure the router to use a technique called **port forwarding**. Search the Internet for that phrase along with the brand and model number of your router hardware to find the relevant instructions.

There are a variety of Web sites that will tell you what your IP address is on the Internet. These include `whatismyipaddress.com`. If your Internet IP address reported differs from the IP address shown by `ifconfig`, it is because the Internet IP address is assigned to your router and the IP address indicated by `ifconfig` is your IP address on the subnetwork.

Connecting to your own computer with `ssh`

Once the `ssh` server software is running, you can also connect *to* your own computer, *from* your own computer. When you do this, your computer is acting as both the server and the client. Connecting to your own computer works even if you aren't connected to a network. This is a bit pointless, in the sense that you are already sitting at your own computer, meaning you don't need `ssh` to gain access to it. However, it provides an opportunity to get experience with `ssh` when you don't have access to a server. It also comes in handy when you are having trouble with `ssh` and want to test parts of your connection to isolate the problem.

Once you have configured your computer as an `ssh` server as described above, try connecting to it using `localhost` as the address, your own username in place of `lucy`, and the same password that you use when logging in to your computer at startup:

```
ssh lucy@localhost
```

After you press ⌷return⌷ and enter your password, you will see the same greeting that you get when you first open a terminal window. This is because `ssh` starts a new shell when it logs into your computer. You can navigate your system and enter commands just as you would if you had not used `ssh`. To leave this `ssh` session, type `exit`, and you will get back to your original login prompt.

SUMMARY

You have learned:

- That computers are found on the Internet with hostnames and IP addresses

- The distinction between clients and servers

- How to gain command-line access to remote machines with `ssh`

- How to package and compress files using `zip` or `tar`, `gzip`, and `gunzip`

- How to transfer files to and from remote computers with `sftp`, `scp`, or graphical interfaces

- How to manage programs using ⌐ctrl⌐ Z, `bg`, `fg`, and the ampersand

- How to terminate processes with ⌐ctrl⌐ C, `kill`, or `sudo kill -9`

- How to determine process identities and CPU usage with `ps` and `top`

- How to keep programs running when the shell closes with `nohup` and `disown`

- Some of the major issues involved with running large parallel analyses on remote, high-performance clusters

- The steps needed to set up a standard computer as a server, with a focus on configuring an `ssh` server

Chapter **21**

INSTALLING SOFTWARE

Nearly everyone has installed software on their computer at some point. Many widely used programs come with double-clickable installers that make it simple to get a new program up and running. Installing more specialized software, including scientific command-line programs, can be a bit more involved. For many users, it is installing software, not using it, that becomes the most frustrating aspect of working within the command-line environment. In this chapter you will be introduced to basic installation procedures and cover background material that will help you understand how to address a range of installation scenarios.

Overview

In the simplest cases, installing command-line software that someone else wrote differs little from the steps you used to get your own programs running in earlier chapters. There are at least three things that have to happen in both cases. First, the program has to be in a language that your computer can understand. The programs you wrote in earlier chapters are in Python and `bash`, languages that can be interpreted by the software on your computer. Second, the program has to be placed in a folder where it can be found by the shell (i.e., in one of the folders specified by the `PATH` variable). Third, the program file has to have file permissions that make it executable. Satisfying these last two criteria should now be familiar; if not, review Chapter 6. In fact, if you have copied any of the scripts from the example folder to your own `~/scripts` folder you have already installed software written by someone else (us).

Installing tools for use at the command line sometimes is this simple. Usually, though, a bit more needs to happen. There are a variety of factors that can complicate the software installation process, which can be a source of frustration for beginners and a big time sink for experienced users. These complications include the following:

- Software isn't always provided in a language that can be understood as-is. Depending on the language, the program may need to be translated (compiled) to a language that can be understood by the computer. In other cases an interpreter may need to be installed.

- A new piece of software often needs other software packages, libraries, or modules to operate. These **dependencies** must be installed before you can use the new program. Dependencies often have their own dependencies, so it is not uncommon to need to install a dozen or more programs just to get the one you want to use running.

- Programs often need certain files and directories to record configurations, store intermediate files, and handle data input and output. Installation can include creating these folders.

- Software often needs to know things that are specific to your computer, such as the configuration of the Internet connection or the preferred program for handling certain file types. Installation scripts can either figure these things out automatically or ask the user to specify them.

- It is often necessary to modify and update existing files, such as .bash_profile and other configuration files, to use a new program. Again, the installation process can take care of this automatically, or you may be required to manually change some settings.

The rest of the chapter will explain some of these issues in more detail, and describe how they are typically approached. The skills you will develop can also enable you to use some software on computers the programmers didn't explicitly support. Specifically, you can often install basic Unix command-line programs under OS X simply by recompiling them on your own system.

Interpreted and compiled programs

The difference between interpreted and compiled programming languages was briefly mentioned in Chapter 7. At that point the distinction may have seemed relatively abstract, but it becomes relevant when installing a new program. Microprocessors, the brains of your computer, can only understand their own native machine language. In order to run, a program must either be interpreted (translated on the fly by another program such as Python or Perl) or already be in this native machine language.

Interpreted programs are stored and executed as text files containing a series of commands that can be read by both humans and the interpreter. The programs you wrote in earlier chapters were all built with interpreted languages, and the interpreting was done by the program indicated at the shebang line.

Programs in the native machine language, on the other hand, are stored and executed as binary files rather than as text files; for this reason, such programs are often simply referred to as **binaries**. Machine languages are designed for the convenience of digital logic circuits rather than of people, and it is rare that anybody directly writes binary code. Usually such files are written in another, more human-friendly language, and then translated into a binary file for subsequent use. The original human-friendly program file is referred to as the **source code**, and this one-time translation to a binary is called **compiling**. We won't describe the computer languages that are commonly used to write source code, but we will describe in this chapter standard procedures for how to compile and install programs from source code. Common compiled computer languages include C, C++, and Fortran. Most GUIs are developed in this way and draw upon platform-specific libraries to handle things like graphics and user-interface elements. This process of compiling from source code is a focus of the present chapter.

A **compiler** is a program that translates source code into an executable binary. There are many compilers, including the widely used `gcc` program on Unix computers. For example, if you create a text file including the short program `rosetta.c` from Appendix 5 (available in the `pcfb/scripts` folder), and then type `gcc rosetta.c` at the command line, this will lead to `gcc` creating an executable file, called `a.out` by default. This `a.out` file can be run by typing `./a.out`. (The reason for the `./` is explained later in this chapter.)

> **OPEN SOURCE SOFTWARE** There is a way to reap the advantages that come with the speed of compiled binaries and the transparency that comes with human-readable computer code—namely, by distributing source code in addition to or instead of binaries. This is the **open source** model, in which users can inspect the source code to see what it is doing, as well as compile the binary executable themselves so that it is optimized for their own computer. Open source programs are also usually licensed in such a way that anybody can use and modify them, as long as they apply the same license when they pass on the software to others. Open source licenses are often applied to interpreted programs as well, even though the program file is the source file. This just makes it clear that it is okay to distribute and modify the program.
>
> Increasingly, software written by scientists is distributed according to the open source model. To the uninitiated, it sometimes comes as a shock and even an outright offense that they are expected to compile their own binary from the source code. It is actually a courtesy.
>
> Some of the most popular repositories for open source software are `sourceforge.net`, `github.com`, and `code.google.com`. Python programs and modules, some of which have special installation requirements, can be found at `pypi.python.org`, and Perl programs are archived at `www.cpan.org`. When you download and install software, you are exposing the guts of your computer to the wilds of the world—do not take security lightly, and only install software from trusted sources.

Using `gcc` directly is the most bare-bones way to create binaries. It does not readily take advantage of other libraries or user preferences during the compilation. To have a more flexible compilation method, programmers typically use the program `make`, which reads from a settings file and handles all of the housekeeping before calling `gcc` with all the dependencies and options specified.

Dependencies and platform-specific aspects are not limited to compiled binary files. Interpreted programs have similar requirements—specifically, a pre-compiled interpreter for the language that the program is written in, and any libraries or modules needed. OS X and Linux both come with interpreters for common languages, but if you are running Windows, you will have to install such interpreters yourself.

Approaches to installing software

This chapter applies mostly to installing binaries of command-line programs that operate in the shell. There are three basic approaches to installing binaries on your computer. In order of increasing complexity, these are: using binaries that have already been compiled; using a package manager system; or compiling the binaries yourself. Usually you will start with the simplest method available for your platform.

Readme.txt *and* Install.txt

Software is often distributed with a file called `Readme.txt` or README. This file usually has a general explanation of what the program does, who wrote it, what kind of license it is distributed under, its history, and, most importantly, any out-of-the-ordinary details you need to install it. It is easy to get blasé about `Readme.txt` and proceed with installation without opening it, but taking a quick look can save a lot of time later. It is also a good place for a beginner to start, since it often gives a step-by-step guide to the installation process, even in cases where nothing is out of the ordinary.

Installation details are sometimes stored separately from `Readme.txt` in a file called `Install.txt` or INSTALL. Often, though, `Install.txt` is a configuration file for installation scripts rather than directions for the would-be user. Open the files in a text editor or view them using the `less` command to find out.

Installing programs from precompiled binaries

Just because a program is open source doesn't necessarily mean that you have to install it from source code; it just means that the source code is available. Many authors of open source software provide compiled binaries for a variety of computers, right alongside the source code. If you don't need to inspect or modify the source code and the binary is available for your operating system, you might as well just download and install the binary. Sometimes the binary is available as a separate download. In many cases the source code is distributed as a compressed folder, and when you expand the file it includes some binaries with the source code files.

From the user perspective there is often little difference between an executable binary and an executable text file. They can be located in the same places in the file system and called from the command line in the exact same way. Installing an executable binary file can be as simple as installing an executable text file. Although the binary does have to be compatible with your computer type, you don't need to have an interpreter installed since it is in machine language that can be directly understood by the microprocessor. The file does, however, need to have executable permissions and be in your path. Refer to Chapter 6 if you need a refresher on this.

Automated installation tools

The second method of installing software uses a variety of tools known as **package management systems**, which can install open source software with as few as two clicks. These programs not only download and compile the source code (or download a binary for your system if it is available), they also install any dependencies and take care of system configuration if need be. This can save a lot of time, and doesn't require much expertise. If you find that you need an open source program that isn't already installed on your computer, it is usually a good idea first to see if it is available via one of these tools rather than do all the work yourself.

On OS X, one of the most widely used systems for making open source programs available is Fink. You can download Fink from the project Web site at `www.finkproject.org`. The first time Fink is launched, it will update the list of available open source programs and you can select which of these you would like to install. Fink can be used from the command line, but also has a convenient GUI called Fink Commander.

Another option for OS X is MacPorts. This system has many advantages, including a large repository of software and automatic handling of dependencies. However, it also can be annoying, due to its insistence on installing its own versions of software that may already be available on your computer. For example, if you ask it to install a Python package, MacPorts will try to install its own instance of Python from scratch, a process that can take many hours. Once you have Python up and running, however, it is simple to add more packages. To install MacPorts, download the appropriate binary files from `www.macports.org`. Operations are done using the `port` command. To search for packages containing a keyword or part of the name, type `port search` *keyword*. Once you find a package, you must install it using the exact name of the package listed, with the command:

```
sudo port install packagename
```

For example, if you search for `biopython`, you will come up with versions that correspond to different versions of Python, such as Python 2.5 versus Python 2.6. This means you cannot just say `port install biopython`. Instead, you must specify `py25-biopython` or `py26-biopython` as the target.

Most Linux distributions come with their own easy-to-use software installation managers and GUIs, some of which include an extensive assortment of scientific applications. In Ubuntu Linux, you can add programs using the Add/Remove item of the Applications menu. You can also try the package manager called Synaptic, available via System ▸ Administration. From the command line, you can also use `sudo apt-get install` to install packages that have been prepared for Ubuntu. For other Linux distributions, you will want to investigate other package managers, including `yum` and `rpm`.

Installing command-line programs from source code

Sometimes, you may discover that a program you want isn't available as a compiled binary or via a package manager such as Fink. Other times, you may discover that you need to modify the program to make it work on your system. In all such cases, the solution is to install from the source code. Because of their Unix underpinnings, both OS X and Linux give you access to many kinds of programs written for general Unix platforms. For command-line programs, you will often actually prefer downloading software written for Unix even when there is an OS X version available. This is because the Unix version is most likely to be updated first and is fully usable when compiled for your OS X computer.

Getting your computer ready

At a later date you may want to check the version of a particular program, modify it, redistribute it, or recompile it, so it is a good idea to hold on to source code. Make a folder called `src` in your home directory to store and build all the software you compile:

```
mkdir ~/src
```

Next, you need to make sure that you have all the software necessary to compile source code. At a minimum, this requires a compiler. Nearly all versions of Linux are distributed with a compiler, but it must be installed separately on OS X and Windows. OS X does come with a set of compiler tools, but they are not installed by default (at least not at the time of writing this book). They are distributed under the name XCode in the Optional Installs section of the installation DVD that came with your Mac, and they can be installed for free from there. If your install disk isn't handy, they can also be downloaded and installed from Apple's Web site at `developer.apple.com/mac/`; free registration is required.

Unarchiving the source code

Almost always, source code is a collection of files rather than a single file. These files are usually distributed as an archive that you can then expand on your own system. To save Internet bandwidth and hard drive space, this archive is usually compressed. Files will typically be a compressed archive—called a tarball—with the extension `.tar.gz` or `.tar.Z`; or they may come compressed in a `.zip` archive. These compression and archiving schemes were discussed in Chapter 20 in the context of remote servers.

We will walk through the process using ExifTool discussed in Chapter 19 as the first example file. With your web browser, download the source archive for the Unix version using the shortcut URL we have created at `tinyurl.com/pcfb -exif`. When you download the `Image-Exif` archive, your web browser may leave the file as a compressed archive; it may automatically uncompress the tarball (leaving you with a `.tar` file); or it may even go on to untar the `.tar` file, leaving you with the `.tar` file plus a folder representing the file contents.

In a Linux GUI, when you double-click a compressed file or archive, it will open a window that looks as through it is displaying the contents of the uncompressed archive. This is actually a preview of the archive's contents, and you have to click the Extract button at the top of the window to complete the operation. If the archive is automatically expanded after downloading, move the folder containing the source code into your ~/src folder. Otherwise, move the archive into your ~/src folder. Then cd into ~/src, uncompress with gzip -d or gunzip, and use tar -xf to separate out the individual files from the archive:[1]

```
mv ~/Downloads/Image-Exif* ~/src  ← Your download location may vary
cd ~/src
guzip Image-Exif*  ← Uncompress
tar -xf Image-Exif*.tar  ← Unarchive
```

Compiling and installing binaries

Now that you have a folder with the source code on your computer in a convenient location, you would typically cd into the source folder and look around with ls. In some cases, the software developer will have distributed several precompiled binaries with the source code. If that is the case and there is one for your particular system, you can just cp it to ~/scripts and you are ready to go. (Remember, programs have to be somewhere listed in the variable PATH for the shell to find them, and we already set up ~/scripts as a place to put executables and added it to PATH. Alternatively, you could create a folder called ~/bin for compiled binaries and add it to your PATH).

In other cases, you will usually find a file called README or INSTALL with specific instructions on how to compile and install the software. The other important file is called Makefile, which is a text file with instructions for the compiler on how to build a binary from the source code files.

In the case of ExifTool, you will have to build the Makefile, and then compile the executable files. After you execute the tar -xf command, you have a folder containing the source files and the configuration files needed to guide its installation. Move into the program's directory and begin the process of configuring, then building (with make), and finally installing (with sudo make install):

```
cd Image-ExifTool-8.23  ← Move into the newly created folder; version number may differ
perl Makefile.PL  ← This program uses a Perl script to write the Makefile; some others don't
make  ← Build (that is, compile) the program using the Makefile generated above
sudo make install  ← Using your special privileges as superuser, install files in their locations
```

By default, the make command expects that the settings file is called Makefile (and remember, character case matters). If the file with instructions is called some-

---

[1] There are many variations on the commands used to uncompress and unarchive files. Some are designed as one-liners using shell commands piped together. You may also be able to uncompress and unarchive in one step, with tar xfz archive.tar.gz. In this case, there is no dash before xfz as there is in tar -xf.

thing else, such as `Makefile.OSX`, you need to rename the file, or tell make the name of the file using the `-f` argument:

```
make -f Makefile.OSX
```

The most common reason for the make file to have a name other than `Makefile` is that there are separate versions for different operating systems and computer architectures. If everything goes well, `make` will parse the `Makefile` and from it determine which compilers to use for which files and how to stitch everything together into a single executable binary.

In many installations, including this one, the `Makefile` includes information not only on how to build the binaries, but where to install them on the system and how to make other modifications necessary for the installation. These post-build processes are executed with the command:

```
sudo make install
```

This command usually takes much less time than the building of the binaries themselves. If this step succeeds, then the binaries are copied from the `~/src` directory into special system folders and each time you log back in you will have the new program available for use.

Variation 1: Off-the-shelf `Makefile`

A very simple compilation and installation procedure is to build the `agrep` tool, discussed in Chapter 16. Download the Unix source from `ftp://ftp.cs.arizona.edu/agrep` (there are several files available, but at the time of printing the relevant one is `agrep-2.04.tar`),[2] then move it to your `~/src` folder and unarchive the file. In this case, a `Makefile` is supplied off-the-shelf and doesn't need to be created with a script. This is common for simple programs that don't require computer-specific configurations. Within the unarchived directory, simply type `make`, and it should build the executable for you.

The supplied `Makefile` doesn't include any information on installing the program after it is compiled, so `make  install` won't do anything. You need to manually place the executable file in the appropriate location on your system.[3] Move the `agrep` executable from the source folder to `~/scripts` or some other place in your PATH:

```
mv ./agrep ~/scripts
```

---

[2] Note that this is an FTP URL, not a Web URL. It will probably download using a different program than your web browser.

[3] Within OS X, to install the man file for `agrep`, copy it from the source folder with this command: `sudo cp agrep.1 /usr/share/man/man1/`.

Variation 2: Generating a `Makefile` with `./configure`

In the ExifTool example earlier, the command `perl  Makefile.PL` generated your settings file for you, but this is not the most common case. If there is no `Makefile` in the source folder that you download, but there *is* a file called `configure`, then that is a separate script which will generate a customized `Makefile` to match your system capabilities. You will first run this command before proceeding with the `make` process. The `README.txt` file will typically explain how and whether the `configure` command is to be used. Sometimes there are options for what features will be enabled in the compiled program which can be specified at the time that you run the `configure` command.

This next installation example compiles the program ImageMagick, which takes advantage of the `configure` command. However, because it has many dependencies, this command may or may not work on your system as specified. If you cannot get it to work with these instructions, you might use this as a chance to test out Fink, MacPorts, or another package manager as the installation mechanism.

You may remember from Chapter 19 that ImageMagick is a powerful image processing system. It is actually installed by default on some systems (type `which  convert` to see if it is already present on yours). In such cases, the procedure described here can be used to update your installation and to install the Unix manual pages.

For OS X or Linux, download the source files for the Unix version of ImageMagick from the project page at `http://sourceforge.net/projects/imagemagick/files`.

The appropriate file will be either a `.zip` or `.tar.gz` file without `windows` in its name. Move the file to your `~/src` folder, uncompress and unarchive as before, and `cd` into the folder that is produced.

As you know, a script will not run unless it can be found. We have talked about adding locations to your `PATH`; however, there are likely to be many configure files on your system, and you don't want to have to move them all into your `~/scripts` folder just to get them to run. Another way to run a script which is not in a folder in your `PATH` is to specify its full absolute path when typing the command. If the script is located in your current working directory, then you can use a shortcut to indicate its absolute location.

Recall that `cd  ..` moves you to the folder enclosing the present folder, because `..` is a shortcut describing "the folder containing this one." Similarly, a single dot indicates the current folder of residence. Thus to specify the full path to the `configure` script in your current working directory when you try to run it, you can type the command preceded by `./`:

```
host:ImageMagick lucy$ ./configure
```

This `./` path specifier is similar to the `~/` shortcut, which inserts the full path to your home folder into the command line, but instead it inserts the full path of the working directory.

From within the ImageMagick source folder, run `./configure`, and it should print out a long list of status messages, ending with some details about what compiler it will use for building the executables. Upon completion, it has written that information into a customized `Makefile`. After it is done, you can proceed with the `make` command to compile the binaries of the required programs. This may take a long time!

For the ImageMagick installation, the instructions are found in the `Install-unix.txt` file. In this case, that file tells you how to enable or disable many optional features. If your installation fails because some of the dependencies cannot be installed or found, you may be able to get through the installation of basic functions by disabling those options, in particular, certain fonts, X11, or PostScript support (achieved with a package called Ghostscript). The choice to disable components is made at the time you run `./configure`, so that your `Makefile` is generated appropriately. If you have to go back and rerun `configure`, first follow the cleanup steps described later in this chapter. Other troubleshooting tips are included in the `Install-unix.txt` file.

Once ImageMagick is installed, you will most commonly use it via the `convert` command, so type `man convert` to read about the program's features and functions.

Installing Python modules

Python modules are often supplied with their own installation scripts. Such a script is usually contained in the source code folder and is called `setup.py`. To install a Python library, run the installation script by changing into the source directory, then typing:

```
sudo python setup.py install
```

This will perform whatever compilation and configuration steps are necessary to get the library working on your system. You don't have to specify `./configure` before `setup.py` because the executable that is being called is the `python` program, which is in your PATH.

As an example, try installing the `pyserial` library used for serial communications. Download the source archive `pyserial-2.5-rc2.tar.gz` (or the most recent version) from `pypi.python.org/pypi/pyserial/`. Move the archive file to your `~/src` folder, then uncompress and unarchive it; remember you can press ⌷tab⌷ to complete filenames at the command line:

```
host:~ lucy$ mv ~/Downloads/pyserial-2.5-rc2.tar.gz ~/src
host:~ lucy$ cd ~/src
host:src lucy$ ls
Image-ExifTool-8.23.tar.gz    agrep-2.04.tar
ImageMagick-6.6.2-8.tar.gz    pyserial-2.5-rc2.tar.gz
host:src lucy$ tar xfz pyserial-2.5-rc2.tar.gz  ← Uncompress and unarchive
host:src lucy$ cd pyserial-2.5-rc2  ← Move into the newly created folder
host:pyserial-2.5-rc2 lucy$ ls
CHANGES.txt MANIFEST.in README.txt  examples     setup.py  ← The installer file
LICENSE.txt PKG-INFO    documentation   serial       test
host:pyserial-2.5-rc2 lucy$ sudo python setup.py install  ← Required for privileges
Password:
running install
running build
running build_py
...
changing mode of /usr/local/bin/miniterm.py to 755
running install_egg_info
Writing /Library/Python/2.5/site-packages/pyserial-2.5_rc2-py2.5.egg- info
host:pyserial-2.5-rc2 lucy$
```

That's all there is to it (when it works). You can see by the status updates that in addition to building the program, the `setup.py` script also copies the files to a central installation folder where they will be accessible to Python programs that you write. In our experience, installing Python libraries using the `setup.py` procedure is generally easier and gives better results even than using a package manager.

Troubleshooting

What to do when software won't compile or installations don't work

Installation is not always as easy as it could be. Most often, the culprit is one or more dependencies—supporting files that are used by the program. Because they deal with graphics, communications, and reading and writing to files, dependencies tend to be more platform-specific than a program designed just to do mathematical calculations. As a result, getting feature-rich graphical programs to install may require a bit of experimentation and Web searching. Here are a few approaches to troubleshooting an installation.

Look inside the `Makefile` If you look inside the `Makefile`, you may find that some lines particular to your operating system need either to be uncommented (remove the # sign), commented out (add a # at the beginning of the line), or otherwise edited. Often—but not always—guiding comments inside the file tell you where these modifications need to be made.

If you do try editing your `Makefile`, save the original version under a different name, so that you can go back to it if necessary. (The reason you have to save the original under a new name, rather than the edited file that you intend to experiment with, is because `Makefile` is a special name that the `make` program looks for to find its settings.) After editing the file, and before attempting to compile again with `make`, you will want to clear out any files generated by your previous compilation attempt. To do this, either try the `make clean` command, or manually delete all the files whose names end with the extension `.o`. You could also delete the entire source code folder for the program and then create a new one from the original archive file you downloaded. This ensures that previous attempts at making the program don't leave files that interfere with your troubleshooting attempts, although you would want to save the `Makefile` that you were editing in the original folder.

Warnings There are important differences between a warning message and an error. Errors are typically fatal to an installation. Many compilation procedures will issue a long stream of `Warning` lines, flagging places where the developer might have taken a few shortcuts or followed a procedure that is not strictly proper for your system. If your compiled program runs in the end, these warnings can usually be ignored without consequence; often they are a matter of style, completeness, and programming rigor. If your binary fails to run, however, the warnings may hold clues to libraries that failed to get linked up with your program. Pay special heed to any lines that talk about `lib` or `dylib` files not being found.

Permissions If you are lucky, then something will fail just because of incorrect permissions. This is relatively easily corrected using the `chmod u+x` command, or by invoking the `make` command with superuser privileges using `sudo make` or especially `sudo make install`.

Troubleshooting dependencies If your program is failing because some required files or libraries are not installing, try to get those to install separately first. This will narrow down the problem to the actual portion of the installation that is causing trouble. Conversely, if you installed the dependencies separately before trying the whole installation, the installer might not be finding your special library files. This can be a problem when using a combination of MacPorts for certain files, and manual installation for others. In these cases, you can try re-installing the dependencies as part of the process of installing your desired package. ImageMagick is an example of where this might happen, since it is dependent upon so many platform-specific graphics libraries.

Technical details of your platform If a binary is compiled for the wrong platform (e.g., Linux versus OS X), when you try to run it, you may get the error message `Cannot execute binary file`. A similar error may occur if a program has been compiled to use multiple processors. Sometimes a program will be compiled for 64-bit operation—this has to do with how blocks of memory are handled during execution—and will fail with a cryptic error when you try to run it on an older system. In any of these cases, you will have to go back and recompile or download the appropriate executable, making sure that you have chosen 32-bit mode. If you are given the choice and you are not sure what your system supports, you can probably just use 32-bit mode without noticeable effect.

Search the Web If your installation continues to fail, there may be a bug or an error in the Makefile. These bugs are often not the programmer's fault, so don't complain too vehemently online. Just because a program works as written on one system does not mean that it will work on yours. Chances are that someone else has probably come across this problem before, so search the Web for the exact error you are receiving (with quotes around it so it is searched as a phrase) and see what solutions others have recommended.

Installing software from source code can be difficult, but it opens up a wide world of programs that you may find useful and which may not otherwise be available to you.

SUMMARY

You have learned:

- Three methods for installing software:

    Using precompiled binaries for your system

    Using package managers such as Synaptic, apt-get, Fink, and MacPorts

    Compiling from source

- Compiling steps:

    Uncompressing and unarchiving with gunzip and tar -xf

    Generating the Makefile with ./configure

    Compiling with make

    Installing with sudo make install

Moving forward

There are many programs specific to subdisciplines in biology that are available only as source code. Download, compile, and install tools that will be useful for your own work.

Chapter 22

ELECTRONICS: INTERACTING WITH THE PHYSICAL WORLD

Many scientific computing tasks require more than reformatting and analyzing data. The data have to get into your computer in the first place, usually through various sorts of automated sensors operating out in the physical world. Although some types of sensors are extremely sophisticated and must be purchased as part of expensive instruments, some of the most common data acquisition tasks, such as monitoring temperature and other environmental variables, can be accomplished with simple systems that you build yourself—and that is the focus of this chapter. The ability to create your own sensor systems and custom electronic devices to interact with the physical world can be one of the most rewarding aspects of scientific computing. It is often far less expensive to do it yourself than to purchase off-the-shelf systems; moreover, you can customize your equipment in ways that would never be possible with prefabricated systems. Lastly, since so many devices communicate with serial ports, we provide background on enabling and troubleshooting serial communication.

Custom electronics in biology

Typical scenarios for custom electronics in biology

There are many situations where building your own electronic devices can greatly facilitate your biological research. For example:

- You are concerned that the temperature of your freezer is unstable, but don't want purchase an overpriced $400 Internet-enabled temperature monitor. You put a temperature probe in the freezer and wire it to a custom circuit attached to a computer adjacent to the freezer. With a few lines of computer code, you program the circuit to send the temperature of the freezer to the computer every minute. You then program the computer to save these data to a database and send a message to your mobile phone via email if the temperature goes outside a specified range.

- You are column-purifying DNA extractions, which requires switching collection tubes every twenty drops. You don't have a fraction collector, and with more than a minute between each drop you would be watching dripping columns all day. You arrange a light-emitting diode and phototransistor so that each drop interrupts the path of light between them, and wire them to a counting circuit. Now you can leave the setup unattended for long intervals of time.

- Your experiments require monitoring solar radiation and oxygen levels during the course of an experiment at a remote field site. You wire several sensors to a self-contained circuit that can convert the sensor data to numbers and store them on a memory card.

- You want to monitor depth and orientation of a marine turtle as it forages, so you create a circuit to log data from a pressure sensor and from an electronic compass, then embed the device into epoxy so it can be attached to the turtle's shell. To download the data afterward, you use a wireless transmitter built into the circuit.

- You need a programmable agitator that can tip a rack of tubes at different intervals at different times of the day. You attach a $10 hobby servo (a special type of motor) to the rack, wire it to an actuator circuit, and program it to turn the servo at the appropriate time. The whole system costs $40, and once you're done using it, all the parts can be used for other things.

We won't cover all the skills needed for each of these examples, but we will get you pointed towards the basic tools. In particular we will focus on general-use microcontroller circuit boards.

Simple circuits with complex microcontrollers

Building custom electronic systems without extensive formal engineering experience is much easier now than it was in preceding decades. Forty years ago anyone building a circuit had to work with basic components such as resistors, transistors, capacitors, inductors, and diodes. This required a detailed knowledge of electrical engineering theory, extensive time commitment, and the experience to assemble tens or hundreds of components. Even then, the overall complexity of the circuit was by necessity limited, since it was built from the ground up. Over the last thirty years, however, a wide variety of microchips has become available to the general public. These microchips contain prefabricated circuits with tens to billions of components, and by containing all the components in one package, they greatly facilitate the construction of complex circuits.

Many microchips are designed to perform a specific task, but the last decade has seen the proliferation of microcontrollers, entire general-purpose computers on a microchip. They are too slow and underpowered for complex data analysis tasks, and most don't even have enough computational power to interface with a screen and keyboard. However, they are very cheap (most are less than $10), use little electricity, are relatively simple to incorporate into circuits with other components and computers, and are designed to be flexible and easy to program. They can be connected to many types of actuators (such as motors and servos) and sensors. Microcontrollers are replacing specialized custom-built circuitry all around you, in washing machines, thermostats, electronic micropipettors, data loggers, flashlights, and just about anything with a battery or plug.

Microcontrollers fit right in with the other technologies we have chosen to include in this book—they are flexible tools that allow you to tackle a variety of challenges (Figure 22.1), and at the same time, working with them builds general skills. They can serve as interfaces to the physical world for your laptop or desktop computer, digitizing sensor data, re-

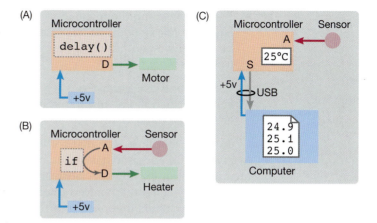

FIGURE 22.1 Common uses for microcontrollers in biological applications On the microcontroller block (orange) "D" represents digital output, "A" indicates an analog port, and "S" indicates a serial port. (A) This stand-alone circuit turns a motor a prescribed number of degrees after delays of specified intervals. No sensor is involved. (B) A stand-alone circuit monitors a temperature probe through an analog input, and uses the microcontroller's own program to determine when to turn on a heating element using the digital output when the temperature falls below a certain threshold. (C) A microcontroller as an external sensor for a computer. The microcontroller receives temperature information from a probe, converts this to numeric values, and sends the data via the USB port to the computer, where it is logged in a file. The USB connector also provides power to the microcontroller.

laying it to the computer (Figure 22.1C), and controlling actuators when instructed to do so. In such cases, they function as fully customizable computer peripherals. Alternatively, they can also be incorporated into stand-alone devices. In that role, they can monitor sensors and record data (e.g., a humidity logger), perform simple tasks based on sensor data (e.g., a drop counter or thermostat, as shown in Figure 22.1B), or serve as controllers for physical devices that need no sensory input (e.g., a tube rocker, as shown in Figure 22.1A).

Building a custom microcontroller-based instrument requires several steps:

- First, envisioning and designing the means of connecting the instrument to the physical world—identifying a sensor and how to connect with it.

- Second, wiring up the microcontroller chip with other components to build the sensor/actuator circuit.

- Third, connecting the microcontroller to a computer and programming it. Since the microcontroller has no screen or keyboard, all the software must be written on another computer and then transferred to the microcontroller memory.

- Fourth, for those situations where the instrument is to function as a computer peripheral, programming the attached computer to read and analyze the data.

In much the same way that software has become more usable and accessible to a wider audience as it has grown more sophisticated, complex microcontrollers are much easier for non-specialists to use than their simpler predecessors. This is because they contain advanced internal circuitry that reduces the complexity of the external circuits that need to be built around them. In the simplest cases, microcontrollers need only a battery and one or two other electronic components to work. In practice, they are most often used with a few more components than that, including components to regulate the power supply and protect input and output pins, a button to reset the software if needed, indicator lights, and connectors to interface with sensors and actuators.

The electronics community has designed simple circuit boards that contain microcontrollers and these other basic components. In less than five minutes such a board can be taken out of its package, connected to a computer, and set up to collect data, without even needing to warm up a soldering iron. The most popular of these standardized boards is the **Arduino** (`www.arduino.cc`), which comes in several sizes and shapes and with different sets of built-in components. The Atmega microcontrollers on Arduino circuit boards come pre-loaded with software that simplifies the programming process. Arduino is "open-source hardware." Like open-source software, all the designs are freely available, and the systems are built with the intention that people will modify them and use them in novel ways. Several manufacturers build boards that comply with the Arduino standards, and all the needed software for programming and using them can be freely downloaded.

Basic electronics

We don't aim to teach you electronics within this one chapter; we only present a few bare-bones essentials necessary to build the most basic microcontroller-based instruments. We strongly suggest that you consult the guides listed at the end of the chapter if you are interested in pursuing electronics and microcontrollers further.

Electricity

Unlike the Internet, an electronic circuit can be usefully thought of as a series of tubes through which fluid is flowing. This "fluid" is electrons, and some of the same intuitive concepts that apply to water in a pipe apply to electrons in a wire. The rate of electron flow in a circuit is called the **current** and is measured in amps (or often in milliamps). The **potential** at any point in a circuit is analogous to pressure in a water pipe; it is measured in **volts**, and this is why potential is often called **voltage**. Power is a measure of the rate of energy flow. It is simply the product of the current times the voltage, and is measured in **watts** (or frequently in milliwatts).

Basic components

Understanding the nature of basic electronic components, and getting a sense of the vocabulary used to describe them, will help you get started not only with building your own circuits, but with interpreting circuits you find online and elsewhere. Table 22.1 introduces some of the more common components that you will see in schematics.

TABLE 22.1 Common electronic components

Symbol	Name	Properties and use
-ʌʌʌ-	Resistor	Restricts flow of electricity along a path
⊣⊢	Capacitor	Temporarily stores charge; dampens changes in voltage
→⊢	Diode	Allows current to flow in one direction; protects circuits; used to "rectify" alternating current to direct current; also used in voltage dividers
⊕	LED	Diode that emits light
⊣|||⊢	Battery	Power supply to the circuit
⏚	Ground	Common reference point for circuit potential voltage; equivalent to the negative terminal
⊕ ⊕ ⊕	Transistor	Electronically switches or amplifies a signal
-ʌʌʌ- ⊕	Variable resistor	Restricts the flow of electricity in a flexible manner; can be controlled either mechanically with a knob or based on a signal like light or temperature
(relay symbol)	Relay	Electromagnetic switch; can use a lower-power circuit to switch on much higher-powered circuits
Vcc 5 / 2 6 / 3 7 / 4 GND	IC	Integrated circuit; packaged components to provide any of several functions, including timers, logic, and processing

A resistor is one of the most important electronic components. As its name implies, it resists the flow of electrons. It is analogous to a narrow section of a pipe. A higher voltage difference passed across the ends of a resistor (like a high pressure difference on either side of a pipe constriction) will lead to higher current through it, with the exact relationship depending on how much resistance is encountered. The unit of resistance is an **ohm**.

Encoding information with electric signals

Sensors are devices that convert a physical property into information encoded in an electronic signal. The properties that can be measured are nearly unlimited, and include temperature, force, pressure, light, orientation, conductivity, and chemical cues such as pH and oxygen concentration. Broadly speaking, the encoding of information into an electronic signal can be **analog** or **digital**. We will explain these terms shortly.

The specifications of data-acquisition devices like microcontrollers include descriptions of how many I/O (input/output) connections are available for each type of encoding. Often the same pin (i.e., connection) can be programmed to serve different functions. For example, a particular digital I/O pin could be configured to support simple ON/OFF input and output, to send text-formatted data via serial communication, or to produce variable output signals through pulse–width modulation.

Analog encoding

Analog signals are the most common starting point for simple sensors. Analog signals represent changes in a physical state as an electric property that varies continuously within a particular range (Figure 22.2A), usually current or voltage. Many sensors are essentially variable resistors that are sensitive to a particular environmental stimulus: the changing resistance is measured as a changing voltage by combining the sensor with resistors of fixed value and passing current through them. For example, in the case of a light sensing circuit, you might detect 0 volts in total darkness, 1.3 volts when a light bulb turns on, and 5.0 volts in full room light. Most sensors will saturate at some point, so in our example, you might continue to receive only 5.0 volts even as the light increases to full sunlight. Several analog-to-digital converters (also called A-to-D converters) are built into most microcontrollers. These convert the continuous range of

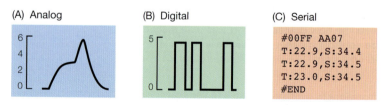

FIGURE 22.2 Common types of electric signals used to encode information (A) Analog signals can vary continuously. (B) A digital signal can only be on or off. (C) Serial signals are a special type of digital signal, where a series of on and off pulses represents more complex data, such as letters and numbers.

voltages into a numeric representation that can then be analyzed with software. For example, integer values between 0–255 might have a linear correspondence to the 0–5.0 volt values produced by your circuit, and that value would correspond in turn to light levels received at the sensor.

Digitally encoded signals

Digital signals have discrete rather than continuous states (Figure 22.2B). Usually these states are binary, that is, either off (zero volts) or on (often 5 volts for many devices). Just as there are many ways to represent information with an analog signal, there are a variety of approaches to encoding information with a digital signal. A few of these are described here.

Toggling When a digital signal is either on or off, it can be used directly to control simple devices. A digital output could be used to turn a single indicator light on or off, close a valve to stop the flow of gas, or open a door to introduce an animal into the experimental arena. Similarly, many types of input only have two states and are therefore readily encoded as a digital electronic signal. Such discrete events might include detecting when a firefly is flashing, whether or not a drop is passing by a sensor, or simply whether a button has been pushed.

All of these events can be controlled or measured with a single digital pin from a microcontroller. The microcontroller can be programmed to monitor when that pin is on or off, or to turn the pin on or off at particular times. Most microcontrollers will have a certain number of pins that can be used for digital I/O (input/output) of this sort. However, while dedicating an entire I/O pin to a particular piece of information works fine for simple applications, it is not practical for controlling devices that need to communicate large volumes of information about many different things. Such tasks are handled by the digital communications protocols described in the next two sections.

Parallel data transmission Most of the data inside your computer is transmitted within and between microchips as **parallel** digital information. A series of wires encodes numbers in binary (see Appendix 6), with each wire specifying a particular digit (1, 2, 4, 8, etc.) and with on corresponding to 1 and off corresponding to 0. These numbers can have literal numeric interpretations (e.g., integers or floats, as described in the earlier programming chapters), or they can stand for other types of information, such as ASCII characters (again, see Appendix 6). The number of wires used to transmit a value depends on how many bits (binary digits) the microprocessor employs in its calculations. In computers, 32-bit and 64-bit processors are standard. Most microcontrollers use only 8 bits, which is plenty for simple control and monitoring circuits, but not as suitable for complicated computing.

Computers used to have parallel ports for connecting external devices, mainly printers. However, this method of connection required one wire for each bit, resulting in bulky sockets and thick cables. The resulting cost and inconvenience has ended the days of parallel ports in modern computers. Thus, although parallel data transmission is regularly used within devices, it is now rarely used between them.

Serial data transmission Serial data transmission is a close relative of parallel data transmission. Information is transferred as binary numbers encoded with on and off signals. Rather than send the signal for each digit in a separate parallel wire, though, all the signals are sent together in the same wire; they are staggered through time as a series of sequential pulses which encode information (Figure 22.2C). Common serial data interfaces include USB (Universal Serial Bus), Ethernet, FireWire, SATA (Serial-ATA, used for hard drives), and RS-232. Wireless serial data protocols include WiFi and Bluetooth. Serial data transmission is everywhere.

Serial data transmission may sound a lot more complicated than parallel data transmission, and in some respects, it is. Data must be packaged into bursts of signals, fed through a single wire, and then unpacked again on the other end. This all requires background processing. In addition, since each digit is sent single-file rather than in parallel, data transmission takes longer than it would with a parallel port with more wires. But these inconveniences have been easily surmounted as electronics have become cheaper and faster, and the huge savings of doing away with large connectors and cables with dozens of wires more than makes up for the other costs.

Sensors and instruments frequently come with circuitry already built in to convert measurements into serial data. Many of these devices use the RS-232 serial data interface, which is old and relatively slow, but also widespread and reliable. Few computers come with RS-232 ports anymore, but cheap RS-232-to-USB adapters are widely available. Serial data from an external port can be received directly from a special kind of terminal window in your computer, and it can also be accessed from within Python and other programs using easily installed libraries and modules, as we will describe shortly.

Pulse-width modulation PWM is another common way of encoding information. PWM is simply a series of pulses that occur at regular intervals, but whose duration is variable. At any point in time the signal is either on or off, but information is encoded as the width of the pulse rather than as binary numbers. If the interval between the pulses is short enough, PWM can be used to vary the average power to a device, just as with a continuous analog signal. For example, a light powered by a PWM circuit will appear brighter when the width of the pulse is longer. PWM is frequently used to control motors and other high-power actuators, since it is more efficient to rapidly turn the power on and off than it is to reduce the voltage when adjusting the power.

Building circuits

Schematics

The language used to communicate an electronic circuit is a two-dimensional diagram called a **schematic** (Figure 22.3A). This represents the wires and components and how they are interconnected. When you are building a circuit, you are translating the schematic diagram into actual connections with your physical components.

The schematic shown in Figure 22.3A is a very simple connection between a battery and a light-emitting diode (LED), including a resistor that is required to restrict the flow of current (without it, the LED would burn out). The schematic shows the three components of the system; in this case the components are labeled with words, but usually they will only be shown by symbols.

Many components have polarity, meaning that if they are put into a circuit backwards they won't work in the intended way. The symbols for polarized components are asymmetric, as is the structure or labeling of the components themselves. It is important to be sure that such components are oriented in the correct direction. LEDs, for instance, have a longer leg for the positive side of the circuit. Other components, such as resistors, have no polarity.

Breadboards

Once you have the design of a circuit, the next step is to translate it into an actual circuit and put together the electronic components. Some of the components you need may already be present on the microcontroller board, but in most cases you will need to supplement the board with at least a few more sensors and parts for your particular task. Electronic circuits are usually made by soldering together wires and components (soldering is an electronics version of welding); however, for testing a prototype and for debugging your circuits, you can use a device called a **breadboard**. This is a plastic rectangle filled with holes (Figure 22.3B,C). Behind the scenes, the holes are interconnected in such a way that wires and the legs of components inserted into them can be connected to other wires without soldering. There are several options for connecting your breadboard circuits to your microcontroller: (1) you can make connections between the microcontroller and breadboard with a few wires; (2) you can plug the entire microcontroller into the breadboard as if it were just another component (this works for small boards, like the Arduino Nano); or (3) you can get a breadboard "shield" that can piggy-back directly onto the microcontroller board.

In the breadboard depicted in Figure 22.3, the two rows of red and green holes along the top and bottom are connected along their length. A wire plugged into a red row will be connected to another wire plugged into any other red hole in that row, but not with the red row on the opposite side of the board. (The holes will not appear colored on an actual breadboard—they are just shown that way in the illustration.) These rows are used to provide power and ground (that is, plus and minus from the battery or power supply) to all the places that power is needed in the board. Simply plug in a wire to one end of row and tap into it by plugging in the end of another wire anywhere else along that row.

The patches of holes near the center of the board are connected in columns, some of which are highlighted in blue in Figure 22.3. Any wire inserted into a hole in that section will be connected electrically to any other wire in that same column. Holes on opposite sides of the central divide are not connected to each other. These holes also allow you to connect to the wires of a chip (an integrated circuit) that is placed with two rows of pins straddling the central divider. Each pin of the chip becomes available to wires inserted in holes of the adjacent column.

Translating a schematic to a breadboard

Look at the breadboard circuit in Figure 22.3C and trace through the connections as they correspond to the schematic. The black wire coming from the negative terminal of the battery (outside the boundary of the image) is connected to a green hole, so anything plugged into the green row—such as the blue wire—is effec-

(A)

(B)

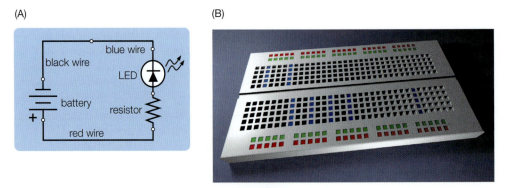

(C)

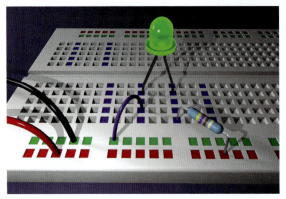

FIGURE 22.3 A schematic, breadboard, and circuit (A) A simple circuit in schematic form. (B) A breadboard showing the patterns of connection between the holes. (C) The circuit depicted in A, but implemented with components placed into a breadboard.

tively connected to the battery. The blue wire then connects to a column in the center of the board, and one leg of the LED is plugged into that same column. The other leg of the LED is inserted into the same row, but in a different column, so it is not directly connected to the blue wire. It is, however, connected to one end of the resistor inserted in the same column. The other end of the resistor is connected indirectly to the red wire (the positive battery terminal), because it is inserted into the same red row.

As shown, this circuit would cause the LED to light up continuously until the battery ran down. If you wanted to add a switch or control to the circuit, a good spot to do it would be in place of the blue wire. Breaking that link would cause the LED to turn off and no current to flow through the circuit.

Serial communication in practice

Unlike the stand-alone circuit above, your data-collection devices will probably be designed to interact with a microcontroller. In those cases, after you have built your electronic circuit and connected it to your microcontroller, you will need to connect the microcontroller to your computer's serial port. This is necessary to program it and may also be needed to operate it if you are building a device that passes data to the computer during normal operation. Some microcontroller boards have USB adapters built in, while others have RS-232 ports. Boards with USB ports can draw power from the same connection, obviating the need for a separate power supply when operating.

Serial communication is so widespread and critical to getting many types of physical devices to interact that here we provide a general background that extends beyond the topics needed to interface with microcontrollers. If you are fortunate, someone will have already built the exact instrument that you need for your research, and it will output your data in text format through a serial connection. With the right setup, you can connect a cable between your instrument and your computer, then use a variety of mechanisms and programs to capture the data stream to your computer for processing, storage, or display. (There are also options like Bluetooth, ZigBee, or XBee, which transmit serial data wirelessly.)

The traditional connector for serial communications to an instrument is a 9-pin RS-232 connector, shown at right, called DB-9 (not to be confused with an analog video connector, which looks similar but has three rows of pins instead of two). Of the nine connections, only three pins are required: two pins are used in communication, one for transmitting signals (TX) and one 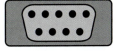 for receiving signals (RX), while the third pin is used for a common ground connection. On older PCs, these connectors are attached to serial port interfaces in the operating system called COM1 or COM2. Newer computers lack these ports, but USB-to-RS232 connectors make them available. When using one of these adapters, you will need to know the port address—that is, the place to send and receive data. With a USB-to-serial adapter installed, there will be what looks like a file in your /dev folder that acts as a socket for connecting to the serial port. On Linux or Mac

OS X, you can list the contents of your /dev folder and look for files named starting with tty.*:

Try /dev/tty*, /dev/ttyS*, or ttyusb0. Your device may be called ttyS0.

```
host:~ lucy$ ls /dev/tty.*
/dev/tty.Bluetooth-Modem        /dev/tty.usbserial-A70064y
/dev/tty.Keyserial1
```

The port for a serial adapter will only show up while it's plugged in, so you might not see any of these devices if you try this command now. You will use the appropriate path in programs or when trying to access your serial port through other data acquisition systems, as described shortly.

Baud rate and other settings

Serial communications protocols have a few options regarding what conventions to use when sending data back and forth. The most important option is the speed at which communication will occur, known as the **baud rate**. If data are being sent at a speed of 9,600 but the computer is set to receive at 19,200 baud, then you might get strange, unrecognizable characters on the screen. A good first step is to match the baud rate to the highest speed that is supported at both ends of the connection, and which can be transmitted the required distance without errors.

Other settings that affect serial communications include those which describe the nature of each data packet that is sent, including the number of bits, whether a parity bit is included (meaning one of the bits is used to detect errors in the other bits), and the number of stop bits (signals sent to mark the end of each series of bits). By far the most common setting for these is 8-N-1 (a standard abbreviation for the above settings), and, unless otherwise instructed, you should start with these options. **Handshaking**, the negotiation between devices about when and how serial data are transmitted, can be done either by the program (called software handshaking), by additional electronic lines (called hardware handshaking), or by built-in buffers on the devices (no handshaking). With this last option, one device sends the data and just assumes the other is listening, and the serial port itself stores as much data as it can until read by a program. Usually you can leave flow control off and rely on the buffers, but in the event that you need to turn it on, software flow control is often called XON/XOFF and hardware flow control is RTS/CTS. Although you will encounter a variety of baud rate settings, the data and flow values are almost always set to the defaults described here.

Null modem

The port on a computer's RS-232 connection will likely be set to use pin 3 of the connector to transmit data (TX) and pin 2 to receive (RX). In contrast, when an instrument is set to talk to a computer, its connector will have the pins in the opposite order: pin 3 will receive data and pin 2 will transmit. This way, all traffic from the computer goes out along pin 3 and all data from the instrument goes out on pin

2 (Figure 22.4A). If you try to connect two devices that both consider themselves to be master devices (or both slave devices), they will clash, with each transmit pin trying to push data to the other (Figure 22.4B). To solve this, you insert a **null modem** connecter between the two devices. This swaps pins 2 and 3 as they pass so that the two computers can talk to each other (Figure 22.4C). This situation sometimes occurs when you create your own device designed to operate by a serial connection or when you try to connect a laboratory instrument to a circuit designed to monitor or control its output.

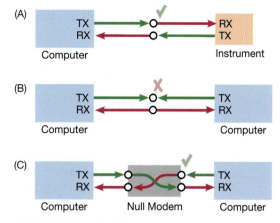

FIGURE 22.4 **Serial connections and the use of a null modem connector**

Software for serial communication

There are a few programs that will allow you to see serial data and send data to a serial device. On Windows, you can download the graphical applications PuTTY or Tera Term, or you can use Hyperterminal, which was installed as a utility prior to Windows 7. On Mac OS X, the most widely used graphical interface for serial communications is the venerable ZTerm. In a shell terminal window within OS X and Linux, you can use kermit or minicom, which may need to be installed.

Serial capabilities are also available in the shell environment of OS X via the built-in screen command. Here is an example session using the screen command in a terminal window; note that it only works if you have a serial-port adapter device plugged in:

```
host:~ lucy$ ls -a /dev/tty.*    ← Gives a list of devices
/dev/tty.Bluetooth-Modem      /dev/tty.Keyserial1
host:~ lucy$ screen /dev/tty.Keyserial1 19200   ← Begin communication at 19,200 baud
```

Now your screen will display any serial data coming through the serial adapter, and items that you type will be sent out through that serial port. (Often, instruments will take simple text commands to change their sampling frequency or other parameters.) To end the screen session, type ⎡ctrl⎤ A followed by ⎡ctrl⎤ \.

You can have a lot of fun with a cheap USB-to-RS-232 adapter and a cable, which will let you talk to instruments, sensors, and even freezers around the laboratory.

Serial comms through Python

Building a microcontroller-powered device is often only part of the mission. You may want to log information from your new device, or for that matter, from another sensor. To get more sophisticated with your serial communications, you can write Python programs to store and display serial data. To use serial communications in your programs, download and install the pyserial library from the link at `pyserial.sourceforge.net`. Uncompress the archive and `cd` into the resultant directory in a terminal window. To install, type:

```
sudo python setup.py install    ← This will ask for your password
```

Note that the module you import into your program after installation is called `serial`, not `pyserial`. In a lot of ways, talking to a serial device is similar to opening a file for reading and then using the corresponding variable (below named `Ser`) as the source of data. Because a serial stream doesn't necessarily have a prescribed length, you will read from it with a `while` loop instead of a `for` loop, and program the loop to exit when a certain condition has been met or when a certain number of lines have been read. The `serialtest.py` program shown here, which is available in the `pcfb/scripts` example folder, reads data from the serial port and prints them to the screen:

```python
#!/usr/bin/env python

# serialtest.py -- a demo of serial comms within python
# requires pyserial to be installed
import serial

# This address will not be the same
# find out your port name by using: ls /dev/tty.* and insert here
MyPort='/dev/tty.usbserial-08HB1916'

# be sure to put the timeout if you are using the readline() function,
#   in case the line is not terminated properly
# the other option is using ser.read(1), which reads one byte (=char)
# 19200 is the baud rate here.
Ser = serial.Serial(MyPort, 19200, timeout=1)
if Ser:
    i = 0
    # count lines and exit after 20
    while (i<20):
        Line = Ser.readline()    # read a '\n' terminated line
        print Line.strip()
        i += 1
    Ser.close
```

Arduino microcontroller boards in practice

Where to start

Arduinos are multipurpose microcontroller boards simple enough to be used in grade school projects, yet powerful enough to create complex and sophisticated robots, instruments, and devices. There are many excellent tutorials, including some that also teach basic principles of electronics. We recommend the official Arduino site `arduino.cc`, and `www.ladyada.net/learn/arduino`. Some books related to Arduino and electronics are also recommended at the end of this chapter.

The easiest way to begin with Arduino is to purchase a starter kit from Adafruit Industries (`www.adafruit.com`) or SparkFun Electronics (`www.sparkfun.com`). These kits include the microcontroller board, breadboards, wires, and sensors you will need to build simple projects. The supplied tutorials walk you through basic circuits and the software needed to run them. Once you have these example circuits running, you can start adding your own sensors and actuators and modifying and adding to the supplied software. This will allow you to build your skills incrementally.

Building circuits with Arduino

Arduino pins can be flexibly configured: Fourteen pins can perform digital input or output. Two of the digital pins can be used as TX and RX in serial communication, and six can generate pulse-width modulation output (simulating analog output). Six additional pins can be used for analog input, using 10 bits of information (see Appendix 6) to encode voltages from 0 to 5 volts as integer values from 0 to 1,023. The uses of these multipurpose boards are wide-ranging. Figure 22.5 shows a simple thermostat circuit, like that diagrammed in Figure 22.1B, which requires a single analog input for reading temperature and a digital output for toggling a relay that turns a heater on and off. Microcontrollers cannot directly supply enough electric power to devices that require large amounts of current, such as heaters or motors. Instead, they can control an intermediary device, such as a relay (see Table 22.1), to switch on and off the power supplied to the device.

There are a number of advanced sensors now available that enable surprisingly sophisticated yet simple-to-build projects. Types of sensors include temperature probes, accelerometers, gyroscopes, light sensors, strain gauges, proximity detectors, radiation sensors, magnetic field detectors, gas sensors, pressure sensors, flex

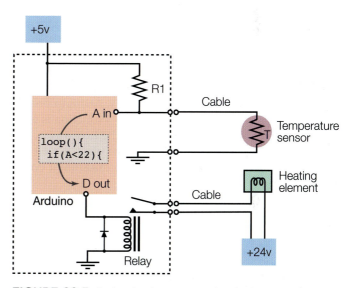

FIGURE 22.5 A simple thermostat circuit A sensor that changes resistance with temperature is combined with a resistor (R1) to convert this information to a changing voltage, which is then monitored with an analog input of the Arduino. A digital output controls a relay, which turns the heater on and off. The heater is supplied from a separate power supply that can handle higher voltage and current than the Arduino is capable of.

sensors, compasses, and microphones. Arduino software is available for most of these sensors, and hardware can be ordered from suppliers like SparkFun Electronics, so building projects with them can be just about plug-and-play.

Some Arduino microcontroller boards, including the Arduino Uno (`arduino.cc/en/Main/ArduinoBoardUno`), are designed to work with shields—premade circuit boards that piggyback on the microcontroller board. The breadboard shield for simple prototyping has already been mentioned. Other shields add functionality for connecting to the Internet via an Ethernet adapter, connecting to other Arduinos wirelessly, logging data to a flash memory card, motor and servo control, self-location with the Global Positioning System (GPS), and other tasks. These shields make a good starting point for many scientific tasks. For compact projects, the Arduino Nano is a fully functional unit with a reduced footprint that plugs directly into a breadboard.

Programming Arduino

An open-source Arduino programming tool (Figure 22.6) is available from the Arduino website (`arduino.cc/en/Main/Software`) for OS X, Linux, and Windows operating systems. This software package is used to compose programs, compile them, and transfer them to the Arduino, as well as monitor traffic through the Arduino's serial port. One of the nice things about Arduino, in addition to broad online support, is that there are many example scripts linked from within the program menus, so you are never far from seeing a working code sample.

The Arduino programming language is based on the language C, but it is not difficult to perform most operations that you have already seen demonstrated within Python. Being aware of a few structural differences will help you get up to speed more quickly. Arduino programs have two main sections: `setup()` and `loop()`. The `setup()` function is where you put commands that you want to run a single time when the board is powered up. Here you will configure how you want to use each of the pins, as well as set them to their initial states. You also need to set up serial communication if you would like to use that in the program:

```
int RedPin = 10;
int GrnPin = 8;
int InputPin = 14;
long previousMillis;

void setup()
{
  pinMode(RedPin, OUTPUT);    // set pin 10 as digital output
  pinMode(GrnPin, OUTPUT);    // set pin  8 as digital output
  pinMode(InputPin,INPUT);    // set pin 14 as digital input

  Serial.begin(19200);   // ...set up the serial output
  Serial.println("Starting...");
```

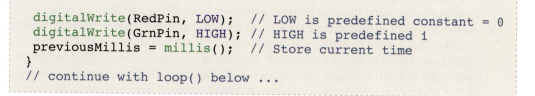

```
digitalWrite(RedPin, LOW);   // LOW is predefined constant = 0
digitalWrite(GrnPin, HIGH);  // HIGH is predefined 1
previousMillis = millis();   // Store current time
}
// continue with loop() below ...
```

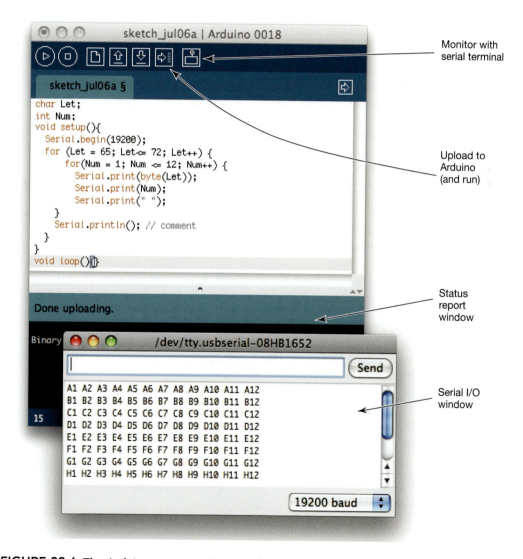

FIGURE 22.6 The Arduino programming interface showing an example program

Some things to notice about the Arduino language: Most lines need to end in a semicolon to mark the end of the command. Comments are indicated by // instead of #. Blocks for functions, loops, or logical expressions are bounded by curly brackets {} before and after the commands. Indentation does not matter, but it is still a good idea to use it to mark off sections of your program. Unlike in Python, you need to tell the system up front what kind of variable each name corresponds to; at that point you can also assign an initial value. If you are using an integer value, for example, you might say:

```
int X = 0;
```

Because of the amount of memory devoted to storing its value, an integer variable of type `int` in an Arduino program can only range from -32,768 to +32,767. If you add 1 to a value of 32,767, it will jump to the value of -32,768. To avoid problems like this, you can use an integer variable of type `long` instead, and this will let your values range above two billion. In Arduino operations, you often will need to monitor time in milliseconds. A regular integer would barely allow you to keep track of time for one minute, so `long` numbers become necessary even for relatively short intervals.

After the initial `setup()` function comes the main program `loop()`. The commands in this block are carried out repeatedly, until the Arduino is powered off. This is very different from the relatively linear Python programs you have been using, which may contain loops, but exit when the commands are completed. This intrinsic looping property makes Arduinos well suited to operations involving monitoring and sensing.

Many operations designed for use in the `loop()` portion of an Arduino program perform an operation once within a defined period (e.g., check the temperature once per 10 seconds). Instead of using a `delay()` function, which halts operation of the program until a prescribed interval has passed, it is better to allow the loop to run free, but check the passage of time and perform an operation when the interval has been exceeded. This basic convention is used in many loop-based systems, including LabVIEW, which we will introduce shortly.

Here is a fragment of code to achieve a timed loop:

```
int Interval = 500;   // time between operations -   doesn't change
long PreviousMillis = 0;   // current milliseconds - use long int

void setup(){
  //use the built-in millis() function to get the system time
  PreviousMillis = millis();
}
void loop()
```

```
{
    //check if an interval has passed. if not, continue looping
  if (millis() - PreviousMillis > Interval) {
    //we have passed the interval. do an operation here
    PreviousMillis = millis(); // redefine the reference time
    // continue the loop
}
```

If you would like to connect your Arduino creation to a graphical interface or a more complex data acquisition program than a simple serial monitor, you have several choices. You can connect to a programming environment called Processing (`processing.org`) that has high-level graphing capabilities; you can use some of the Python interface protocols (`arduino.cc/playground/Interfacing/Python`); or you can connect to just about any other data acquisition environment through a serial connection, supported by most Arduino devices.

Other options for data acquisition

MATLAB has built-in support for collecting data from a serial port (type `doc serial` from the MATLAB prompt), so it is possible to write your data capture, analysis, and presentation pipeline all within that environment.

LabVIEW by National Instruments (`ni.com`) is a powerful but expensive option for data acquisition. NI, as the company is known, sells many cards and interfaces for doing analog, digital, serial, and other types of data acquisition. Instead of using text like most languages, LabVIEW uses diagrams to wire together programming elements, as shown in Figure 22.7. Each variable (for example user input, output, a measured value, or a predefined constant) is represented by a box, and

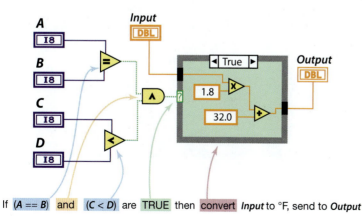

If (A == B) and (C < D) are TRUE then convert Input to °F, send to Output

FIGURE 22.7 A fragment of LabVIEW code with an accompanying translation into more familiar programming terms

wires are used to represent connections between the variables, mathematical and logical operators, loops, and if statements.

The LabVIEW programming environment is a unique and capable way to rapidly develop a graphical interface for your instrument, including built-in modules for creating knobs and buttons, sliders, graphs, and indicators (Figure 22.8). As

There are a few key concepts that differ for developing LabVIEW programs. As with the Arduino, most programs run continuously within a loop. In LabVIEW, loops are represented by a box, and everything in the box is executed repeatedly. LabVIEW itself figures out the sequence of events for your wired diagram, so you don't have to plan the order of operations. Values needed for later calculations are automatically obtained in the proper sequence. Global variables are also not usually used: instead, special operators called shift registers are placed at the left and right edges of the while loop. Variables wired to a shift register on the right side are passed back to the corresponding representation on the left, thus persisting through the next iteration of the loop.

Although LabVIEW hardware and the basic software are expensive (more than $1,000), there are academic licenses available to students and teachers which make the software available for less than $100. Because LabVIEW programs are inherently graphical in nature, this is a relatively easy way to incorporate a user-friendly interface into your data acquisition and control tasks, if that is a priority. Many

FIGURE 22.8 The graphical interface of a custom designed LabVIEW virtual instrument Each item on the front panel can be dragged from a palette, and then the elements are wired up on a diagram to provide input and output variables.

example programs (called VIs, or virtual instruments) can be downloaded from the online support community at `forums.ni.com`.

Common sources of confusion

When you work through an electronics book or tutorial, you may encounter a few common sources of confusion.

Measuring voltage

Like difference in pressures, voltage must be measured relative to a standard. The standard is usually called the **ground**, after the historical practice of using the Earth as a reference point for electric potential. Within a circuit, the ground, indicated diagrammatically by the symbol ⏚, is your reference point—the electronic equivalent of sea level. It corresponds to the negative terminal of a battery or power supply. The convention is that current flows in electrical circuits from the positive terminal to ground (but see the next section for the physical actuality that this convention simplifies).

You will often see apparently dead-end connections to the ground symbol all over the place in a circuit schematic. This doesn't mean those wires or terminals are unconnected at one end, or that there are several different grounds; it is just that they are all connected back to the negative power side. Showing all these wires would complicate the presentation of the schematic.

Current flow versus electron flow

It is sometimes confusing to think about the way that current flows through a circuit. In an electronics sense, it is said that positive charge flows from plus to minus—a convention dating back at least to Benjamin Franklin. However, in the physics sense, current is actually made up of electrons flowing from minus to plus, with positive "holes" (a lack of negative charges, like a double negative) flowing as a counter-current. As long as current is described in a manner which is internally consistent, this won't affect your circuit diagrams. The practical result, though, is that arrows in the schematic symbols for diodes, transistors, and other components may sometimes look like they are pointing in a direction opposite to the way you imagine current to be flowing (i.e., from positive to ground).

Pull-up and pull-down resistors

When you create a circuit, for example with an Arduino, you might assume that an unconnected input, since it has no voltage applied to it, would be off and have a logical value of zero (`LOW`). However, if you leave an input floating like this, it actually has an undetermined state. It will certainly go on (`HIGH`) when you apply a positive voltage, but it might also be high in this undetermined state.

To avoid this problem, you can connect the input to either +5 volts or the ground through a large resistance (typically 4,700 to 10,000 ohms), so that very little current flows through this part of the circuit. If you use a **pull-up resistor** (Figure 22.9A),

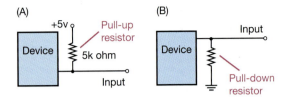

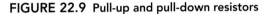

FIGURE 22.9 Pull-up and pull-down resistors

the input will be normally high, and you will have to apply a low signal (connect to ground) to switch the input. If instead you tie the input to ground through the resistor, (Figure 22.9B) it is known as a **pull-down resistor**, and the input will be held low until a sufficient positive signal is applied. In either case, because the resistance is so large, the signal from the pull-up or pull-down line will be overwhelmed by the sensor voltage when it is applied, and very little current will be drained from your circuit through this path.

SUMMARY

You have learned:

- How physical properties can be measured through the use of electronic sensors

- How information can be encoded as electric signals

- How a breadboard works to prototype a circuit

- The basic principles of serial communication

- That microcontroller boards like the Arduino can be used to build scientific instruments

- That LabVIEW is a powerful but expensive approach to data gathering

- How to avoid some common sources of confusion when starting out with electronics

Moving forward

- Get yourself an Arduino board from Adafruit Industries (www.adafruit.com) or SparkFun Electronics (www.sparkfun.com) and work through the tutorials online.

- Read about XBee and other wireless communication systems that can provide a wireless serial connection, sometimes over distances of more than a kilometer.

- Compare the specifications and capabilities of other data-acquisition systems, such as the SerIO board and PICaxe microcontrollers. These could be more suited to the demands of your project, such as having the ability to conserve power by sleeping and waking when needed.

- To test your circuits, get a multimeter, which measures resistance, voltage, and current. An oscilloscope, either as a stand-alone instrument or a USB-attached computer peripheral, will let you see plots of voltage over time with resolution of microseconds. This is very useful for determining what is going on behind the scenes in your circuits.

- Think about what measurements or experiments you perform that could be automated, or what sensors would make your field work easier. Work with a technician or try to come up with an electronics solution to the problem.

Recommended reading

References are listed with general texts at the top, and more specialized books at the bottom.

Mims, Forrest M. *Getting Started in Electronics*. Master Publishing, 2003.
    A classic text that provides a friendly introduction to electronics, components, and circuits.

Mims, Forrest M. *Science and Communication Circuits & Projects*. Master Publishing, 2004.

Mims, Forrest M. *Electronic Sensor Circuits & Projects*. Master Publishing, 2004.
    Both pamphlets by Forrest Mims include creative solutions to designing sensors and environmental monitors. Although these are now a bit old, they have clear explanations and still contain a great deal of useful information and clever ideas.

Scherz, Paul. *Practical Electronics for Inventors*. McGraw-Hill, 2006.
    Provides a more in-depth and comprehensive treatments of electronics. (We did not just select this because we felt kinship with the title.) Be sure to get at least the second edition (2006) or later, as the first edition is reported to have a fair number of errors.

Margolis, Michael. *Arduino Cookbook*. O'Reilly Media, 2010.
    This book contains a wealth of useful examples presented in Problem:Solution format. Each solution includes circuit diagrams and an example program, so that it can be rapidly implemented.

Igoe, Tom, and Dan O'Sullivan. *Physical Computing: Sensing and Controlling the Physical World with Computers*. Course Technology PTR, 2004.
    Although this book does not talk about Arduino in particular, it covers a lot of the principles and concepts used when developing sensor devices and networks, and when working with microcontrollers.

APPENDICES

Appendix *1*

WORKING WITH OTHER OPERATING SYSTEMS

This first appendix provides details on how to get set up if you have a computer with Microsoft Windows or Linux, rather than Mac OS X. As described in *Before You Begin*, Linux and OS X are both based on Unix. They are much more similar to each other than to Windows. Because of this, it will be easier to follow along with this book if you are using Linux than if you are using Windows. The main difficulties for using this book with Windows will be related to the command line, introduced in Chapters 4–6. There are work-arounds for that section, and it should not be much of a problem to work through the rest of the book.

Most of the important tools used in this book are in fact available for all three major operating systems—OS X, Linux, and Windows. Some of these tools, such as regular expressions, are entirely platform independent, and can be used within many programs and languages. Certain others, such as specific brands of text editors, do depend on the operating system, but are available for each OS with more or less similar features. Still other tools, however, have only approximate alternatives for the different operating systems.

The instructions in this appendix can't anticipate all of the possible variations of systems, so if you get stuck, check the content at `practicalcomputing.org`.

Microsoft Windows

Should I work in Windows or install Linux?

If you have a computer running Microsoft Windows, you have two options. The first is to install programs, languages, and environments within Windows that add most of the functionality needed to develop and use the skills we present—for example, jEdit, Python, and Cygwin. The second and more comprehensive option is to install Linux on your computer alongside Windows, giving you the chance to learn your way around an operating system based on Unix. Robust and freely available virtual machine software lets you do this without needing to reformat your hard drive or reboot your computer to switch between operating systems.

We'll cover both approaches in this appendix, starting with the first—installing programs within Windows for added functionality.

Text editors for text editing and regular expressions in Chapters 1–3

Using jEdit At this time, our two recommendations for Windows text editors are Notepad++, described below, and jEdit. jEdit is a full-featured open-source text editor that is available for all major operating systems. It has many of the same features and advantages as TextWrangler, the editor for OS X that is used throughout the main text of this book, and like TextWrangler, supports regular expressions. You can obtain jEdit from www.jEdit.org. Click the download link on the Web site and select the Windows Installer for the stable version. Follow the instructions when you launch the installer. You will need to have Java Runtime Environment (JRE) 1.4 or greater installed, which you can find at www.java.com/en/download/. Once you've installed the editor, select Find... from the Search menu and make sure that Regular expressions is checked.

There are some differences from TextWrangler in the way jEdit handles regular expressions. One is that when you capture text with parentheses, the text is still held in numbered variables, but these are preceded by a $ rather than a \. For instance, the replacement term \1\t\2\t in TextWrangler is written as $1\t$2\t in jEdit. Note that not all backslashes are replaced with dollar signs, only those used to designate that a number represents a bit of captured text.

Another difference is that in regular expression searches, jEdit uses \n as the universal end-of-line character; this is different from TextWrangler. To change this character in an open document (whether to carriage return, line feed, or both), click the Utilities menu and select Buffer Options... to open a dialog box. One of the options in this dialog is to modify the Line separator, which is just another term for line ending.

Using Notepad++ One of the most popular text and script editors for Windows is Notepad++, which supports many features for programmers (Figure A1.1). It can be downloaded from www.sourceforge.net/projects/notepad-plus/. With this editor, in Chapters 2 and 3 you will use a Replace dialog box (ctrl H) rather than a Find dialog, and you will need to make sure that regular expression is checked at the bottom of the dialog box.

Notepad++ doesn't support all of the regular expressions presented in this book. Because of this, when working through Chapters 2 and 3, you might want to use jEdit instead. See Appendix 2 for more on regular expressions and compatibility. Notepad++ does not support the modifier ? to change the greediness of quantifiers. It uses the same convention as TextWrangler, \1, for using captured text in substitutions. Support for end-of-line characters in searches may require selecting Extended mode instead of Regular expression.

For programming, however, Notepad++ will probably give you a better user experience. One minor advantage Notepad++ has over jEdit is that it does not

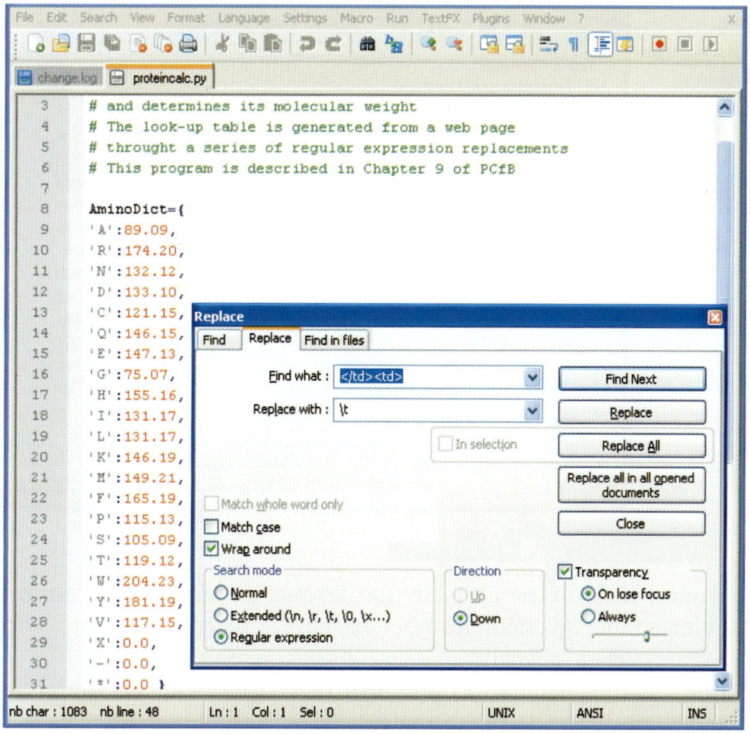

FIGURE A1.1 The Notepad++ editor, showing syntax coloring and the regular expression search box

require Java or other ancillary programs, so file opening and saving dialog boxes have a more familiar look and feel.

Cygwin for emulating Unix shell operations in Chapters 4–6

The DOS and PowerShell command-line interfaces available in Windows are very different from the Unix command line, and they cannot be used in place of it when working through the examples in this book, nor in real life. To get a proper Unix command line, you can install a full Linux system, via the instructions described later in this appendix. However, it is also possible to partially emulate a Unix command line with the Cygwin package. Cygwin can be downloaded from `cygwin.com`, which provides installation instructions as well.

Many of the programs described in the shell chapters, such as `cd`, `ls`, `mv`, and `cp`, are installed by default with Cygwin. Other useful programs which aren't installed by default can be added from within the `setup.exe` installer, ideally at the time you originally install Cygwin. Programs you should be sure to add in this manner include:

- Editors: `nano`

- Interpreters: `python`

- Net: `curl`

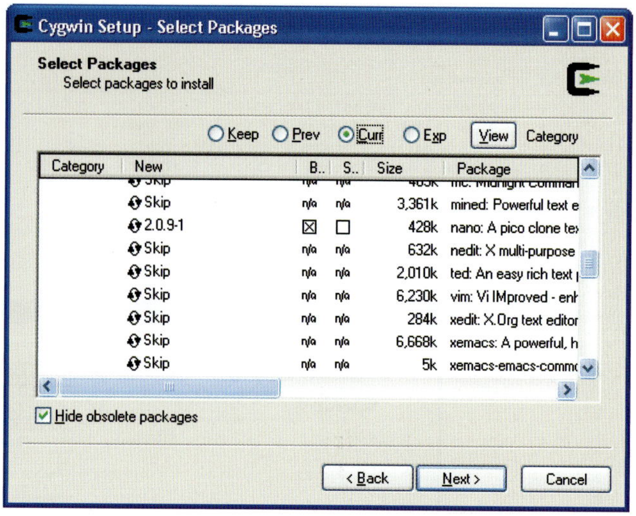

FIGURE A1.2 The installer for Cygwin, a Unix subsystem for Windows Be sure to install the optional packages as indicated in the text.

To select these packages, click on the little circular arrows under the New column heading in the installer, so that a version number appears in place of the word Skip (Figure A1.2).

More information on the installation and use of Cygwin can be found at `www.physionet.org/physiotools/cygwin/`. When you launch the Cygwin environment, you will be presented with a window showing the bash shell. You will be able to operate using Unix commands as described in the text, somewhat equivalent to running a terminal program in a true Unix system.

The root directory within Cygwin (the directory specified by / as described in Chapters 4–6) is not the Windows root directory, `c:\`, but corresponds instead to the Windows directory `c:\cygwin`. This means, for instance, that the file `/home/Administrator/.bash_profile` in the Cygwin environment is the same file as `c:\cygwin\home\Administrator\.bash_profile` in Windows. Conversely, the Windows root directory `c:\` is available from within the Cygwin environment as the directory `/cygdrive/c`. This allows you to access files from Windows without having to copy them within `c:\cygwin`.

When you launch the Cygwin environment it puts you within the default home directory. Depending on your system and installation, this can be `/home/Administrator` or, particularly confusing, `/cygdrive/c/ Documents and Settings/lucy`. This directory corresponds to the example home directory we use throughout the book, `/Users/lucy`, and at the command line you can always use the `~/` shortcut within Cygwin to refer to your home directory, rather than typing out the whole thing.

If you are using Cygwin, you should create your scripts and examples folders as subdirectories within whatever directory you are in upon first launching Cygwin. You can find this initial directory by typing either pwd at the initial prompt,

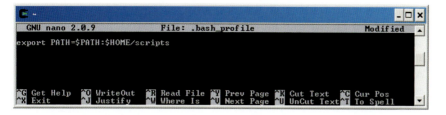

FIGURE A1.3 Using the editor **nano** in Cygwin to modify the **PATH** variable in the **.bash_profile** settings file

or echo $HOME. The instructions for editing the $PATH variable in your .bash_profile, as described in Chapter 6, should apply here as well; this is because the $HOME variable is used to make a reference which is portable between systems (Figure A1.3).

One drawback of Cygwin is that you cannot copy and paste from within the terminal window by default. There is a convoluted way to enable this by right-clicking on the Cygwin tab in the Windows taskbar and in the preferences enabling QuickEdit. Now you can select text, copy it by pressing ⟨return⟩, and paste into another terminal window.

Another catch is that ⟨ctrl⟩C, the usual Windows shortcut for copy, is also a widely used Unix and Cygwin shortcut for interrupting a process, such as when you have a runaway program. Similarly, ⟨ctrl⟩Z, which typically means "undo" in Windows, is the Unix and Cygwin command to stop (suspend the operation of) a running process. You will have to retrain your fingers to use the right shortcuts and avoid the wrong ones for each environment you work in.

Python on Windows for Chapters 8–12

If you are going to use Windows with a virtual Unix environment—either Cygwin or Linux—then follow the Linux or Cygwin instructions for Python later in this Appendix. Otherwise, to run Python in Windows, download Python for Windows from www.python.org/download. For compatibility with this book, choose the most recent version of Python 2.x, not any of the 3.x versions. You will need administrator privileges to complete the installation process. Install for All Users, and note the name of the folder name where the installation is placed. You might also want to install the Python Win32 Extensions, available at sourceforge.net/projects/pywin32. Among other things, this will give you Unicode support for use of extended characters.

Integrated Development Environments, or IDEs, are available for Python, and these too can run inside Windows. A typical IDE serves as a text editor, code debugger, and language reference source all in one. Popular free IDEs for Python include IDLE, which is included in the default Python installation, as well as the more advanced PyDev, which is available at pydev.org. From within these IDEs, you can edit and run the Python programs described in this book. In Windows, the commands for reading and writing files will vary slightly from our examples; this is because you will be using paths such as 'C:\scripts' rather than Unix or OS X paths written as '~/Documents/scripts'. Even so the underlying idea is the same: make a new folder called scripts on your main drive, for example C:\scripts, and use this as your storage location for all working scripts.

In Windows, running Python scripts from the Command Prompt is more complicated than running them from within an IDE such as IDLE. You will have to inform the Windows command-line environment of where to find both the Python program and your scripts. There is a chance that the installer has made the first modification for you already, but proceed with these modifications to make sure that your scripts folder can also be found. First you must determine where python.exe was installed on your computer—typically this will be something like C:\Python27. At the same time, you can also let the operating system know about the scripts directory you created above. In the Windows Control Panel, click on the System icon, then select Advanced system settings. In the dialog that appears, click the Environment Variables button near the bottom (Figure A1.4).

Scroll down among the list of Environment Variables until you see the Path variable, and double-click to edit this. At the end of the existing string value, type a semicolon, followed by the name of the folder that contains python.exe and the location of your scripts folder. For example, you might type something like this: ;C:\Python27; C:\scripts;.

Click OK to save your new path. Thanks to adding python.exe to your Path variable, you will now be able to launch the Python interactive prompt just by

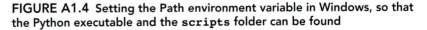

FIGURE A1.4 Setting the Path environment variable in Windows, so that the Python executable and the scripts folder can be found

typing `python` at the Command Prompt. In Windows you use ⌷ctrl⌷Z instead of ⌷ctrl⌷D to exit the Python environment. Note that because the Python installation process will have already associated `.py` files with Python, you do not strictly need to use a shebang line in your scripts as you must on OS X and Linux. You should still include it, though, so that your programs can work on these other platforms as well.

To execute scripts stored in the scripts directory, just type their name at the Command Prompt. You can get the prompt window by clicking the Start button in the Task bar and selecting Programs ▸ Accessories ▸ Command Prompt. Remember that the Command Prompt is not the same as the Unix command-line environment, and will therefore not have the same commands available. You can also run a script by double-clicking its icon on the desktop or in an Explorer window, but this is only useful if the program doesn't require command-line parameters, and also doesn't need to read or write files based on the working directory, rather than just the folder in which the icon happens to be located.

Python under Cygwin If you install Cygwin as described above, then during the installation process you should also install Python as one of the optional packages. You can use the Unix instructions in the text with little modification. Editing your `.bash_profile` for setting the scripts path will use your Cygwin home directory instead of `/Users/lucy`, as set by the `$HOME` variable. In your programs, for indicating paths to files you can actually use normal slashes, for example `c:/Users/lucy`, to avoid backslashes being misinterpreted in a special way. Also, open files using the mode `'rU'` instead of simply `'r'`, to avoid problems with end-of-line characters.

Working with MySQL on Windows for Chapter 15

In Windows, if you haven't configured MySQL to autostart during the installation, first get a Command Prompt, then type `cd c:\mysql\bin`. In that directory, start the database server with the command `mysqld`. At this point, you can launch `mysql` and follow along as described in the chapter. File paths will need to be specified using the `c:\` notation instead of the Unix-style notation of paths like `/Users/lucy`.

To configure the server to autostart, see the detailed instructions (and helpful user comments) at `dev.mysql.com/doc/refman/5.1/en/windows-start-service.html`.

Working with vector and pixel art in Windows for Chapters 17–19

In Windows, you can use the free vector and pixel editors Inkscape, GIMP, and ImageJ, or the commercial Adobe Illustrator and Photoshop. The experience will be the same as described in these graphics chapters except that some key combinations will be a bit different. Specifically, the ⌷alt⌷ key is used in place of ⌷option⌷, and ⌷ctrl⌷ is used where Command (⌘) would be used in OS X.

Linux

Installing Linux

There are many options for installing Linux. The first decision you need to make is which of the many Linux distributions to select. We recommend starting with Ubuntu (www.ubuntu.com) for the simple reason that it works well with most consumer computers. It also has a very wide user base, in part because it is so accessible to beginners. The instructions here were written for Ubuntu 10.04, and some details may change with future versions. Consult www.ubuntu.com for updated installers or practicalcomputing.org for additional instructions.

If you are running Windows, we recommend that you install Ubuntu within a virtual machine, as described below. This is the simplest way to get Linux and Windows going on the same computer, and to easily switch between them as you work. A virtual machine is a program that runs within a **host** operating system and simulates an entire computer running a **guest** operating system. In this case the host operating system is Windows, and the guest operating system is Ubuntu Linux. You can run your virtual machine right alongside other Windows programs, without any need to restart your computer to switch between the host and guest operating systems. You can even suspend the guest computer if you want to close the virtual machine program, and then restart it later right where you left off.

If you would like to erase Windows and install Linux alone on your computer, or install Linux on a separate disk partition that you can boot Linux from when you restart the computer, you can follow the standard installations instructions on the Ubuntu web site. A dual-boot system that can start up as either Windows or Linux can be quickly configured using Wubi, found at wubi-installer.org. These standard installations, though, don't give the benefits of using Windows programs when you need them, alongside the Linux system.

There are virtual machine programs that work with just about any combination of host and guest operating systems, except that OS X cannot be installed as a guest operating system on any virtual machine. We recommend the free, full-featured virtual machine VirtualBox by Oracle. It works within many host operating systems, including Windows, OS X, and Linux. To get up and running, you will need to install VirtualBox, create a virtual computer within it, install Ubuntu on the virtual computer, and then configure the system (Figure A1.5).

Installing VirtualBox If you have a standard PC running Windows, download and install the x86 version of VirtualBox for Windows hosts. It is available at www.virtualbox.org/wiki/Downloads. The default installer settings should be fine. Once VirtualBox is installed, launch it and click the New button to create a new virtual computer that you will later install the guest operating system on. The setup wizard will guide through this process. You can give the virtual machine any name you like—Ubuntu would do just fine. Select Linux for the operating system, Ubuntu for the version, and select a minimum of 512 MB base memory. You will then need to create a virtual hard disk. In the Virtual Hard Disk window, click New... and select the option to create a dynamically expanding storage disk.

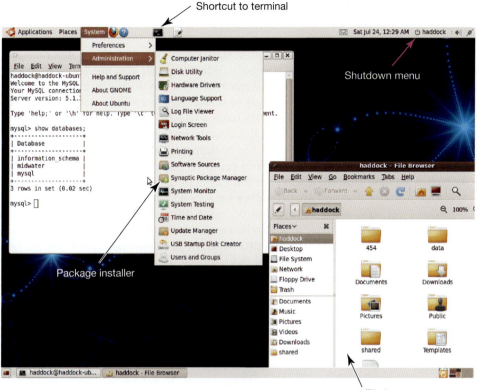

FIGURE A1.5 The Ubuntu Linux environment running in a VirtualBox window Note the convenient terminal shortcut in the VirtualBox window bar.

You will then need to select a size for the virtual hard disk. 8 GB is sufficient for most purposes.

Installing Ubuntu in VirtualBox Once you have installed VirtualBox and created a virtual computer, as described above, you'll need to install Ubuntu on the virtual computer. Follow the links for downloading Ubuntu from www.ubuntu.com and download the 32-bit Desktop edition. This is a large file, and may take several hours if you have a slow Internet connection. Start the virtual computer you created in VirtualBox by double-clicking it in the left side of the VirtualBox window. This will launch the First Run Wizard. Select CD/DVD–ROM Device for the Media type, click on the folder icon under Media Source, click the Add button at the top of the window that pops up, and then browse to and select the Ubuntu file you downloaded. Once you have finished the First Run Wizard, your virtual computer will start up and Ubuntu will guide you through the installation process. If you have any questions or difficulties with the Ubuntu installation process, consult Ubuntu's Web site for help.

When the Ubuntu installer asks to prepare disk space, it is only looking at the virtual disk and not your entire hard drive. You should follow the default settings, which will dedicate the entire virtual disk to Ubuntu. Once the installation is complete you will get a message to restart your computer. Since this message is within your virtual machine, it is only indicating that you need to restart that virtual computer and not the physical machine. You can do this by clicking Restart in the message window. In the process of restarting, you will be asked to remove the installation CD. You don't actually have a CD, of course, but you do need to disconnect the virtual machine from the Ubuntu installation file you downloaded. To do this, click on the CD icon at the bottom of the virtual machine window, and select Unmount CD/DVD ROM. You can then complete the restart, which will launch your new Ubuntu guest operating system. It is a good idea to install any software updates that you are prompted for when Ubuntu starts.

Whenever you want to use Linux, open VirtualBox and launch Ubuntu. VirtualBox will capture your mouse when you are using Ubuntu, which will keep you from moving it outside the Ubuntu window. To release the mouse, hit the key indicated at the lower right corner of the window. There are many books and on-line resources for learning your way around Ubuntu and around Linux in general. We encourage you to seek out these resources if you are interested in becoming more familiar with the specifics of this computing environment.

Installing Guest Additions After the basic installation is working, you will probably want to do two things: first, increase the screen resolution, and second, configure shared folders to easily transfer files between the host operating system, Windows, and the guest operating system, Linux. To accomplish these tasks, you will install Guest Additions, which are enhancements to the guest operating system that help it interact more fluidly with the host operating system. Follow these instructions:

1. From the VirtualBox menu bar on your host operating systems, go to the Devices menu and choose Install Guest Additions. This won't actually install the Guest Additions, it just makes the installation files available from within the guest operating system.

2. At the top of the Ubuntu screen, click on the Places menu and select Computer. This will open a window with several icons, including one for a CD the name of which will start with VBOXADDITIONS. Double-click on the CD icon to see the Guest Additions installation files.

3. You will need to do the installation via the command line, which requires opening a terminal window. Click on the Applications menu at the top of the Ubuntu screen, select Accessories, and then click Terminal.

4. Type the word sudo with a space after it at the command line. Then drag the file VBoxLinuxAdditions-x86.run from the installation file window you opened earlier to the terminal window. This just inserts the full name of the file with its path.

5. Press ⌐return⌐. sudo will ask you for your Ubuntu password, and then launch the installation.

6. When the installation is done, restart Ubuntu.

Now that the Guest Additions are installed and are ready to configure shared folders, you can resize the Ubuntu Desktop just by resizing the VirtualBox window.

Setting up shared folders Configuring shared folders will allow you to easily transfer files between operating systems, but still requires a few more steps. Unfortunately, configuring the shared folder is a bit complicated and isn't particularly well documented, even though this is one of the first things that most people installing VirtualBox will want to do. If you don't successfully complete the optional shared folder instructions below you can still transfer files to and from Ubuntu over the web or with disks. If you have trouble, consult the Ubuntu page on VirtualBox at help.ubuntu.com/community/VirtualBox, or the VirtualBox documentation at www.virtualbox.org/manual/ch04.html#sharedfolders.

Setting up a shared folder requires multiple steps. You must create a folder on the host (Windows) operating system, create a folder on the guest (Ubuntu) operating system as well, configure VirtualBox to share the host folder with the guest, and then configure the guest operating system to connect the two folders each time it boots up.

1. On the Windows desktop (or somewhere else on the Windows filesystem, if you prefer), create the folder you will share with the guest operating system. For the sake of this example configuration let's call it UbuntuShare and put it on the desktop.

2. Open a terminal window in Ubuntu, and create a folder on the Ubuntu side. We'll call it hostshare:

```
mkdir ~/hostshare
```

From the VirtualBox menu bar on your host operating systems, go to Devices menu and select Shared folders... Click the folder icon with a plus to add a shared folder, and browse to the UbuntuShare folder you created above. Make sure that the Make Permanent option is selected, and click OK twice to exit.

3. Restart Ubuntu using the menu at the right side of the upper menu bar.

At this point you have created the needed folders and configured the VirtualBox program to share the host folder with the guest operating system. You still need to configure the guest operating system to mount the host folder, which is the most complicated part. First, you'll gather the information you need mount the folder. Then you'll edit a configuration file to mount the folder each time Ubuntu starts. All of the following commands are issued in the Ubuntu terminal window.

4. Open a terminal window and type in the following commands shown in bold:

```
lucy@lucy-ubuntu:~$ cd ~/hostshare
lucy@lucy-ubuntu:~/hostshare$ pwd
/home/lucy/hostshare
```

5. The output of the above commands, e.g., /home/lucy/hostshare, is the path of the Ubuntu shared folder. The middle part of the path is your username. In this example the username is lucy, though yours will probably be something else. You will need this path and username in later steps.

6. The following steps will automatically mount the shared file each time Ubuntu starts up. Edit your system configuration file fstab in the terminal window by typing:

```
sudo nano /etc/fstab
```

7. Add the line below to the fstab file, right below the line that starts with: # <file system> <mount point>...:

```
UbuntuShare /home/lucy/hostshare vboxsf uid=lucy 0 2
```

where /home/lucy/hostshare is the path you got with the pwd command in step 4 and lucy is your username (the middle part of the path). Once you have finished, type ⌗ctrl⌗ X, then follow the prompts to save changes and exit the nano program.

8. Restart Ubuntu. Try transferring some files to make sure that the shared folders are working properly.

If you cannot get it to work, try this additional step, just to see whether the shared folders can be mounted at all.

Use the following command in the terminal window to mount the folder. UbuntuShare is the name of the folder on the host operating system and /home/lucy/hostshare is the path to the folder on the guest operating system that you got in the previous step. (This path will be something other than the example used here). This command will require your Ubuntu password:

```
sudo mount -t vboxsf -o uid=1000,gid=1000 UbuntuShare /home/lucy/hostshare
```

Without restarting, you should now be able to share files between the operating systems by moving them to the shared folder. Test that the sharing worked by moving some files in and out of the hostshare folders in Ubuntu and the UbuntuShare folder in Windows. If you get error messages or can't transfer files, check to make sure that the folder names are correct and retry the command. If you have further problems consult the Web sites mentioned in the introduction to sharing. If it does work, you are almost there. Recheck the `fstab` line and make sure that your directory names correspond. The `mount` command typed above is only a temporary fix that lasts until that terminal session is ended.

You can learn more about using VirtualBox from the user manual and from the forums at `practicalcomputing.org`. The manual is bundled with the VirtualBox installer and is also available at `http://download.virtualbox.org/virtualbox/UserManual.pdf`.

Installing software in Linux An important advantage of Linux is that it comes with a wider range or pre-installed software than other operating systems, and all of it is free. Right out of the box it will have many of the tools, from Python modules to full-functioned office productivity software, you will need on a daily basis. Chances are, though, that you will soon want to install additional software, including some of the more specialized tools described in this book.

There are several ways to install software on Linux. While you can download software packages from the Web and install them yourself, Ubuntu Linux comes with a few tools that streamline the process. If the software you want is available via these tools, it is a good idea to use them, since they simplify the installation process and ensure that everything is configured properly for your system. At the command line in Ubuntu, use `apt-get` for downloading and installing software. There are also a couple of GUI software installer interfaces. If you know the program you are looking for, use the Synaptic Package Manager (found in the System ▸ Administration menu). If you know what type of program you want but aren't sure which particular software packages are available, Ubuntu has a curated list of programs arranged by categories. This makes it simple to browse and install new software. This tool is available at Applications ▸ Ubuntu Software Center. Other versions of Linux have their own package managers, including the command-line tools `yum` and `rpm`.

Text editing and regular expressions with jEdit *for Chapters 1–3*

Ubuntu comes with several text editors pre-installed, and makes additional editors available via its Synaptic Package Manager. We suggest that you install jEdit. The Synaptic Package Manager makes installing not only jEdit but other programs a simple process. Select the System menu at the top of the Ubuntu screen, then the Administration submenu, and then click on the Synaptic Package Manager. Search for jEdit in the Quick search box. Click the checkbox next to jEdit in the Applications list, select Mark for Installation from the pop-up menu, and click Apply at the top of the window. Once the installation is complete, jEdit can be launched by selecting the Programming submenu from the Applications menu, then clicking the jEdit icon.

Basic instructions for using jEdit with regular expressions and for changing the end-of-line character can be found in the Windows jEdit section earlier in this appendix. If you prefer a native program rather than a Java application, try gedit, which is installed by default on Ubuntu. It is found in the Accessories submenu of the main Applications menu. You should also install the developer plug-ins (launchpad.net/gdp) to get regular expressions and other useful features. There are some reports that gedit cannot save directly to shared folders.

Using the Linux shell for shell operations in Chapters 4–6

To start using the command line in Ubuntu, launch the Terminal program from within the Accessories submenu of Applications. Since you will be using Terminal frequently, you might as well also drag its icon from Applications to the status bar at the top of the Ubuntu screen (see Figure A1.5).

Both OS X and Ubuntu use the bash shell by default, and nearly all the command-line programs used in this book that come with OS X also come with Ubuntu (install curl separately). This makes it simple to apply the lessons from the main text of the book directly to the Ubuntu command line. There are a few minor differences that could be confusing without some explanation, though.

The primary difference between OS X and Ubuntu at the command line is that the filesystems are arranged slightly differently. Home directories in Ubuntu are located in the /home folder, not the /Users folder employed by OS X. /Users/lucy on OS X would therefore correspond to /home/lucy on Ubuntu. On either type of system, ~/ is still the relative path to the home directory of the current user, even though those directories have different absolute paths. If you are using Cygwin on Windows, ~/ will also work, but there will be some additional path elements added before the home directory name.

If you are using Linux, or if you are having problems getting the instructions from Chapter 6 to work, check to make sure that you have the bash shell running, rather than tcsh or some other shell variant. To do this, in a terminal window, type:

```
echo $SHELL
```

It will return the name of the program that currently runs within your terminal window, which should be /bin/bash. If your system reports another shell, such as tcsh, ksh, or csh, you have the option to change the default shell over to bash using the command chsh (short for **ch**ange **sh**ell).

First, determine the full path pointing to the bash program using the which command:

```
which bash
/bin/bash
```

Then run chsh with the -s option, followed by the location of bash returned by that command:

```
chsh -s /bin/bash
```

The system will probably ask you for your login password to confirm this operation.

Setting up your `.bash_profile` Your `bash` settings might be stored in a different location than described in Chapter 6. On Ubuntu and some other Linux variants, the settings for `bash` are stored in the file `~/.bashrc`, rather than the `~/.bash_profile` file used by OS X. The default Ubuntu installation already has a `~/.bashrc` file, so you won't need to create it, you can just modify it as indicated in the text. You can add additional lines after the end of existing lines in the file.

If you are using `tcsh` or `csh` shells, then the command to modify your path should go into the file `.cshrc`, and you should use this command instead:[1]

```
setenv PATH $PATH:$HOME/scripts
```

Setting up aliases in `tcsh` is also slightly different from `bash`. Instead of the equals sign, you just put spaces between the shortcut and its definition:

```
alias ll 'ls -l'
```

Other shells might use `.profile` as the settings file. If you are restricted to using a shell besides those above, do a search to find the location of the user login file on your system.

Python on Linux for Chapters 8–12

Python is included as part of the basic Ubuntu installation, so the instructions included in the main text of this book should also apply to your system. In some Linux installations, the path to the `python` folder or the `env` program used in the `#!` line of a script may be different. You can check this by typing `which env` in a terminal window, and using that location in place of `#!/usr/bin/env python` at the beginning of your scripts.

Many third-party Python modules, including NumPy, are installed by default in Ubuntu. Others, such as Biopython, can be easily installed through the Synaptic Package Manager or your Linux distribution's package installer.

Working with MySQL for Chapter 15

In Linux, the user experience for MySQL should be the same as described in the text. You will not have the same MySQL Preference Pane to control launching the MySQL database server each time your computer starts up.

---

[1] Many hidden shell settings files end with `rc`, so storing `csh` settings in a file called `.cshrc` is not as odd as it may sound.

You can install MySQL at the command line or using the GUI through System ▶ Administration ▶ Synaptic Package Manager. If you do install MySQL with `apt-get` at the command line, it is a good idea to update your installer information first with the command:

```
sudo apt-get update
```

Then type this command in a terminal window:

```
sudo apt-get install mysql-server
```

It will ask you a few times if you want to set the `root` password. You can leave it blank for the purposes of this demo and change it later if you wish to make your databases work over the Web.

Working with vector and pixel art in Linux for Chapters 17–19

The open source graphics programs discussed in this book (Inkscape, ImageJ, and GIMP) can be installed on Ubuntu through the Synaptic Package Manager. Select the System menu at the top of the Ubuntu screen, then the Administration submenu, and click on the Synaptic Package Manager. Search for the programs you want to install, mark them for installation, and click the apply button at the top of the screen.

Appendix 2

REGULAR EXPRESSION SEARCH TERMS

Regular expressions—ways to perform adaptive searches and replacements—are described in Chapters 2 and 3. Here we provide a quick reference to some of the more common regular expression terms. This table and the text of the book itself do not encompass the entire range of regular expressions. There are many other useful constructs, for example, embedding miniature scripts into your replacement terms, and searching for A or B in a string using the syntax (sword|jelly)fish. If you would like to delve deeper, there are many online references, and there is even an in-depth reference guide built into the Help menu of TextWrangler.

There is some variation in the terms supported from program to program and from language to language. The most widespread terms, which can be used almost anywhere that regular expressions are supported, are the POSIX Extended Regular Expressions. These include ., *, +, {}, (), [ ], [^], ^, $, ?, and |. While quite a bit can be accomplished with the POSIX terms, in many implementations the language has been supplemented with some nonstandard terms. Most of these nonstandard terms are based on Perl regular expressions. These include many of the character class wildcards listed in the tables below, such as \d, \w, and \n. These extra wildcards make it easier to write clear regular expressions. Lack of support for Perl-like regular expressions is one of the most common causes of confusion when moving to a new programming context.

If you are using regular expressions in a new context but find that they don't behave as expected, or that they generate errors, check to see which regular expressions are supported by the tool you are using. POSIX does define its own set of wildcards, but the syntax is different from the Perl-style \w format that we use in this book. These wildcards include [:digit:] in place of \d and [:alpha:] instead of \w that we use in this book (though not including the digits). These POSIX character classes can be used in some contexts where Perl classes aren't available, including SQL queries and the command-line tool grep. If you don't want to switch between wildcard types, a more universal solution is to replace character class wildcards with an explicit character range, such as [0-9] or [A-Z].

Wildcards	
\w	Letters, numbers and _
.	Any character except \n \r
\d	Numerical digits
\t	Tab
\r	Return character. Also used as the generic end-of-line character in TextWrangler
\n	Line-feed character. Also used as the generic end-of-line character in Notepad++
\s	Space, tab, or end of line
[A-Z]	A single character of the ranges indicated in square brackets
[^A-Z]	A single character including all characters *not* in the brackets. Note that this will include \n unless otherwise specified, and may cause you to match across lines
\	Used to escape punctuation characters so they are searched for as themselves, not interpreted as wildcards or special symbols
\\	The \ symbol itself, escaped
Boundaries	
^	Match the start of the line, i.e., the position before the first character
$	Match the last position before the end-of-line character

Quantifiers, used in combination with characters and wildcards

+	Look for the longest possible match of one or more occurrences of the character, wildcard, or bracketed character range immediately preceding. The match will extend as far as it can while still allowing the entire expression to match.
*	As above, matches as many of the previous character to occur, but allows for the character not to occur at all if the match still succeeds
?	Modifies greediness of + or * to match the shortest possible match instead of longest
{}	Specify a range of numbers to repeat the match of the previous character. For example: \d{2,4} matches between 2 and 4 digits in a row [AC]{4,} matches 4 or more of the letter A or C in a row

Capturing and replacing

( )	Capture the search results between the parentheses for use in the replacement term
\1 $1	Substitute the contents of the matched into the replacement term, in numerical order. Syntax depends on the text editor or language that you are using.

Appendix 3

SHELL COMMANDS

Terminal operations are described in Chapters 4–6, 16, and 20. Many of the built-in `bash` shell commands are summarized here for quick reference. To get more information about a command and its options, type `man`, followed by the name of the command. If you are not sure which command applies, you can also search the contents of the help files using `man -k` followed by a keyword term.

Command	Description		Usage
`ls`	List the files in a directory		`ls -la`
	Parameters that follow can be folder names (use * as a wildcard)		`ls -1 *.txt`
			`ls -FG scripts`
			`ls ~/Documents`
	`-a`	Show hidden files	`ls /etc`
	`-l`	Show dates and permissions	
	`-1`	List the file names on separate lines. Useful as a starting point for regexp into a list of commands	
	`-G`	Enable color-coding of file types	
	`-F`	Show a slash after directory names	
`cd`	Change directory		
	Without a slash, names are relative to the current directory		`cd scripts`
	With a preceding slash (/) names start at the root level		`cd /User`
	Tilde (~/) starts at the user's home directory		`cd ~/scripts`
	Two dots (..) goes "up" to the enclosing directory		`cd My\ Documents`
	One dot refers to the current directory		`cd 'My Documents'`
	Minus sign goes to the previously occupied directory		`cd ../..`
	Use ⌷tab⌷ key (see below) to auto-complete partially typed paths		`cd ..`
			`cd -`
	Use backslash before spaces or strange characters in the directory name, or put the whole name in quotes		

Command	Description	Usage
pwd	Print the working directory (the path to the folder you are in)	
↑	↑ key to step back through previously typed commands The cursor can be repositioned with the ← and → keys, and commands can then be edited Press return from anywhere in the line to re-execute. On OS X you can also reposition by option-clicking at a cursor location	
tab	Auto-complete file, folder, or script names at the command line	cd ~/Doc tab
less	Show contents of a file, page by page These commands also apply to viewing the results of man While less is running:	less data.txt
	q Quit viewing	
	space Next page	
	b Back a page	
	15 g Go to line 15	
	G Go to the end	
	↑ or ↓ Move up or down a line	
	/abc Search file for text abc	
	n After an initial search, find next occurrence of the search item	
	? Find previous occurrence of the search item	
	h Show help for less	
mkdir	Make a new directory (a new folder)	mkdir scripts
rmdir	Remove a directory (folder must be empty)	rmdir ~/scripts
rm	Remove file or files Use the -f flag to delete without confirmation (careful!) Use the -r flag to recursively delete the files in a directory and then the directory itself	rm test.txt rm -f *_temp.dat
man	Show the manual pages for a Unix command Use -k to search for a term within all the manuals The result is displayed using the less command above, so the same shortcuts allow you to navigate through	man mkdir man -k date man chmod

Command	Description	Usage

cp	Copy file, leaving original intact
	Does not work on folders themselves
	Single period as destination copies file to current directory, using same name
		`cp test1.txt test1.dat`
		`cp temp ../temp`
		`cp ../test.py .`

mv	Move file or folder, renaming or relocating it Unlike cp, this does work on directories	`mv test1.txt test1.dat` `mv temp ../temp2`
|	Pipe output of one command to the input of another command	`history | grep lucy`
>	Send output of a command to a file, overwriting existing files Do not use a destination file that matches a wildcard on the left side	`ls -1 *.py > files.txt`
>>	Send output of a command to a file, appending to existing files	`echo "#Last line" >> data.txt`
<	Send contents of a file into command that supports its contents as input	`mysql -u root midwater < data.sql`

./	Represents the current directory in a path—the same location as pwd
	Trailing slash is optional
	Can execute a file in the current directory even when the file directory is not included in the PATH
		`cp ../*.txt ./`
		
		`./myscript.py`

cat	Concatenate (join together) files without any breaks. Streams the contents of the file list across the screen	`cat README` `cat *.fta > fasta.txt`
head	Show the first lines of a file or command Use the -n flag to specify the number of lines	`head -n 3 *.fasta` `ls *.txt | head`
tail	Show the last lines of a file or output stream Use the -n flag to specify the number of lines to show With a plus sign, skip that number of lines and show to the end. Use -n +2 to show from the second line of the file to the end, skipping one header line	`tail -n 20 *.fta` `tail -n +3 data.txt`
wc	Count lines, words, and characters in an output stream or file	`wc data.txt` `ls *.txt | wc`
which	Show the location of executable files in the system path	`which man`

Command	Description	Usage
`grep`	Search for phrase in a list of files or pipe and show matching lines: `grep -E "`*`searchterm`*`" `*`filelist`* Often used in conjunction with piped output: `command	grep searchterm` Use quotes around search terms, especially spaces or punctuation like >, &, #, and others To search for tab characters, type `ctrl`V followed by `tab` inside the quotes Optional flags:	
	`-c` Show only a count of the results in the file	
	`-v` Invert the search and show only lines that do not match	
	`-i` Match without regard to case	

	`-E` Use full regular expressions
	       Terms should be enclosed in quotes. Use `[ ]` to indicate a character range rather than the wildcards of Chapters 2 and 3
	       General wildcard equivalents:
	       `\s  [[:space:]]`
	       `\w  [[:alpha:]]`
	       `\d  [[:digit:]]`	

	`-l` List only the filenames containing matches	
	`-n` Show the line numbers of the match	
	`-h` Hide the filenames in the output	
`agrep`	Search for approximate matches, allowing insertions, deletions, or mismatched characters. (Must be installed separately.) See Chapter 21 Optional flags include:	`agrep -d "\>" -B -y ATG seqs.fta` `agrep -3 siphonafore taxa.txt`
	`-d ","` Use comma as delimiter between records	
	`-2` Return results with up to 2 mismatches. Maximum is 8 mismatches	
	`-B -y` Return the best match without specifying a number of mismatches	
	`-l` Only list file names containing matches	
	`-i` Match without regard to case	

`chmod`	Change access permissions on a file (usually to make a script executable or Web accessible)
	First option is one of `u`, `g`, `o` for user, group, other
	Second option after the plus or minus is `r`, `w`, or `x`, for read, write, or execute. Can also use binary encoding as explained in Appendix 6
		`chmod u+x file.pl`
		`chmod 644 myfile.txt`
		`chmod 755 myscript.py`

Command	Description	Usage
set	Show environmental variables, including functions that have been defined	
$HOME	The environmental variable containing the path user's home directory	`echo $HOME` `cd $HOME`
$PATH	The user's PATH variable, where the directories to search for commands are stored	`export PATH=$PATH:/usr/local/bin`
nano	Invoke the text editor. Control key sequences include:	`nano filename.txt`
	ctrl X Exit nano (will be prompted to save)	
	ctrl O Save file without exiting	
	ctrl Y Scroll up a page	
	ctrl V Scroll down a page	
	ctrl C Cancel operation	
	ctrl G Show help and list of commands	
ctrl C	Interrupt the current process	

sort	Sort lines of a file	`sort -k 3 data.txt`
		`sort -k 2 -t "," F1.csv`
		`sort -nr numbers.txt`
		`sort A.txt > A_sort.txt`

	–k *N* Sort using column number *N* instead of starting at the first character. Columns are delimited by a series of white space characters	
	–t "," In conjunction with –k, use commas as the delimiter to define columns	
	–n Sort by numerical value instead of alphabetical	
	–r Sort in reverse order	
	–u Return only one unique representative from a series of identical sorted lines	
uniq	Return a single line for each consecutive instance of that line in a file or output stream. To remove all duplicates from anywhere in the file, it must be sorted before being piped to the uniq command Use –c flag to return a count along with the repeated element	`uniq -c records.txt` `sort names	uniq -c`

Command	Description		Usage
cut	Extract one or more columns of data from a file		`cut -c 5-15 data.txt`
			`cut -f 1,6 data.csv`
	`-f 1,3`	Return columns 1 and 3, delimited by tabs	`cut -f2 -d ":" > Hr.txt`
	`-d ","`	Use commas as the field delimiter instead of tabs. Used in combination with –f	
	`-c 3-8`	Return characters 3 through 8 from the file or stream of data	
curl	Retrieve the contents of a URL from over the network. URL should be placed in quotes. Without additional parameters, will stream contents to the screen		`curl "www.myloc.edu" > myloc.html`
	For some Linux versions, wget offers similar functionality		`curl "http://www.nasa.gov/weather[01-12]{1999,2000}" -m 30 -o weather#1_#2.dat`
	See man `curl` for ways to send user login information at the same time		
	`-o`	Set the name of the output file to save individual files for the data. See #1 below	
	`-m 30`	Set a time out of 30 seconds	
	`[01-25]`	In the URL, substitute two digit numbers from 01 to 25 into the address in succession	
	`{22,33}` `{A,C,E}`	Substitute items in brackets into URL	
	`#1`	The substituted value, for use in generating the filename	
sudo	Run the command that follows as a superuser with privileges to write to system files		`sudo python setup.py install`
			`sudo nano /etc/hosts`
alias	Define a shortcut for use at the command line. To make persistent, add to startup settings file `.bash_profile` or equivalent		`alias cx='chmod u+x'`
function	Create a shell function—like a small script		`myfunction() {`
	`$1` is the first user argument supplied after the command is typed		`    # insert commands here`
	`$@` is all the parameters—useful for loops as below		`    echo $1`
	Variable names are defined with the format `NAME=` with no spaces. They are retrieved with `$NAME`		`}`
	Save it in `.bash_profile` to make it permanent		
;	In a command or script, equivalent to pressing [return] and starting a new line		`date; ls`

Command	Description	Usage

`for`	Perform a `for` loop in the shell. Can be useful in the context of a function	<pre>for ITEM in *.txt; do
		    echo $ITEM
		done</pre>

`if`	An `if` statement in a shell function:
	<pre>if [test condition]
	then
	   # insert commands
	else
	   # alternate command
	fi</pre>
	Comparison operators are eq for equals, `lt` for less than and `gt` for greater than
		<pre>if [$# -lt 1]
		then
		   echo "Less than"
		else
		   echo "greater than 1"
		fi</pre>

`` ` ` ``	Backtick symbols surrounding a command cause the command to be executed and then substitute the output into that place in the shell command or script	<pre>cd `which python`/..
		nano `which script.py`</pre>

`host`	Return IP number associated with a hostname, or the hostname associated with an IP address, if available	<pre>host www.sinauer.com
		host 127.0.0.1</pre>

`ssh`	Start a secure remote shell connection	`ssh lucy@pcfb.org`

`scp`	Securely copy files to or from a remote location	<pre>scp localfile user@host/path/remotefile
		scp user@host/home/file.txt localfile.txt</pre>

`sftp`	Start a file transfer connection to a remote site. The prompt changes to an ftp prompt, at which the following commands can be used: `open` From the prompt, open a new `sftp` connection `get` Bring a remote file to the local server `put` Place a local file on the remote system `cd` Change directory on the remote server `lcd` Change directory on the local machine `quit` Exit the `sftp` connection	`sftp user@remotemachine`

`gzip`
`gunzip`
`zip`
`unzip`
	Compress and uncompress files	<pre>gzip files.tar
		gunzip files.tar.gz
		unzip archive.zip</pre>

`tar`	Create or expand an archive containing files or folders
	`-cf` Create
	`-xvf` Expand
	`-xvfz` Expand and uncompress `gzip`
		<pre>tar -cf archive.tar ~/scripts
		tar -xvfz arch.tar.gz</pre>

Command	Description	Usage
&	When placed at the end of a command, runs it in the background	
ps	Show currently running processes. Flags controlling the output vary greatly by system. Usually a good starting point is -ax. See man ps for more	ps -ax | grep lucy
top	Show current processes sorted by various parameters, most useful of which is processor usage -u	top -u
kill -9	Terminate a process emphatically, using its process ID. Retrieve PID from the ps or top command	kill -9 5567
killall	Terminate processes by name	killall Firefox
nohup	Run command in background and don't terminate it when logging out or closing the shell window Use in this odd format shown, to prevent program output to cause the command to quit	nohup command 2> /dev/null < /dev/null &
ctrl Z	Suspend the operation to move it into the background or perform other operations	
jobs	Show backgrounded or suspended jobs, won't show normal active processes	
bg	Move a suspended process into the background. Optional number after it in the format %1 will specify the job number	
apt-get yum rpm port	Package installers for various Unix distributions. Search for and install remote software packages. Typically used with sudo	sudo apt-get install agrep yum search imagemagick

Appendix 4

PYTHON QUICK REFERENCE

Conventions for this appendix

In the examples below, italicized terms are not real variable or function names, but are stand-ins for an actual name. If a function name is shown as `.function()` then the dot means it is used as a method, coming after the variable name, as in `MyString.upper()`.

Format, syntax, and punctuation in Python

- Indented lines define blocks of statements that are executed in loops, decisions, and functions.

- Comments are marked by # and extend from that symbol to the end of the line. Multi-line comments can be bracketed on both sides by three quote marks.

- To continue a statement on the next line, use the \ character at the end of a line.

- Parentheses ( ) pass parameters to functions.[1]

- Square brackets [ ] define lists and retrieve subsets of values from strings, lists, dictionaries, and other types.

- Curly brackets { } define dictionary entries.

Python scripts begin with the shebang line, and can include an optional line to enable support of Unicode characters:

```
#! /usr/bin/env python
# coding: utf-8
```

---

[1] They also are used to define tuples, non-changeable list-like variables that we don't address in this book.

The command-line interpreter

Start by typing `python` at the command line. Cycle up through history of previous Python commands using ↑. Use `quit()` or `ctrl` D to exit (`ctrl` Z in Windows).

You should be able to paste entire programs into the interpreter, but sometimes the indented block of a loop or conditional statement might not be carried over properly. Pasting commands at the Python prompt also does not work well for things involving user input or reading and writing files. In addition, the buffer of your terminal program may not keep up with large pasted blocks, resulting in errors on the text pasted.

Command summary

Variable types and statistics

Changing variable types and getting information	
Convert numbers and other types to strings This conversion is required for the `.write()` function used with a file or the `sys.stderr.write()` function	`str()`
Convert integers or strings to floating point	`float()`

Can specify the base in alternate base systems. To specify the number in hex, use `int(MyString,16)`	`int(3.14)`
	`int("3")`
	`int("4F",16)`

Give the length of a string, list, or dictionary	`len("ABCD")`
	`len([1,2,4,8])`
	`len(Diction)`

Strings

Defining and formatting strings	

Strings are defined by pairs of single (') or double (") quotation marks, not curly quotes ("")	`Location = "Hawai'i"`
	`Region  = "3'-polyA"`
	`Genus   = 'Gymnopraia'`

Multi-line strings are defined by three quote marks in a row	`MultiString = """`
	`   Triple-quoted strings`
	`   can span several lines.`
	`   They also act like comments`
	`"""`

Convert from number to string	`str(100.5)`
Find the ASCII code for a string character with `ord()`	`ord('A')`

Manipulating strings

Change case with `.upper()` and `.lower()`	`MyString.upper()` `MyString.lower()`
Join two strings with +	`MyString + YourString` `'Value' + str(MyValue) + '\n'`
Repeat a string with *	`print '='*30` `==============================`
Literal substitution (not using wildcards or regular expressions) with `.replace()`	`MyString.replace('jellyfish','medusa')`
Count occurrences of `'A'` in MyString with `.count()`	`MyString.count('A')`
Remove all white space from rightmost end of string with `.rstrip()`	`MyString.rstrip()`
Remove only linefeeds, not tabs	`MyString.rstrip('\n')`
Strip all white space from both sides of string with `.strip()`	`MyString.strip()`

See *Working with lists* in this appendix for converting strings or characters to lists and *Searching with regular expressions,* also in this appendix, for advanced search and replace techniques.

Gathering user input

Get user input during execution of program	`raw_input("Enter a value:")`
Get space-separated parameters given when program is run at the command line. You can pass parameters with wildcards, like `dive*.csv`	`import sys` `sys.argv`
The script or program name, using the zeroth parameter	`sys.argv[0]`
All subsequent command-line arguments	`sys.argv[1:]`
Determine how many command-line parameters were provided, via the `len()` function	`if len(sys.argv) > 1:`

Building strings

Printing strings	
Print variables separated by a space	`print MyString, MyNumber`
Print variables not separated by space	`print MyString + str(MyNumber)`

Generating strings with the formatting operator, `%`:

```
MyString = '%s %.2f %d' % ("Value",4.1666,256)
```
↳ Substitution points   ↳ Values to insert

This creates the string: `'Value 4.17 256'`

Given the string `s = '%x' % (4.13)` **where** `%x` **is a placeholder listed below:**		
Placeholder	**Type**	**Result**
`%s`	String variable	`'four'`
`%d`	Integer digits	`'4'`
`%5d`	Integer padded to at least five spaces	`'    4'`
`%f`	Floating point	`'4.130000'`
`%.2f`	Float with precision of two decimal points	`'4.13'`
`%5.1f`	Float with one decimal, padded to at least five total spaces (includes decimal point)	`'  4.1'`

Comparisons and logical operators

Comparison operators[a]	
Comparison	**Is `True` if...**
`x == y`	x is equal to y
`x != y`	x is not equal y
`x > y`	x is greater than y
`x < y`	x is less than y
`x >= y`	x is greater than or equal to y
`x <= y`	x is less than or equal to y

[a]These operators return `True` (1) or `False` (0) based on the result of the comparison.

Logical operators[a]	
Logical operator	**Is True if...**
A and B	Both A and B are True
A or B	Either A or B is True
not B	B is False (inverts the value of B)
(not A) or B	A is False or B is True
not (A or B)	A and B are both False

[a]In this table, A and B represent a True/False comparison like those listed in the previous table.

Note that in Python, when an expression involving logical operators is found to be true, the value returned is that of the first true item being tested, not True itself.

```
>>> 1 and 2
2
>>> 3 or 4
3
```

Math operators

Normal order of precedence applies. Operations involving only integers produce only integers, even at the expense of accuracy.

Addition	+
Subtraction	–
Multiplication	*
Division	/
Modulo (remainder after division)	% 7 % 2 → 1
Power	** 2**8=256
Truncated division (result without remainder)	// 7//2.0 = 3.0
Increment a variable by a value	+= X += 2

Decisions

The if, elif, and else commands control the flow of a program according to logical tests. Statements built on these commands end with a colon. Below is a description of each, with example code on the right.

```
if logical1:
    # do indented lines
    # if logical1 is True

elif logical2:
    # if logical1 is False
    # and logical2 is True

else:
    # do if all tests
    # above are False
```

```
A=5
if A < 0:
        print "Negative number"

elif A > 0:
        print "Zero or positive number"

else:
        print "Zero"
```

Loops

For and while loop definitions end with a colon. Use for loops to step through ranges and lists. Below are a series of loop examples, with code shown on the right.

for loop using range()

```
for Num in range(10):
    print Num * 10
```

for loop with a list

```
for Item in MyList:
    print Item
```

for loop with a string

```
for Letter in "FEDCBA":
    print Letter
```

while loop

```
X=0
while X < 11:
        print X
        X = X + 2
```

Searching with regular expressions

Regexp to find matching subsets in a string

Use regexp within your program to extract and substitute portions of a string. The basic format is:

```
Results = re.search(query,string)
```

The query is a text string containing the regular expressions pattern that you would enter into a Find dialog box.

Import the module	`import re`
Define a search query, using raw string	`MyRe = r"(\w)(\w+)"`
String to search	`MyString = "Agalma elegans"`
Search and save matches	`MyResult = re.search(MyRe, MyString)`
All the matches together	`MyResult.group(0)`
The first captured match	`MyResult.group(1)`
All matches as separate items	`MyResult.groups()`

Regexp to substitute into a string

The basic format is:

```
re.sub(query, replacement, string)
```

When used in a program, this is the same as a Replace All command for that string.

Import the module	`import re`
Define a search query, using a raw string	`MyRe = r"(\w)(\w+) (.*)"`
Define the replacement term, using \1, \2, etc., to represent entities captured with parentheses	`MySub = r"\1. \3"`
String to search	`MyString = "Agalma elegans"`
Search and save matches	`NewString = re.sub(MyRe, MySub, MyString)`
The result saved in `NewString`	`"A. elegans"`

Working with lists

Lists are ordered collections of objects. Items in a list can be of any type, including other lists and heterogeneous mixes of variable types. The first element has an index of 0; so, for example, a list with five members does not have an item at index 5.

Creating lists	
Create a list from string or other variable type If the variable is a string, the list elements will be each character of the string	`list(MyString)`
Define with square brackets	`MyList = [1,2,3]` `OtherList = [[2,4,6],[3,5,7]]`
Define an empty list; required before the list can be appended to	`MyList=[]`

Define numerical lists with the `range()` function
 The left element is included in the retrieval, the right index is not
 Given one parameter, `range(N)` creates `N` elements, from 0 to `N-1`. A third parameter optionally sets the step size between elements, positive or negative
	Function **Result**
	 `range(5)` `[0, 1, 2, 3, 4]`
	 `range(1,8,2)` `[1, 3, 5, 7]`
	 `range(5,0,-1)` `[5, 4, 3, 2, 1]`

Parse strings into lists with `.split()` Default delimiter is any amount of white space, or specify delimiter character in the `()`	`MyList = MyString.split()`
Add elements with `.append()`	`MyList.append(10)`

Insert elements with a single index repeated on both sides of the colon	`MyList=range(5)`
	 `MyList[3:3]=[9,8,7]`
	 `>>> MyList`
	 `[0, 1, 2, 9, 8, 7, 3, 4]`

Delete elements from list with `del` Assign `=[]` to delete indexed elements	`del MyList[2:5]` `MyList[2:5]=[]`

Accessing list elements

Extract elements with `[ ]`	`MyList[Start:Finish]`
Index range: Start element is retrieved, finish element is not	
Indices can count from either the beginning, or, using negative numbers, the end of the list	`MyList[begin:end+1:step]`
Skip first element of a list	`MyList[1:]`
All but last element	`MyList[:-1]`
Return list elements in reverse order, leaving the original list unchanged	`MyList[::-1]`
Sort list in place, modify original	`MyList.reverse()`

Extract even or odd elements	`MyList=range(8)`
	`MyList [1::2]`
	`[1, 3, 5, 7]`
	`MyList[0::2]`
	`[0, 2, 4, 6]`

Unpacking two or more values at once	`a,b=MyList[0,1]`

List information and conversions

Convert lists of strings to strings with `.join()`	`''.join(MyList)`

The `.join()` method works a bit backwards, acting on the character used to join, with the list as a parameter	`MyList = ['A', 'B', 'C', 'D']`
	`print '-'.join(MyList)`
	`A-B-C-D`

Test if an item is in a list with the `in` operator	`print 'A' in MyList` `True`

Create a list of unique elements of a list with `set()`	`MyList=list('aabbbcdaa')`
	`print list(set(Mylist))`
	`['a','b','c','d']`

Sort lists Return a sorted list, leaving original list unaltered	`NewList=MyList.sorted()`
Sort in place, modifying original list	`MyList.sort()`
	`Keys=Diction.keys()` `Keys.sort()`
Retrieve elements and their indices together, using `enumerate()`	`Ind, Elem = enumerate(MyList)`

List comprehension

Performs an operation on each item in a list, and returns a list of the results. List comprehensions are very useful for manipulating lists in Python.

```
Squares = [Val**2 for Val in MyList]
Strings = [str(Val) for Val in MyList]
```

Dictionaries

Dictionaries are somewhat like lists, except that instead of values being accessed by sequential numerical keys (indexes), they are accessed by non-sequential keys defined as you wish. Keys and values can be of many types, including numbers, strings, or lists, and they can occur together in one dictionary. Only one instance of a key is allowed in a dictionary, but values can occur repeatedly; that is, it is keys that are required to be unique, not values. Dictionaries have no intrinsic order to their contents, and values are returned only by key, not by position or order of entry.

Defining dictionaries	

Define entries within curly brackets with the format {key: value}
Key–value pairs are separated by commas
Between the brackets, the definition can span several lines and indentation is not important
	`Diction = {1:'a', 2:'b'}`
	`Diction={`
	`'Lilyopsis' :3, 'Resomia'  :2,`
	`'Rhizophysa':1, 'Gymnopraia':3 }`

A list of keys and a list of values having the same number of elements can be zipped together to form a dictionary	`SiphKeys = ['Lilyopsis','Rhizophysa',`
	`   'Resomia','Gymnopraia']`
	`SiphVals = [3,1,2,3]`
	`Diction = dict(zip(SiphKeys,SiphVals))`

Add entries using indexed values with square brackets Requires a pre-existing dictionary, which can have no entries	`Diction={}` `Diction['Marrus'] = 2`
Delete dictionary entries with `del` The method used to clear list elements by assigning to [ ] does not work with dictionaries. The key will still exist	`del Diction['Marrus']`

Extracting values from a dictionary	
Index with square brackets `[ ]` and the key	`print Diction['Resomia']` `2`
If the key is not present, results in an error	`print Diction ['Erenna']` `...KeyError: 'Erenna'`

Retrieve with `.get()`
Optionally, provide a value to return if the key is not present
	`print Diction.get('Resomia')`
	`2`
	`print Diction.get('Erenna',-99)`
	`-99`

Information about a dictionary	

Get a list of keys or values with `.keys()` and `.values()`, but not in any predictable order
The order, however, will be internally consistent between the two lists
	`Diction.keys()`
	`['Resomia','Lilyopsis',`
	`'Gymnopraia','Rhizophysa' ]`
	`Diction.values()`
	`[2, 3, 3, 1]`

Number of entries in a dictionary	`len(Diction)`

Creating functions

Define the function in the program before it is used, or in an external file which is imported. Functions can be generated with or without additional parameters, and parameters can be assigned default values.

```
def function_name(Parameter = Defaultvalue):
    # insert statements that calculate values
    return Result   # send back the result
```

Call the function from within the program, passing values in parentheses:

```
MyValue = function_name(200)
```

Working with files

Reading from a file	
Open the connection to the file	`InFile = open(FileName, 'rU')`
Read lines in succession	`for Line in InFile:` `    # perform operation on Lines`
Alternatively, read all lines into a list at once. (This can't be used after the command above since InFile is already at the end of the file)	`AllLines = InFile.readlines()`
Close the file connection	`InFile.close()`

An example of a short file-reading program in action:

```
FileName="/Users/lucy/pcfb/examples/FPexcerpt.fta"
InFile = open(FileName, 'rU')
for Line in InFile:
    MyLine = Line.strip()
    if MyLine[0]==">":
        print MyLine[1:]
InFile.close()
```

Getting information about files	
Use the os module	`import os`
Check if string is path to a file; fails if it is not found or if it is a folder rather than a file	`os.path.isfile('/Users/lucy/pcfb/')`
Check if a folder or file exists	`os.path.exists('/Users/lucy/pcfb/')` `True`
Fails with ~/ as part of path	`os.path.exists('~/pcfb/')` `False`
Get a list of files matching the parameter, using * as a wildcard	`import glob` `FileList = glob.glob('pcfb/*.txt')`

Writing to a file	
Open file stream, overwriting existing file if it exists	`OutFile = open(FileName, 'w')`
Open file stream, appending to the end of a file if it already exists	`OutFile = open(FileName, 'a')`
Write a string to the specified `OutFile` Line endings are not automatically appended, and numbers must be converted to strings beforehand, using the `str()` function or the format operator `%`	`OutFile.write('Text\n')`
Close the `OutFile` when done writing	`OutFile.close()`

Using modules and functions

First import the module, then call the function, usually followed by parentheses.

Ways to import functions from a module	
Import all the functions and use them thereafter by appending the function name to the module	`import themodule` `themodule.thefunction()`
Import a module, but use a different name for it within the program	`import longmodulename as shortname` `shortname.thefunction()`
Import all the functions from a module, and use them with only the function name	`from themodule import *` `thefunction()`
Import a particular function, and use it with just its name	`from themodule import thefunction` `thefunction()`
To see a list of commands in the module, after importing in the Python interactive environment	`dir(modulename)` `help(modulename)`

To create your own modules, use `def` to define functions as indicated above, place them in their own file, and save with a filename ending in `.py` somewhere in your PATH. Import them into your script using the filename without the `.py` extension.

Some built-in modules	
`random`	Random sampling and random number generation
`urllib`	Downloading and interacting with Web resources
`time`	Information related to the current time and elapsed time
`math`	Some basic trigonometric functions and constants
`os`	Items related to the file system
`sys`	System-level commands, such as command-line arguments
`re`	The regular expressions library for search and replace
`datetime`	Date conversion and calculation functions
`xml`	Reading and writing XML files
`csv`	Read in a comma-delimited file using the function `csv.reader()`

Other installable modules	
`MySQLdb`	Interact with a `mysql` database
PySerial	Connect through the serial port to external devices. Use with `import serial`
`matplotlib`	MATLAB-like plotting functionality
`numpy`, `Scipy`	Large package of numerical and statistical capabilities
Biopython	Functions for dealing with molecular sequence files and searches. Use with `import Bio` or `from Bio import Seq`

Miscellaneous Python operations

Presenting warnings and feedback

`sys.stderr.write()`

Sends output to screen (but does not send output to a file when a redirect such as >> is used).

Catching errors

Statements indented under a `try:` function will be executed until an error occurs. If there is an error, then the block of code indented under a subsequent `except:` statement will be executed.

Shell operations within Python

`os.popen("rmdir sandbox")`

The shell command specified in parentheses is executed. If you want to read the results the command would usually print to the screen, append `.read()`:

`Contents = os.popen("ls -l").read()`

For example, `os.popen(pwd)` will try to operate whether or not there is printed feedback.

Reference and getting help

- From the `python` command line, use `dir(item)` to see functions within a variable or imported module. Use `type(item)` to get a simple statement of the variable type.

- Depending on the variable, `help(item)` may give you the information pages related to a function or a variable, showing you information pertinent to its type.

- Consult Web sites such as `diveintopython.org` when stuck.

Appendix 5

TEMPLATE PROGRAMS

To give you an idea of the flavor of different languages, in this appendix we present the same small piece of code written in several popular languages. These all generate the output seen in Chapter 9 where we discuss the range() function.

The source code for each example (which you can copy and paste into separate program files to try out) is available in the file ~/pcfb/scripts/rosetta.txt. Expected output from each program is:

```
A1 A2 A3 A4 A5 A6 A7 A8 A9 A10 A11 A12
B1 B2 B3 B4 B5 B6 B7 B8 B9 B10 B11 B12
C1 C2 C3 C4 C5 C6 C7 C8 C9 C10 C11 C12
D1 D2 D3 D4 D5 D6 D7 D8 D9 D10 D11 D12
E1 E2 E3 E4 E5 E6 E7 E8 E9 E10 E11 E12
F1 F2 F3 F4 F5 F6 F7 F8 F9 F10 F11 F12
G1 G2 G3 G4 G5 G6 G7 G8 G9 G10 G11 G12
H1 H2 H3 H4 H5 H6 H7 H8 H9 H10 H11 H12
```

The programs take two basic approaches to generate the output. All of them use nested loops which cycle through ASCII values with the variable Letter, and numerical values with the variable Number. Some (C, C++, bash, JavaScript, Java, and Arduino) print the character pairs as they appear, and then print an end-of-line character each time the letter changes. Others (Python, Perl, MATLAB, PHP, and Ruby) build up a long string by appending character pairs, and then print that entire line when the letters increment. Either approach could made to work with any of the programming languages. Notice how the syntax of the for loops varies, how the comment characters differ, and how different languages merge strings and generate printed output.

The point of this section is to show that once you understand the basic building blocks used in programming, you will be able to adapt to many different languages by just learning their particular syntax and style.

Python 2.7 or earlier

```
#!/usr/bin/env python
for Letter in range(65,73): # step character 65 to 72
    Labels=''
    for Number in range(1,13):
        Labels += chr(Letter) + str(Number) + ' '
    print Labels    # print the whole line
```

Python 3

The next version of Python, Python 3, has already been released. Several important modules have not yet been updated to be compatible with Python 3, so we have focused on Python 2.x for this book. There is a utility called 2to3.py which can attempt to convert Python 2.x scripts into Python 3 format. The two versions of the language are very similar, but there are important differences—for example, some functions have been moved to other modules or removed altogether, and the print function operates differently, as illustrated here:

```
#!/usr/bin/env python
for Letter in range(65,73): # step character 65 to 72
    Labels=''
    for Number in range(1,13):
        Labels += chr(Letter) + str(Number) + ' '
    print(Labels)      # print the whole line
```

Perl

Perl has been a popular language for text manipulation and bioinformatics, though it is now more common to see Python used for tasks that would have been approached with Perl in previous years. You will still come across many Perl utilities and the BioPerl libraries. Perl is installed on Unix systems by default, and using scripts is handled in the same way that Python scripts are executed.

```
#!/usr/bin/env perl
for ($Letter = 65; $Letter < 73; $Letter++) { # step character 65 to 72
    $Labels = "";
    for ($Number = 1; $Number < 13; $Number++) {
        $Labels .= chr($Letter) . $Number . " " ;
    }
    print $Labels . "\n "  # print the whole line
}
```

bash shell

The bash shell is a user environment, but it also has a fully functional scripting language. It is especially useful for working within the filesystem (e.g., moving, renaming, and logging information) but less so as a stand-alone data processing environment. Note that there can be no spaces on either side of the equal sign when defining a variable.

```bash
#! /bin/bash
for Letter in {A..H}
do
    LABEL=" "
    for Number in {1..12}
    do
        LABEL="$LABEL $Letter$Number"
    done
    echo $LABEL   # print the whole line
done
```

C

C and its derivatives (Objective-C, C++) are some of the most popular languages for developing sophisticated software that runs in a graphical environment on your computer.

```c
#include <stdio.h>
char letter;
int number;

int main(void){
    for (letter = 65; letter <= 72; letter++) {
        for (number = 1; number <= 12; number++) {
            printf("%c%d ",letter,number);
        }
        printf("\n");  // print the line break
    }
    return 0;
}
```

C is a compiled rather than interpreted language, so in order to run this you will have to save it as rosetta.c, then compile it with the terminal command:

```
gcc rosetta.c
```

This assumes that you have the C compiler `gcc` installed. The executable produced by this command will be called `a.out`, and if its directory is not in your path, it can be run from its directory by typing:

```
./a.out
```

C++

Like Objective-C, C++ is a heavily object-oriented descendent of the C language. You can use straight C code in C++, but you can also do more using an object-based approach, which is a fundamental principle of modern computing.

```cpp
#include <iostream>
using namespace std;

char letter;
int number;

int main(){
   for (letter = 65; letter <= 72; letter++) {
      for (number = 1; number <= 12; number++) {
         cout << letter << number <<" ";
      }
      cout << endl; // print the end of line char
   }
}
```

Note that to run this program, you will have to save it as `rosetta.cpp`, then compile it with the terminal command:

```
g++ rosetta.cpp
```

As with our C example, this assumes that you have a compiler installed, in this case the C++ compiler g++. Upon success, the executable will be called `a.out`, and if the directory where `a.out` resides is not in your path, it can be run from its directory by typing:

```
./a.out
```

Java

Java is a compiled language that has the advantage of being cross-platform in most respects. A program written for a particular operating system or instrument should be able to run on other platforms.

```java
class Rosetta {
  public static void main(String[] args){
    //Step through the ASCII values
    for(int letter=65; letter<73; letter++){
      for(int number=1; number<13; number++){
        //print w/o LineFeed
        System.out.print((char)letter + (number + " "));
      }
      System.out.println();
    }
  }
}
```

Save the program as `rosetta.java`. To run it, you have to first compile it using the `javac` compiler:

```
javac rosetta.java
```

This will generate a `.class` file. To run it, you just use the base name of the program (which is actually determined by the program's class statement and not by its filename) along with the `java` command:

```
java Rosetta    ← Uppercase because it refers to the .class, not the .java file
```

JavaScript

JavaScript is the language used to make web pages dynamic. It is interpreted by your web browser, so save this script in a file with the extension `.html`, and drag it into a browser window to see the output:

```html
<html> <body>
<code>
Table of values<br><br>
<script type="text/javascript">
var letter, number;
document.write("<br>");
    for (letter = 65; letter<= 72; letter++) {
        for (number = 1; number <= 12; number++) {
            document.write(String.fromCharCode(letter) + number + " ");
        }
        document.write("<br>"); // print an html line break
    }
</script>
</code>
</body> </html>
```

PHP

PHP is also an interpreted language for web pages, but it is executed by the server just before the web page is sent out to the browser. This gives a bit more privacy because people cannot see the source of your php code—only the rendered result. To test this script, you can use your own computer as a web server. Turn on Web Sharing from System Preferences ▸ Sharing. Then save the file as rosetta.php in your ~/Sites/ folder, and access the URL http://localhost/~lucy/rosetta.php (where lucy is replaced by your username).

```php
<html> <body>
<code>
Table of values<br><br>
<?php
for($Letter = 65; $Letter <=72; $Letter++) {
    $Label='';
    for($Number = 1; $Number <=12; $Number++) {
        $Label .= chr($Letter) . strval($Number) ." ";
    }
    echo $Label . "<br>"; // print an html line break
}
?>
</code>
</body> </html>
```

Ruby

Ruby is a popular object-oriented scripting language that is used behind the scenes in some web applications.

```ruby
#! /usr/bin/env ruby

# note, in ruby, uppercase variable names mean that they are constant
# here we can either use lowercase, or start with an underscore

for letter in 65..72 # step character 65 to 72
  label = ''
  for number in 1..12
    label += letter.chr + "#{number}" + " "
  end  # end for number
  puts(label) # print the whole line
end  # end for Letter
```

MATLAB

As a powerful programming environment, MATLAB can be used for calculations, signal and image processing, data acquisition, simulations, and developing graphical user interfaces. It is extremely useful and well designed for analyses (nearly every variable can be a multi-dimensional array), but licensing and maintenance fees can be prohibitively expensive if you are not able to get an academic rate.

```matlab
for Letter = 65:72    % step through character 65 to 72
   Labels = '';
   for Number  = 1:13
      Labels = [Labels char(Letter) Number2str(Number) ' ' ];
   end % for Number
   disp(Labels)  % print the whole line
end % for Letter
```

R

The R programming and analysis environment is powerful yet free. We recommend you learn it for numerical analysis and plotting. The program below calls a special `Rscript` handler to take care of running the program at the command line instead of in the R console window.

```r
#! /usr/bin/env Rscript
for (Letter in 65:72) {    # step through character 65 to 72
   Lines = ''
   for (Number in 1:12) {
      cat(intToUtf8(Letter),Number,' ',sep="")
   } # end for Number
   cat("\n")  # print the end-of-line character
} # end for Letter
```

Arduino

This program is written to be uploaded to an Arduino device, which will then generate the desired output and sent it out along the serial port. Connect to the Arduino, select the proper serial port, monitor the serial connection, and restart the device to see the output in a serial monitor window:

```
char Letter;
int Number;

void setup(){
   Serial.begin(19200);
   for (Letter = 65; Letter<= 72; Letter++) {
      for (Number = 1; Number <= 12; Number++) {
         Serial.print(byte(Letter));
         Serial.print(Number);
         Serial.print(" ");
      }
      Serial.println(); // print a line break
   }
}
void loop(){} // the program needs a loop function, even if empty
```

Appendix 6

BINARY, HEX, AND ASCII

Alternate base systems

When counting upwards from zero, it seems completely natural to go from the single digit 9 to the two digits of 10. Nor do we pause to think when counting from 19 to 20, even though we switch from incrementing only the rightmost digit to incrementing the one next to it as well. This is because we know intuitively that the different columns or positions in a number represent ten times the value of the column to their right, and that when you count past the highest value of a column, you reset the column back to zero and tally one more on the next higher position. However, although we perform these increments in multiples of ten, there is nothing intrinsically special about the number ten itself which makes this convention any more natural or real than if each position represented eight times its neighbor. If we were born with four fingers on each hand, chances are that is the system that we would use instead.

In a given numerical system the multiplier used for successively higher positions is known as the **base**. We operate in a base 10 system (**decimal**), and our four-fingered friends would operate in base 8 (octal). In an octal counting system, the digits 8 and 9 don't exist. Counting goes 6, 7, 10, 11, with 10 (pronounced one-zero, not ten) equal to a decimal value of 8. Confusing? Perhaps. But non-decimal systems are nonetheless very useful, especially in computing.

One system which is commonly used is base 2 (**binary**). This is because it is very easy for a computer or circuit to represent the states of 0 (off) and 1 (on), in electronic memory as well as in calculations. The sequence of binary numbers is shown in the table at the end of this appendix. A particular column can have a value of zero or one, so if you add 1 to 1, you have run out of digits and have to move to the next column to obtain the binary value 10, representing decimal 2. Add 1 again to get 11 and once more for 100, which is the binary representation of decimal 4.

As you can see, binary numbers can quickly grow to a very large number of digits (called **bits**), but determining the decimal value of a binary number like 1110 is not as difficult as it might seem. You use the same basic approach you use with decimal numbers, although in base 10 you may not realize you're doing it. To get the value of a decimal number like 254, you multiply the digit in each position by the value of that position, and add them together: $2 \times 100 + 5 \times 10 + 4 \times 1 = 254$. Just as the positions in a decimal number represent increasing powers of ten (1, 10, 100, 1000), the digits of a binary number represent increasing powers of two—each is double its neighbor (1, 2, 4, 8, 16, 32). So the decimal value for binary 1110 is 8 + 4 + 2 + 0 = 14.

BINARY ENCODINGS AND chmod Imagine that you want to represent the status of three light switches in a room. Each switch is either on or off, so by using three binary digits you can quickly represent any situation—for instance, when only switch 2 is on (010) or when only switch 3 is off (110). Now, what is a quick way to communicate the switch status to your computer-savvy friend? You can just give them the decimal representation of the binary digits. The first state is 2 and the second is 6, and all values can be represented with a number between 0 and 7.

Computers use similar encodings to represent a range of binary flags. For example, recall from Chapter 6 the output of the shell command ls -l, which shows the read, write, and execute permissions associated with a file. If you imagine that these permissions are your three light switches, you can represent each combination using a single decimal number. Read and write on, but execute off, will be binary 110 or decimal 6, while read-only will be 100. All permissions on will be decimal 7. The chmod command, which you used in the form of chmod u+x, actually accepts these decimal equivalents as input. Remember that there are three groups of rwx permissions you can set: first for the user (you), next for the group, and finally for other people. These three sets of permissions can be represented together by three numbers. For a file called **data.txt**, to give yourself read and write permission (6), but the others read-only (4), you would use the command **chmod 644 data.txt**. To allow yourself and others to execute the script **myscript.py**, but only allow yourself to write to the file, you would use **chmod 755 myscript.py**. While the parameter u+x changes a single element, this syntax lets you set nine bits of information at once.

Computers allocate a certain number of binary digits to store information. These are typically in successive powers of two, starting with 8 bits (called a byte), and then 16, 32, or 64 bits. As a result, certain numbers keep recurring in programming contexts. When an item is represented by 8 bits (one byte), it can take on values from 0–255 (binary 11111111). If you add one to this value, it will roll over to zero. You will see values up to 255 in everything from character encoding (see the section in this appendix, *ASCII and Unicode characters*) to image colors (see *Images and color* in this appendix). Integers are often allotted 16 bits, so they can range from 0 to 65,535 (or plus or minus half of that, up to 32,767). If a file is said to take up 2 K of memory, that indicates it is using 2,000 bytes of information (in most cases, actually 2,048 bytes), or an equivalent of 16,000 on/ off values. Because values in these 8-bit increments occur so frequently, another counting system, called hexadecimal, is employed to encode them in a straight-forward manner.

Hexadecimal

Counting systems are not limited to using fewer than 10 numerals; they can also use more. After binary, base 16 (**hexadecimal**, or hex) is the next most commonly used numbering system in computing. This system uses the letters A to F as digits to represent decimal numbers 10 to 15, similar to how Jack, Queen, King can represent the values 11, 12, or 13 points in a deck of cards. Counting in hex you will go 1, 2, 3, 4, 5, 6, 7, 8, 9, A, B, C, D, E, F, 10, where hex 10 has a decimal value of 16. From there you will get 11, on up to 1A, eventually A1, and FF. To clarify which numerical system is being used, hex values are often written with a preceding 0x (that is, zero with x for hex), so 0x10 is decimal 16. Hex values are also often padded with an extra zero for single digits such as 0x0A for hex A. Values of hexadecimal digits can be determined in the same way you used for binary: multiply each position by its value and add them all together. For example, 0x10 is 1 × 16, 0x1A is 16 + 10, and 0xF1 is 15 × 16 + 1 = 241.

Why use base 16? It makes it very easy to encode binary information in an efficient and readable format. The maximum value of 4 binary digits is 0x0F, and the 8 bits of one byte correspond exactly to two hex digits. So decimal 255 = binary 11111111 = 0xFF. This one-to-two correspondence between bytes and hexadecimal numbers is part of what makes them so useful: unlike decimal values, where the positions increment out of sync with the position of binary values, one hex digit corresponds exactly to 4 bits, as shown in Figure A6.1. This means that they can be translated individually and then be concatenated to maintain the correct value.

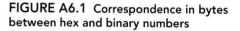

FIGURE A6.1 Correspondence in bytes between hex and binary numbers

ASCII and Unicode characters

When a computer represents text, there needs be some universal convention for storing the characters in memory, so that another program can understand the digital representation. One of the oldest and most widespread of these character encodings is called ASCII. This convention uses 7 bits of memory to store each character, so there are 128 (including 0) possible characters in an ASCII system. (Characters are often stored in a byte of memory, with the leftmost digit a zero.) Characters encoded in ASCII include upper and lowercase letters, numbers, punctuation, tab, space, end-of-line characters, Escape, ⌗ctrl⌗ C, and even a beep.[1] ASCII characters and their corresponding numerical values are also shown in the table at the end of this appendix.

You used the ASCII decimal values, along with Python's chr() function, in Chapter 9 and Appendix 5 to loop through the first few letters of the alphabet.

---

[1] Try typing echo ⌗ctrl⌗ G or just ⌗ctrl⌗ G in a terminal window, or printing ASCII value 07 in some programming environments.

In addition to programs, another common place to see ASCII values is in Web addresses and in the source of web pages. Web addresses do not properly interpret certain characters, so these are commonly replaced by a percent sign followed by the ASCII hexadecimal values. For example, the ASCII value for the space is decimal 32 (which is represented as `20` in hex). If a space occurs as part of a Web address (a practice to avoid), the link will have the value `%20` inserted instead. You performed this substitution in Chapter 6 when working with `curl` and the CrossRef database. Other common values to substitute are `%2F` (a slash), `%3F` (a question mark) and `%3D` (the equals sign). You can find these values in the ASCII table, and many text editors will have an option to encode symbols for the Web.

There are only 26 letters in the alphabet, so the 128 values of ASCII system should be able to cover almost anything, right? Not exactly. Other alphabets, and even common English usage, may require many more characters and symbols— even curly quotes are not in the basic ASCII character set. This has led to the development of other standards, including Unicode's UTF-8 and UTF-16, which allocate more bits in order to encode tens of thousands of characters. UTF-8 uses from one to four bytes to encode characters, with the size dependent on the character set. The first 128 values encode the same symbols in the ASCII character set (fortunately, using the same values). UTF-8 then uses two bytes for nearly two thousand additional symbols and punctuation marks. These additional characters include some common and important punctuation marks and accented characters (é, ü, ø, å, °, "", ∫, ß, Σ, π), which may show up in your datasets.

Failing to support Unicode encodings can prevent a web page from properly rendering many non-English letters, and of particular relevance to this book, it can also cause your programs to fail unexpectedly. Certain shell programs like `tr` can't cope with non-ASCII characters, and you may need to sanitize your dataset before running `bash` scripts. To add UTF-8 support to your Python programs, you can start by adding an extra line beneath your shebang line:

```
#! /usr/bin/env python
# encoding: utf-8   ← Case-sensitive
```

Now your programs will be able to read and print from the extended character set. The Python function `unichr()` translates values to Unicode characters in the same way that `chr()` translates to ASCII. An example script called `asciihexbin.py` shows a brief demonstration of how to generate a table like the one in this appendix.

Images and color

Another situation when you will find yourself trying to speak hexadecimal is when working with images. For many image formats, each pixel is encoded using a set of values which range from 0–255. In a grayscale image there will be one such value, with 0 corresponding to all black, and 255 meaning white. In a so-called 8-bit RGB image, each red, green, and blue value will be allocated one byte. A fourth byte often indicates the degree of transparency for that pixel. (If you allocate 16 bits to each color, it does not let you make them brighter, but rather, enables more levels of gradation between black and full color, increasing your dynamic range.) Using a two-digit hexadecimal number per color, for a total of three pairs, is a convenient way to represent the three bytes describing the color of each pixel. When written out, these values are concatenated, so that the first pair encodes red, the second pair green, and third pair blue. A value of FF0000 means fully saturated red, while 000088 is a dark blue and FF9900 will be a pale orange (red plus green equals shades of yellow).

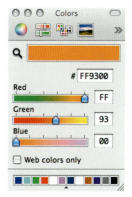

FIGURE A6.2 A typical color-picker in a graphics editing program, showing colors represented via hex numbers

These color definitions are used in creating web content, and are included in style sheets used to set the design of a Web site. In this context, they are often preceded by a number sign, as in color="#A788D2". You can play with these values using the color pickers available in most graphics programs and in web page and style sheet editors (Figure A6.2). Many programming languages and analysis tools also frequently use six hex digits to describe RGB colors. In some languages, two additional digits can be appended (resulting in 8 digits total) to specify the transparency or alpha level. In R, for example, FF at this position means the color is fully opaque, 00 is fully transparent, and 80 would be 50% transparent.

Decimal, hex, binary, and ASCII values

Decimal	Hex	Binary	ASCII character or keystroke	Decimal	Hex	Binary	ASCII character or keystroke
0	0x00	0000000	^@ – Null character	12	0x0C	0001100	
1	0x01	0000001	^A	13	0x0D	0001101	^M – Carriage Return
2	0x02	0000010	^B	14	0x0E	0001110	
3	0x03	0000011	^C – Interrupt	15	0x0F	0001111	
4	0x04	0000100	^D	16	0x10	0010000	
5	0x05	0000101	^E	17	0x11	0010001	
6	0x06	0000110	^F	18	0x12	0010010	
7	0x07	0000111	^G – BELL	19	0x13	0010011	
8	0x08	0001000	^H – backspace	20	0x14	0010100	
9	0x09	0001001	^I – tab	21	0x15	0010101	
10	0x0A	0001010	^J – Line Feed	22	0x16	0010110	
11	0x0B	0001011		23	0x17	0010111	

Decimal	Hex	Binary	ASCII character or keystroke
24	0x18	0011000	
25	0x19	0011001	
26	0x1A	0011010	
27	0x1B	0011011	^[ESCape
28	0x1C	0011100	
29	0x1D	0011101	
30	0x1E	0011110	
31	0x1F	0011111	
32	0x20	0100000	space
33	0x21	0100001	!
34	0x22	0100010	"
35	0x23	0100011	#
36	0x24	0100100	$
37	0x25	0100101	%
38	0x26	0100110	&
39	0x27	0100111	'
40	0x28	0101000	(
41	0x29	0101001	)
42	0x2A	0101010	*
43	0x2B	0101011	+
44	0x2C	0101100	,
45	0x2D	0101101	-
46	0x2E	0101110	.
47	0x2F	0101111	/
48	0x30	0110000	0
49	0x31	0110001	1
50	0x32	0110010	2
51	0x33	0110011	3
52	0x34	0110100	4
53	0x35	0110101	5
54	0x36	0110110	6

Decimal	Hex	Binary	ASCII character or keystroke
55	0x37	0110111	7
56	0x38	0111000	8
57	0x39	0111001	9
58	0x3A	0111010	:
59	0x3B	0111011	;
60	0x3C	0111100	<
61	0x3D	0111101	=
62	0x3E	0111110	>
63	0x3F	0111111	?
64	0x40	1000000	@
65	0x41	1000001	A
66	0x42	1000010	B
67	0x43	1000011	C
68	0x44	1000100	D
69	0x45	1000101	E
70	0x46	1000110	F
71	0x47	1000111	G
72	0x48	1001000	H
73	0x49	1001001	I
74	0x4A	1001010	J
75	0x4B	1001011	K
76	0x4C	1001100	L
77	0x4D	1001101	M
78	0x4E	1001110	N
79	0x4F	1001111	O
80	0x50	1010000	P
81	0x51	1010001	Q
82	0x52	1010010	R
83	0x53	1010011	S
84	0x54	1010100	T
85	0x55	1010101	U

Decimal	Hex	Binary	ASCII character or keystroke
86	0x56	1010110	V
87	0x57	1010111	W
88	0x58	1011000	X
89	0x59	1011001	Y
90	0x5A	1011010	Z
91	0x5B	1011011	[
92	0x5C	1011100	\
93	0x5D	1011101	]
94	0x5E	1011110	^
95	0x5F	1011111	_
96	0x60	1100000	`
97	0x61	1100001	a
98	0x62	1100010	b
99	0x63	1100011	c
100	0x64	1100100	d
101	0x65	1100101	e
102	0x66	1100110	f
103	0x67	1100111	g
104	0x68	1101000	h
105	0x69	1101001	i
106	0x6A	1101010	j

Decimal	Hex	Binary	ASCII character or keystroke
107	0x6B	1101011	k
108	0x6C	1101100	l
109	0x6D	1101101	m
110	0x6E	1101110	n
111	0x6F	1101111	o
112	0x70	1110000	p
113	0x71	1110001	q
114	0x72	1110010	r
115	0x73	1110011	s
116	0x74	1110100	t
117	0x75	1110101	u
118	0x76	1110110	v
119	0x77	1110111	w
120	0x78	1111000	x
121	0x79	1111001	y
122	0x7A	1111010	z
123	0x7B	1111011	{
124	0x7C	1111100	|
125	0x7D	1111101	}
127	0x7F	1111111	^? — delete

SQL COMMANDS

SQL, short for Structured Query Language, is the language used to interact with relational databases, as discussed in Chapter 15. Although our specific examples are drawn from MySQL, learning the basics of SQL can help you work with nearly any database system. MySQL has excellent online references, tutorials, and examples. Many are at the site: `dev.mysql.com/doc/refman/5.1/en/`.

Installing MySQL is described in Chapter 15. The commands listed in the tables below would be entered at the `mysql>` prompt, launched using the command:

```
mysql -u root
```

See Appendix 1 for installation and launching instructions.

If you have assigned a password to the `root` account, the command above should end with `-p`. You can also log in as a user other than `root` if you have configured other users.

Databases are organized into tables containing fields (corresponding to columns), which in turn contain values of related information organized into rows.

Working at the MySQL prompt	
Purpose	**Example**

Entering commands
 	
Commands can span several lines. They are only executed when the line is terminated with a semicolon. Indentation and capitalization are just for readability and are not interpreted
	`SELECT genus FROM specimens`
	`   WHERE vehicle LIKE 'Tib%'`
	`   AND depth > 100`
	`;`

Interrupt a command or cancel a partially typed command. Do not type ctrl C, which will end your entire `mysql` session	`\c` return
Quit MySQL	`EXIT;` `\q` return

Get general help, or help on a command or topic	`HELP`
	`HELP SELECT`
	`HELP LOAD DATA`

Selected MySQL data types

Data type	Description
INTEGER	An integer. Also abbreviated as INT
FLOAT	A floating point number, including scientific notation
DATE	A date in 'YYYY-MM-DD' format
DATETIME	A date and time in 'YYYY-MM-DD HH:MM:SS' format
TEXT	A string containing up to 65,535 characters
TINYTEXT	A string containing up to 255 characters
BLOB	A binary object, including images or other non-text data

Creating databases and tables

Make a new blank database	`CREATE DATABASE databasename;`
Select a database as the target of subsequent commands	`USE databasename;`
Make a new table containing field type definitions	`CREATE TABLE tablename` `    (fieldname1 TYPE, fieldname2 TYPE2);`

Make a new table with an autoincrementing primary key, then other column definitions	`CREATE TABLE tablename`
	`    (primarykeyname INTEGER`
	`    NOT NULL AUTO_INCREMENT`
	`    PRIMARY KEY,`
	`    nextfield TYPE, anotherfield TYPE);`

Adding data into table fields

Import formatted text data whose columns correspond exactly to predefined table fields	`LOAD DATA LOCAL INFILE` `    'path/to/infile';`
Add a row of values to a table in the order that matches the predefined fields	`INSERT INTO tablename VALUES` `    (1,"Beroe",5.2,"1865-12-18");`
Redefine values based on another criterion	`UPDATE tablename SET values = x` `    WHERE othervalues = y;`

Database and table information

List the names of the databases or tables	`SHOW DATABASES;` `SHOW TABLES;`
Show name, type, and other information about the fields of a table	`DESCRIBE tablename;`
Show the number of entries in the table	`SELECT COUNT(*) FROM tablename;`

Extracting data from tables with `SELECT`

List all the rows in all columns of a table. The rows retrieved can be refined with `WHERE` statements at the end of the line	`SELECT * FROM tablename;`
Show the values of the listed columns from the table	`SELECT vehicle,date` `    FROM specimens;`
Show the unique values of a named column	`SELECT DISTINCT vehicle` `    FROM specimens;`
Show a count of the values in a named table	`SELECT COUNT(*)` `    FROM specimens;`

Show a count of the values in a named field, clustered by the unique values of that field. Like `SELECT DISTINCT`, but with counts	`SELECT vehicle,COUNT(*)`
	`    FROM specimens`
	`    GROUP BY vehicle;`

Qualifying which rows to retrieve using `WHERE`

`WHERE` refines the records (rows) retrieved from a `SELECT` command. Criteria include comparisons like greater than and less than, or comparisons of equality, which can apply to numbers or strings. Use `!=` for not equal	`SELECT vehicle FROM specimens`
	`    WHERE depth > 500`
	`    AND dive < 600 ;`

Find approximate matches, using `%` as a wildcard of any characters	`WHERE vehicle LIKE "Tib%"`

Find matches using regular expressions. Wildcards are not all supported, but beginning and end of line, `. [ ] +` are supported	`WHERE field REGEXP query`
	`WHERE vehicle REGEXP "^T"`
	`WHERE species REGEXP "galma$"`

Combine criteria with logical operators
Use parentheses to group logical entities
	`SELECT vehicle from specimens`
	`    WHERE (vehicle LIKE "Ven%")`
	`    OR (vehicle LIKE "JSL%");`

Mathematical and statistical operators

Basic math operators	`+, -, *,  , /`
Basic comparisons	`<, >, =, !=`
Average of the values	`AVG()`
Count of the values	`COUNT()`
Maximum value	`MAX()`
Minimum value	`MIN()`
Standard deviation	`STD()`
Sum of the values	`SUM()`

Deleting entries and tables

Clear all entries from a table	`DELETE FROM tablename;`
Clear entries matching `WHERE` criteria	`DELETE FROM tablename WHERE` `    vehicle LIKE "Tib%";`
Delete an entire table. Use with caution. Can't undo it	`DROP tablename;`

Saving to a file

Save the results from a query into a tab-delimited file	`SELECT * FROM midwater`
	`    INTO OUTFILE '/export.txt'`
	`    FIELDS TERMINATED BY '\t'`
	`    LINES TERMINATED BY '\n'`
	`;`

Export the entire database to an archive. This command is run at the shell prompt, not the `mysql` prompt. The resulting file has all the commands necessary to recreate the original database tables	`mysqldump -u root databasename >` `    datafile.sql`
Read back in a database created via dump Read in a file of SQL commands This command is also run at the `bash` prompt, and the target database must already exist	`mysql -u root targetdb < mw.sql`

User management[a]

Set the password for the current user (from the `mysql` prompt). Remember the equal sign	`SET PASSWORD = PASSWORD('mypass');`
	`SET PASSWORD`
	`    FOR 'python_user'@'localhost' =`
	`    PASSWORD('newpass')`
	`    OLD_PASSWORD('oldpass');`

Add a new user with defined addresses that they can connect from and a preset password	`CREATE USER 'newuser'@'localhost'` `    IDENTIFIED BY 'newpassword';`

Give a user privileges. The capabilities, database and tables, and user and host are specified. Host IP ranges use % as the wildcard character	`GRANT SELECT, INSERT, UPDATE,`
	`    CREATE, DELETE ON midwater.* TO`
	`    'newuser'@'localhost';`

Log in with password (from the shell prompt)	`mysql -u newuser -p`

[a]These commands can also be accomplished from within the Dashboard or SQuirrelSQL GUIs.

INDEX

ABOUT THE BOOK

Editor: Andrew D. Sinauer
Project Editor: Azelie Aquadro
Copy Editor: Randy Burgess
Production Manager: Christopher Small
Book Design and Layout: Joan Gemme
Cover Design: Steven H. D. Haddock